Advances in Delay-tolerant Networks (DTNs)

Woodhead Publishing Series in Electronic and Optical Materials

Advances in Delay-tolerant Networks (DTNs)

Architecture and Enhanced Performance

Second Edition

Edited by

Joel J. P. C. Rodrigues

Federal University of Piauí (UFPI), Teresina - PI, Brazil; Instituto de Telecomunicações, Portugal

ELSEVIER

WP
WOODHEAD
PUBLISHING
An imprint of Elsevier

Woodhead Publishing is an imprint of Elsevier
The Officers' Mess Business Centre, Royston Road, Duxford, CB22 4QH, United Kingdom
50 Hampshire Street, 5th Floor, Cambridge, MA 02139, United States
The Boulevard, Langford Lane, Kidlington, OX5 1GB, United Kingdom

Library of Congress Cataloging-in-Publication Data
A catalog record for this book is available from the Library of Congress

British Library Cataloguing-in-Publication Data
A catalogue record for this book is available from the British Library

ISBN: 978-0-08-102793-6 (print)
ISBN: 978-0-08-102794-3 (online)

For information on all Woodhead publications
visit our website at https://www.elsevier.com/books-and-journals

Publisher: Matthew Deans
Acquisitions Editor: Kayla Dos Santos
Editorial Project Manager: John Leonard
Production Project Manager: Vijayaraj Purushothaman
Cover Designer: Victoria Pearson

Typeset by SPi Global, India

Working together
to grow libraries in
developing countries

www.elsevier.com • www.bookaid.org

Contents

Part Two   Improving the performance of delay-tolerant networks (DTNs)

Contributors

A. Mohammed Ahmed Dalian University of Technology, Dalian, China

M. Asplund Linköping University, Linköping, Sweden

Antoine Auger ISAE-SUPAERO, University of Toulouse, Toulouse, France

Gwilherm Baudic ISAE-SUPAERO, University of Toulouse, Toulouse, France

C. Caini University of Bologna, Bologna, Italy

Shengbo Chen Henan University, Kaifeng, China

H. Chenji Texas A&M University, College Station, TX, United States

N.L. Clarke Plymouth University, Plymouth, United Kingdom

Tarun Dhankhar Division of Information Technology, Netaji Subhas Institute of Technology, University of Delhi, New Delhi, India

Sanjay K. Dhurandher Department of Information Technology, Netaji Subhas University of Technology, New Delhi, India

W.M. Eddy MTI Systems, Greenbelt, MD, United States

M.R. Frater University of New South Wales, Canberra, ACT, Australia

S.M. Furnell Plymouth University, Plymouth, United Kingdom

B.V. Ghita Plymouth University, Plymouth, United Kingdom

B. Jedari Dalian University of Technology, Dalian, China

J. Lacan Université de Toulouse, Toulouse, France

J. Leguay Thales Communications, Colombes, France

Emmanuel Lochin ENAC, University of Toulouse, Toulouse, France

Jahanavi Mishra Division of Information Technology, Netaji Subhas Institute of Technology, University of Delhi, New Delhi, India

J. Morgenroth Technische Universität Braunschweig, Braunschweig, Germany

S. Nadjm-Tehrani Linköping University, Linköping, Sweden

Y. Najaflou Dalian University of Technology, Dalian, China

P. Pirozmand Dalian University of Technology, Dalian, China

W.-B. Pöttner Technische Universität Braunschweig, Braunschweig, Germany

R.H. Rahman University of New South Wales, Canberra, ACT, Australia

Victor Ramiro ISAE-SUPAERO, University of Toulouse, Toulouse, France

Ju Ren Central South University, Changsha, China

Joel J. P. C. Rodrigues Federal University of Piauí (UFPI), Teresina - PI, Brazil; Instituto de Telecomunicações, Portugal

S. Schildt Technische Universität Braunschweig, Braunschweig, Germany

Jagdeep Singh Division of Information Technology, Netaji Subhas Institute of Technology, University of Delhi, New Delhi, India

Vasco N.G.J. Soares Instituto de Telecomunicações, Polytechnic Institute of Castelo Branco, Castelo Branco, Portugal

R. Stoleru Texas A&M University, College Station, TX, United States

P.-U. Tournoux Université de la Réunion, Saint-Denis Réunion, France

A.G. Voyiatzis Industrial Systems Institute, RC "Athena", Patras, Greece

L. Wolf Technische Universität Braunschweig, Braunschweig, Germany

Isaac Woungang Department of Computer Science, Ryerson University, Toronto, ON, Canada

F. Xia Dalian University of Technology, Dalian, China

Zhensheng Zhang University of California, Los Angeles, CA, United States

Preface

Delay-tolerant networks (DTNs) address the problem of intermittent connectivity in a network with long delays between sending and receiving messages, or periods of disconnection. The DTN principle allows the network nodes to store messages and forward these messages later when connectivity is restored. This book reviews the technology involved, its important applications, and the prospects for improving performance.

This book is formed of 15 chapters and starts with an introduction to delay and disruption tolerant networks. DTNs are considered an overlay network that aims to enable communications in disruptive network conditions due to environmental or operational factors. DTNs present a big research potential and social impact contributing to enable services and applications in a variety of environments including disaster-recovery/rescue operations, vehicular communications, military battlefields, habitat monitoring, deep-space communications, underwater networks, social networks, and noninteractive Internet access connectivity in rural and developing areas.

After this introductory chapter, the book (Chapter 1) is organized into two main parts. The first focuses on the different types of DTNs, presenting several reviews considering DTN applications on satellite and deep-space communications, vehicular and underwater communications, and during large-scale disasters. Improving the performance of DTNs includes the chapters of the second part of the book. A brief overview of each chapter is presented below.

Chapter 2 evaluates the potential advantages of DTNs when applied to satellite communications for both geosynchronous (GEO) and low earth orbit (LEO) constellations. Continuing the application of DTNs to space communications, Chapter 3 considers the requirements associated with acquiring data collected by satellites in a deep-space context. It presents an architecture that protects a wide range of DTN-based and non-DTN-based threats to which the system is vulnerable.

Chapter 4 considers the contributions of DTNs for the proposal of a DTN-based network architecture for vehicular communications, called vehicular delay-tolerant networks (VDTNs). The VDTN seeks novel and effective solutions for communicating in vehicular environments where continuous end-to-end connectivity cannot be assumed. This network architecture adopts a store-carry-and-forward paradigm combined with an IP over the VDTN approach and out-of-band signaling with control and data plane separation. Finally, an overview of research conducted in the VDTN architecture is also provided, and open research issues are highlighted.

Underwater communications using DTNs are addressed in Chapter 5. It presents the current state-of-the-art solutions in underwater DTNs. The author mentions that the current solutions either rely on message replication or make some assumptions about the underlying network topology or mobility patterns (contact schedules). Chapter 6 focuses on DTNs for emergent communications. This chapter resumes the use of DTNs during a disaster recovery process. Starting with a motivating scenario based on a natural disaster that occurs over a large geographical area, the industry state of the art is reviewed. After, the literature review of the state of the art is also considered taking into account the main issues related to this DTN application.

Chapter 7 discusses the energy efficiency issues in OppNets related to geocasting and security. It also focuses on various geocasting techniques such as Geocasting for OppNets (GeoOpp), Expected Visiting Rate (EVR), and Floating Content to counter problems related to energy efficiency. Finally, using the Opportunistic Network Environment (ONE) simulator, the author compares the energy-efficient protocols E-PRoPHET, ProWait, PORON, E-EDR, and E-ATDTN, and observed that E-ATDTN outperforms these protocols in terms of average residual energy, number of message drops, and number of dead nodes.

Part Two starts with Chapter 8 that evaluates the Bundle Protocol (BP) and alternative approaches to data bundling in DTNs. Based on the presented study, the authors propose interesting trends for further work. Opportunistic routing in mobile ad hoc DTNs is considered in Chapter 9. This chapter discusses some of the key challenges in designing an efficient opportunistic routing (OR) protocol, views some of the representative OR protocols, and points out a few open issues associated with OR.

Chapter 10 presents some potential solutions to enable stored and live streaming over DTNs. Compared with stored data streaming, which has several solutions, live data streaming remains a challenging problem, and the use of appropriate recovery mechanisms is mandatory. After an overview of the challenges and available solutions to enable a stored data streaming, the author discusses how an on-the-fly coding scheme allows live data streaming to be performed over a DTN network.

The rapid selection and dissemination of urgent messages over DTNs are considered in Chapter 11. It provides a brief overview of the existing research in the area as well as a more in-depth study on how to achieve a higher level of predictability of the message dissemination latency. The authors present the Random Walk Gossip (RWG) protocol, which is a manycast protocol for partition-tolerant networks tailored for disaster area networks. This work extends the design of the protocol by differentiating messages based on their deadline and progress so far, considering several performance studies to evaluate the proposal.

Chapter 12 studies the social characteristics of mobile users, which have been extensively utilized to improve the performance of protocols in DTNs considering three social aspects of DTNs. First, some important social network analysis concepts and metrics such as a social graph, node centrality, tie strength, community structure, etc. are investigated. After, well-known social-based mobility models are introduced, since simulation and evaluation of existing forwarding protocols in DTNs rely heavily on the reality of mobility models. Finally, the state of the art of social-based data forwarding in DTNs with respect to multicasting, user selfishness, and incentive schemes are presented, and the most important features are outlined.

Performance issues and design choices in DTN algorithms and protocols are addressed in Chapter 13. It studies several areas relevant to the DTN performance and takes a look at the main factors impacting the performance of a DTN system. Examples from real-world experiments and deployments are presented. As well as suggestions for implementers based on the author's experience with several DTN systems and protocol implementations. However, the points examined here are not only applicable to specific implementations or even the bundle protocol. The discussed issues are general and relevant to all implementations and deployments of DTN-like systems.

The development of opportunistic applications, i.e., applications running over opportunistic networks, is still in early stages. This is due to lack of tools to support the process in such uncertain conditions. Indeed, many tools have been introduced to study and characterize opportunistic networks, but none of them is focused on helping developers to conceive opportunistic applications. Chapter 14 shows that the gap between opportunistic applications development and network characterization can be filled with network emulation. First, it points out important challenges about the development of opportunistic applications. Then, to cope with these challenges, it details a set of requirements that an emulator should meet to allow the testing of such applications.

Finally, Chapter 15 focuses on the challenge to that is the proposal of the key application that unveils its potential and results in a wide adoption. This chapter surveys the breadth of applications in which DTN is already experimented, solving actual, real-world problems related to intermittent connectivity and harsh operational environments around the earth.

Advances in Delay-Tolerant Networks examines the current state-of-the-art and the importance of this technology. It will be a valuable resource for researchers in electronics, computer engineering, telecommunications, and networking; for R&D managers in the communications industries; and those involved in disaster management.

Joel J. P. C. Rodrigues
Federal University of Piauí (UFPI), Teresina - PI, Brazil
Instituto de Telecomunicações, Portugal

An introduction to delay and disruption tolerant networks (DTNs)

1

Joel J. P. C. Rodrigues[a,b] and Vasco N.G.J. Soares[c]
[a]Federal University of Piauí (UFPI), Teresina - PI, Brazil, [b]Instituto de Telecomunicações, Portugal, [c]Instituto de Telecomunicações, Polytechnic Institute of Castelo Branco, Castelo Branco, Portugal

1.1 Introduction

The Internet Protocol (IP) suite, commonly known as TCP/IP (the well-known Transmission Control Protocol/Internet Protocol), makes implicit assumptions of continuous, bi-directional end-to-end paths, short round-trip times, high transmission reliability, and symmetric data rates (Socolofsky and Kale, 1991). However, a wide range of emerging networks (outside the Internet) usually referred to as opportunistic networks, intermittently connected networks, or episodic networks violate these assumptions. These networks fall into the general category of delay/disruption-tolerant networks (DTNs) (Cerf et al., 2007). DTNs experience any combination of the following: sparse connectivity, frequent partitioning, intermittent connectivity, large or variable delays, asymmetric data rates, and low transmission reliability. More importantly, an end-to-end connection cannot be assumed to be available in these networks. Table 1.1 summarizes the main differences between traditional networks (Internet) and DTN networks.

The TCP/IP stack does not properly handle such connectivity challenges. Firstly, the performance of TCP is severely limited by high latency and moderate to high loss rates. Secondly, the performance of the network layer is affected by the loss of fragments. Furthermore, the high latency also causes traditional routing protocols to incorrectly label links as nonoperational. This motivated the proposal of a new network architecture that was designed to enable communication under stressed and unreliable conditions.

The work on Interplanetary Internet Architecture, later generalized to the DTN architecture, began in the late 1990s (Burleigh et al., 2003). DTN is a network research topic focused on the design, construction, performance evaluation, and application of architectures, services, and protocols that intend to enable data communication among heterogeneous networks in extreme environments (Cerf et al., 2007; Scott and Burleigh, 2007; Fall and Farrell, 2008; Fall, 2003). To answer these challenges the DTN Research Group (DTNRG) (2002), which was chartered as part of the Internet Research Task Force (IRTF) (2013), proposed an architecture (i.e., RFC 4838) (Cerf et al., 2007) and a communication protocol (i.e., RFC 5050) (Scott and Burleigh, 2007) for DTNs.

Advances in Delay-tolerant Networks (DTNs). https://doi.org/10.1016/B978-0-08-102793-6.00001-1

Table 1.1 Main differences between the assumptions of traditional
and delay-tolerant networks.

	Traditional (Internet-like)	**DTN**
End-to-end connectivity	Continuous	Frequent disconnections
Propagation delay	Short	Long
Transmission reliability	High	Low
Link data rate	Symmetric	Asymmetric

This chapter provides an introduction to delay and disruption tolerant networks and it is organized as follows. Section 1.2 reviews the DTN architecture and its key concepts. Next, application scenarios for these networks are presented in Section 1.3. The most relevant well-known routing protocols for DTN-based networks are discussed in Section 1.4. Finally, Section 1.5 concludes the chapter presenting a summary of the review.

1.2   Delay-tolerant network architecture

The DTN architecture (Cerf et al., 2007), illustrated in Fig. 1.1, introduces a store-carry-and-forward paradigm by overlaying a protocol layer, called bundle layer, below the application layer, which provides internetworking on heterogeneous networks (regions) operating on different transmission media. DTN proposes a new

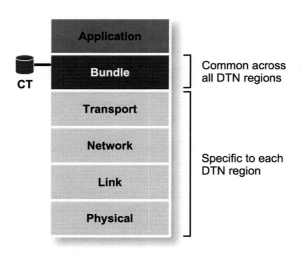

Fig. 1.1 Illustration of DTN layered architecture.

communication paradigm that breaks an end-to-end communication path into hop-by-hop sessions, enabling asynchronous messages (i.e., bundle) delivery over a physically delayed, or disrupted network environments. The bundle protocol (Scott and Burleigh, 2007) is end-to-end, strongly asynchronous, bundle oriented.

In a DTN, the application data units are aggregated into one or more variable-length protocol data units called "bundles," at the bundle layer. The idea is to "bundle" together all the information required for a transaction (i.e., entire blocks of application-program data and metadata/control information). This minimizes the number of round-trip exchanges, which is useful when the round-trip time is very large. To help to route and schedule decisions, bundles contain an originating timestamp, a useful life indicator (i.e., time-to-live), a class of service assignment for bundle priorities (that can be set to expedite, normal, or bulk in order of decreasing priority), and a length indicator.

The bundle protocol also offers an optional hop-by-hop transfer of reliable delivery responsibility, called bundle custody transfer, and an optional end-to-end acknowledgment functionality (i.e., "return receipt") (Fall et al., 2003). When nodes accept custody of a bundle, they commit to retaining a copy of the bundle until such responsibility is transferred to another node. Custody transfer and return receipt functionalities are illustrated in Fig. 1.2. Moreover, the bundle layer also implements security services (Farrell et al., 2009; Symington et al., 2011) and a flexible naming scheme with late binding (Fall et al., 2009) that allows bundles destined to a descriptive name to be resolved progressively until they are delivered to one or several recipients.

The protocols used in the layers below the bundle layer might be diverse and are chosen according to the communication requirements of each region (e.g., terrestrial Internet, interplanetary networks, or sparse wireless sensor networks). DTN gateways are responsible for forwarding bundles between two or more DTN regions, as illustrated in Fig. 1.3. Therefore, they must map data from the lower-layer protocols used in each region they span (Fall, 2003). This approach allows data to traverse multiple regions, via region gateways, to reach a destination region and, finally, a host within that region.

Protocols below the bundle layer of different regions may provide different semantics, thus protocol-specific convergence layer adapters are required to provide the functions necessary to carry the bundles on each of the corresponding protocols.

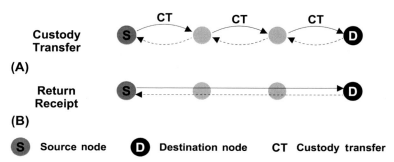

Fig. 1.2 Illustration of DTN (A) custody transfer and (B) return receipt functionalities.

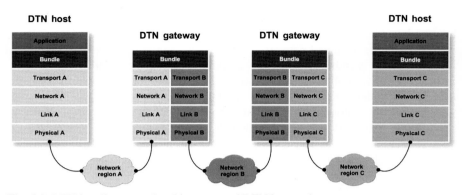

Fig. 1.3 DTN bundle protocol architecture, and DTN host and gateway concepts.

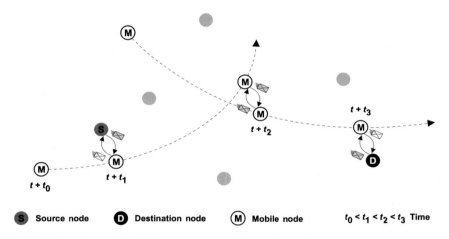

Fig. 1.4 Illustration of the DTN store-carry-and-forward paradigm.

An example of a work in progress as an Internet-draft for a TCP-based convergence layer protocol can be found in Demmer and Ott (2008). Another example is provided by Atkins and Guo (2008) where a vehicle transmission protocol and its correspondent convergence layer for DTN are discussed.

The store-carry-and-forward paradigm avoids the need for constant connectivity. It is used to move bundles across a region, exploiting node mobility. This paradigm, which is illustrated in Fig. 1.4, can be described as follows. A source node that originates bundle stores it using some form of persistent storage (such as a hard disk) and physically carries it while waiting until a communication opportunity becomes available. When a contact opportunity occurs (i.e., two nodes are in range), the bundle is forwarded to an intermediate node, according to a hop-by-hop forwarding/routing scheme. Then, this process is repeated and the bundle will be relayed hop-by-hop until eventually reaching its destination node.

A contact is defined as a period during which two network nodes have the opportunity to communicate and it depends on the application area (Jones and Ward, 2012). DTN architecture (Cerf et al., 2007) defines different types of contacts that can be classified as persistent, on-demand, scheduled, predicted, or opportunistic. In persistent contacts, links are always available and no action is required to instantiate such a contact. On-demand contacts are similar to persistent contacts, but require some action to instantiate. In scheduled contacts, it is assumed that nodes move along predictable paths. Therefore, it is possible to predict or receive time schedules for their future positions. Thus, communication sessions will be scheduled, i.e., the contact is established at a particular time, for a particular duration. Predicted contacts require analyzing previously observed contacts (or some other information) to predict future opportunities (i.e., the contact times and durations) to transmit data. In opportunistic contacts, communication opportunities happen unexpectedly, without any prior knowledge.

There may be some situations where contacts between network nodes are of such a short duration that a bundle can be inevitably too large to be sent in one piece. This results in incomplete bundle transmissions. DTN architecture considers the use of fragmentation and reassembly to ensure that contact volumes are fully utilized, thus avoiding the retransmission of partially transmitted bundles. Two types of fragmentation/ reassembly are proposed: proactive and reactive (Cerf et al., 2007). In proactive fragmentation, a node splits a bundle into smaller fragments before a transmission attempt and then transmits each fragment as an independent bundle over the DTN network. The fragmentation decision may be based on knowledge of the link availability (ahead of time) or account for buffer limitations on the next node, among others.

In reactive fragmentation, the fragmentation process is executed after an attempted transmission has occurred. In this case, a node may learn via lower-layer convergence protocols that only a portion of the entire bundle was transmitted to the next node, for instance, due to a sudden link failure. Then, both nodes can cooperatively reconcile the remaining and the already received portions into valid (fragmented) bundles, which can be sent at new contact opportunities.

For both fragmentation types, the fragments are only reassembled into the original larger bundle at the destination node. Fragments may be further fragmented, either proactively or reactively. A study on the effects of fragmentation on the bundle delivery success in DTNs is presented in Pitkänen et al. (2008).

1.3 DTN application scenarios

The DTN concept was initially designed for communicating with spacecraft, to compensate for disconnections over interplanetary distances (Burleigh et al., 2003; Akyildiz et al., 2003). Interplanetary networking (IPN) is characterized by high intermittent connectivity, extremely long propagation delay (due to the finite speed of light), low transmission reliability (due to the positioning inaccuracy and limited visibility), and low and asymmetric data rate. Several projects have been carried out on this topic by several organizations, such as the InterPlaNetary Internet Project

(Internet Society IPN Special Interest Group, 2011), Defense Advanced Research Projects Agency (DARPA) (2013), NASA Jet Propulsion Laboratory (2013a), and MITRE—Applying Systems Engineering and Advanced Technology to Critical National Problems (MITRE, 1997).

However, over the years, researchers have identified numerous terrestrial environments where DTN concepts may be employed. For example, underwater networks make use of the DTN paradigm to cope with the problems caused by intermittent connectivity, mobility, sparse deployment, high propagation delay, high transmission cost, low asymmetric data rate, and poor transmission reliability (due to positioning inaccuracy and high attenuation). These networks enable applications for oceanographic data collection, pollution monitoring, offshore exploration, disaster prevention, assisted navigation, and tactical surveillance applications (Partan et al., 2006; Katz, 2007). Examples of projects in this area include Underwater Acoustic Sensor Networks (UW-ASN) (Broadband and Wireless Networking Laboratory (BWN LAB), 2004), Underwater Acoustic Network (UAN) (2007), and SiPLABoratory (2008).

Wildlife tracking networks, which are designed for biology research, may consider a DTN approach to face the problems resulting from intermittent connectivity, mobility, sparsity, energy constraints, large end-to-end delay, and asymmetric data rate. These networks allow monitoring the long-term behaviors of wild animals sparsely distributed over a large area. Examples of projects for animal tracking are ZebraNet (Juang et al., 2002; Zhang et al., 2004), SWIM (Small and Haas, 2003), and TurtleNet (UMass Diverse Outdoor Mobile Environment (DOME) Project, 2007).

Sparse wireless sensor networks (e.g., space, terrestrial, and airborne) can also apply DTN technology to deal with the problems caused by intermittent connectivity, sparse deployment, limited power (and also limited memory and CPU capability), and low and asymmetric data rate (Shah et al., 2003; Jain et al., 2006; Sherwood and Chien, 2007). These networks are usually employed to monitor science and hazard events, like earthquakes, volcanoes, flooding, forest fire, sea ice formation and breakup, lake freezing and thawing, and environmental monitoring. Some examples of projects in this area are Volcano Sensorweb (NASA Jet Propulsion Laboratory, 2013b) and Sensor Networking with Delay Tolerance (SeNDT) (Trinity College Dublin, 2006).

The problem of providing data communications to remote and underdeveloped rural communities in developing countries has been addressed by several projects with approaches that focus on asynchronous (i.e., disconnected) messaging by transportation systems. Vehicles are used as data mules, carrying data to remote villages and regions where there is no network infrastructure. Such an approach reduces the cost of connectivity and allows dealing with intermittent connectivity, mobility, sparse deployment, high propagation delay, and asymmetric data rate issues (Seth et al., 2006; Vallina-Rodriguez et al., 2009). Examples of projects in this area are the following: DakNet (Pentland et al., 2004), Saami Network Connectivity (SNC) (Doria et al., 2002), Wizzy Digital Courier (Wizzy Digital Courier, 2003), Message Ferry (Zhao and Ammar, 2003; Zhao et al., 2004), Networking for Communications Challenged Communities (N4C) (2008), First Mile Solutions (United Villages, Inc, 2005),

KioskNet (Tetherless Computing Lab—University of Waterloo, n.d.; Guo et al., 2007), and Affordable Internet Access Project for Underdeveloped Communities (CCSU, 2007).

Recently, store-carry-and-forward DTN concepts have been proposed for maintaining communications in sparsely connected vehicular network environments (Shao et al., 2009; Pereira et al., 2012; Franck and Gil-Castineira, 2007), enabling the development of emerging vehicular applications such as, but not limited to, road safety, traffic monitoring, driving assistance, and infotainment. Transmissions over these networks are subject to frequent and unpredictable disconnection caused by a dynamic topology, high-speed mobility, variable node density, short contact durations, limited transmission ranges, radio obstacles, and interferences. An example of a network architecture designed according to the store-carry-and-forward operating principle, called vehicular delay-tolerant network, is presented in Soares et al. (2009).

People networks, also called pocket switched networks and social networks, explore transfer opportunities between mobile wireless devices carried by humans (Erramilli et al., 2007; Chen et al., 2007; Chaintreau et al., 2006; Hui et al., 2005, 2008; Glance et al., 2001). These networks experience the problems imposed by intermittent connectivity, mobility, energy limitation, heterogeneous, and asymmetric data rate. They enable applications that forward data based on people's social interests (e.g., news, music, movies, and arts). The Haggle project (Haggle, 2006) is an example in this area.

Integrating DTN concepts to military tactical networks can ease communications in hostile environments (battlefields) where network infrastructure is unavailable (Olajide and Washington, 2008; Jormakka et al., 2008; DARPA, 2013; Defense Advanced Research Projects Agency (DARPA), n.d.). These networks suffer from problems of high intermittent connectivity, mobility, destruction, noise, attack, interference, low transmission reliability (due to position inaccuracy and limited visibility), and low data rate. The Marine Corps Command and Control, On-the-Move, Network, Digital, Over-the-horizon Relay (CONDOR) (Scott, 2005) program is an example in this area.

DTN principles can also be considered for disaster recovery networks. These networks can support communications between emergency responders in catastrophe-hit areas lacking a functioning communication infrastructure (Asplund et al., 2009; Nelson et al., 2009; Dong et al., 2009; Nelson, 2008).

1.4 DTN routing protocols

Routing in DTNs is a challenging issue because of the lack of contemporaneous end-to-end paths. Furthermore, information and resource shortages accentuate this challenge. It is important to note the importance of node mobility, which is exploited to carry data around the network and thus to overcome network partitions.

In these networks, routing consists of a sequence of independent, local forwarding decisions that make bundles "progress in steps" toward their destination. The source of knowledge that is used to make these decisions often differs, and can be used to classify routing protocols. While some routing approaches assume that there is not any knowledge available, others consider and eventually combine information about historical data (e.g., recent encounters, contact time, contact frequency, or contact location), location (e.g., past, present, future location data), or movement patterns.

DTN routing strategies can also be classified as single-copy schemes (i.e., forwarding-based) or multiple-copy schemes (i.e., flooding-based) (Spyropoulos et al., 2008a,b; Balasubramanian et al., 2007). Single-copy schemes maintain a single copy of a bundle in the network that is forwarded between network nodes. These routing schemes have low resource requirements (e.g., storage, bandwidth, energy), however, they suffer from low delivery ratios and large delays. On the contrary, multiple-copy schemes replicate bundles at contact opportunities. Copies of the same bundle can be routed independently to increase security (Burgess et al., 2007) and robustness (i.e., the chances of delivery via different paths). Bundle replication improves the probability of delivery and minimizes delivery latency. The downside is that it consumes a high amount of energy, and increases the contention for network resources like bandwidth and storage. Therefore, it potentially can lead to poor overall network performance, as discussed in Balasubramanian et al. (2007) and Haas and Small (2006). These shortcomings often make multiple-copy routing strategies improper for energy-constrained and bandwidth-constrained DTN applications.

The next subsections present examples of single-copy and multiple-copy routing protocols, which are aimed at generic application scenarios. Therefore, they can be potentially used in any DTN-based network. Detailed theoretical background and surveys of DTN routing protocols may be found in Zhang (2006), Spyropoulos et al. (2008a,b), Ramanathan et al. (2007), Jain et al. (2004), Jones and Ward (2012), and Shen et al. (2008).

1.4.1 Single-copy routing protocols

Direct Delivery/transmission (Spyropoulos et al., 2004) and First Contact (Jain et al., 2004) are two examples of simple single-copy DTN routing protocols. These protocols do not use any knowledge about the network to make forwarding decisions. In Direct Delivery (Spyropoulos et al., 2004), the source node carries the bundle until it meets the destination node. Thus, although this protocol has minimal overhead, it may incur in very long delays for a bundle delivery. Its operation is illustrated in Fig. 1.5.

First Contact (Jain et al., 2004) performs routing by forwarding bundles randomly, as illustrated in Fig. 1.6. Nodes forward bundles to the first node they encounter. This results in a random search for the destination node. Moreover, bundles may oscillate among a set of nodes or be delivered to a dead end.

These basic routing approaches can be enhanced considering a utility function that evaluates the capability of the encountering nodes to deliver a bundle to its destination. One possibility would be to incorporate location information to assist in making forwarding decisions. Such an algorithm belongs to a class of routing protocols known as geographic routing protocols (also known as location-based or position-based routing protocols).

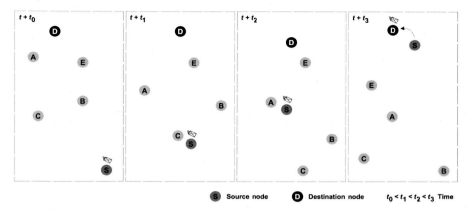

Fig. 1.5 Illustration of the Direct Delivery routing protocol operation.

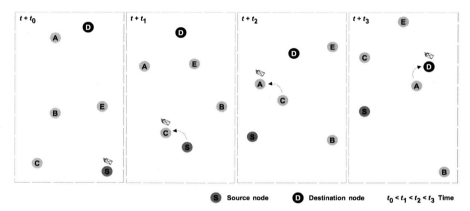

Fig. 1.6 Illustration of the First Contact routing protocol operation.

Geographic routing relies mainly on location information and other mobility parameters provided by positioning devices such as global positioning systems (GPS) (Hofmann-Wellenhof et al., 2001). In this class of routing protocols, routing decisions are made with the goal of, at each step, progressively reducing the geographic distance to the destination node(s). Hence, it is assumed that nodes know their geographical location and the geographical location of the destination node(s).

The increasing availability of vehicle navigation systems (NS) has sparkled the development of geographic routing approaches to vehicular networks. An NS features location hardware (typically a GPS), a roadmap database containing several information items, such as speed limit and average speed, and the shortest path algorithm. With this plethora of information, it is possible to estimate the arrival time to a specific location (namely, of a given destination node).

Although several geographic routing protocols have been proposed for vehicular communications, most approaches do not apply to sparse (i.e., low node density) scenarios. Examples include the position-based routing strategies for vehicular ad hoc networks (VANETs) presented in Lochert et al. (2003), Fiore et al. (2006), and Schwingenschlogl and Kosch (2002), which are not able to deal with intermittency, disruption, or frequent network partitions that can last for a long period. On the contrary, Geographical Opportunistic Routing (GeOpps) (Leontiadis and Mascolo, 2007) is an example of a routing protocol that follows the store-carry-and-forward paradigm to cope with these issues.

GeOpps is a forwarding routing protocol that maintains a single copy of each bundle in the network, and its routing decisions are made as follows. A vehicle moving along a suggested route (determined in function of its destination) uses its navigation system to determine the nearest point (NP) on its route to a location (D) where a data bundle must be delivered. Fig. 1.7 shows an example of NP calculation for three vehicles A, B, and C. Then, the navigation system is used to estimate the time of arrival (ETA) of the vehicle to the NP, and to determine the ETA needed to go from NP to D. The sum of these values is called the minimum estimated time of delivery (METD) as shown in Eq. (1.1), and it is used as a utility function to make routing decisions:

$$METD = ETA \text{ to } NP + ETA \text{ from } NP \text{ to } D \qquad (1.1)$$

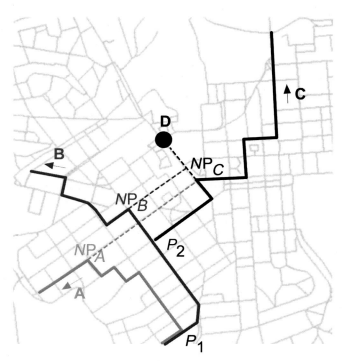

Fig. 1.7 Example of GeOpps calculation of the nearest point (NP) from the bundle's destination (D), for vehicles A, B, and C.

During vehicle travel, a bundle is forwarded to an encountered vehicle, only if the *METD* required by the encountered vehicle to deliver the bundle is lower than the *METD* of the vehicle that currently carries the bundle. This would mean that the encountered vehicle is likely to move closer and/or faster to the bundle's destination. This process is repeated until the bundle reaches its destination or its time-to-live expires.

Fig. 1.7 shows an example where a vehicle A, which is carrying a bundle to be delivered to D, meets a vehicle B at location P_1. The *NP* calculation for A and B allows concluding that B's *METD* value is lower than A's *METD*. This happens because the time required to go from P_1 to NP_B and then to D is lower than the time required to go from P_1 to NP_A and then to D. Hence, A forwards the bundle to B. Afterwards, during its travel, the vehicle B encounters vehicle C at location P_2. Since C is going to pass closer to the destination of the bundle, it has a lower *METD* than B. Thus, vehicle B forwards the bundle to C.

1.4.2 Multiple-copy routing protocols

Some examples of well-known and widely investigated multiple-copy DTN routing protocols are Epidemic (Vahdat and Becker, 2000; Zhang et al., 2007), Spray and Wait (Spyropoulos et al., 2005), PRoPHET (Lindgren et al., 2004, 2011), and MaxProp (Burgess et al., 2006). These protocols are aimed at generic application scenarios. Therefore, they can be potentially used in any DTN-based network that delivers data using a store-carry-and-forward paradigm. These routing protocols make different assumptions about the knowledge available to network nodes (e.g., absence of knowledge or history of node encounters), as discussed in the following.

Epidemic protocol (Vahdat and Becker, 2000; Zhang et al., 2007) does not require any prior knowledge about the network. Under this routing protocol, each node maintains a list of the bundles it carries that have not been delivered. At node encounters, network nodes exchange all bundles that they don't have in common. Using this strategy, all bundles are eventually spread to all nodes, including their destination. Fig. 1.8 clarifies the protocol operation.

The epidemic is shown to be effective but suffers from the disadvantages of flooding as the node density increases. It creates lots of contention for buffer space and required bandwidth, resulting in many bundle drops and retransmissions in resource-constrained network environments. In an environment with infinite buffer resources and bandwidth, this protocol provides an optimal solution, since it delivers all the bundles that can be delivered in the minimum amount of time. For this reason, it is considered "unbeatable" and it is used as a benchmark to compare with other routing protocols (Ramanathan et al., 2007).

Spray and Wait (Spyropoulos et al., 2005) limits the number of bundle replicas (i.e., copies) per bundle allowed in the network to control flooding. This routing protocol assumes two main phases. In the "spray phase," for each bundle originated at a source node, L bundle copies are spread to L distinct nodes. If the destination node is not found during the "spray phase," then at the "wait phase" direct transmission is performed. Hence, it waits until one of the L relays (i.e., nodes) finds the destination node.

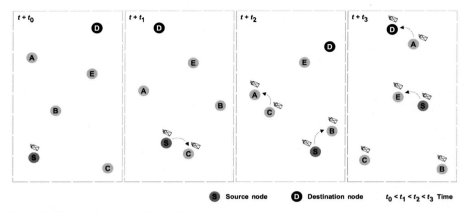

Fig. 1.8 Illustration of the Epidemic routing protocol operation.

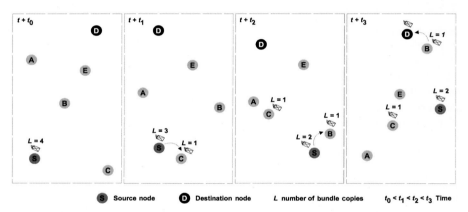

Fig. 1.9 Illustration of the Source Spray and Wait routing protocol operation.

Two different spraying schemes are proposed for the "spray phase," *source spray* (also called *normal spray*), and *binary spray*. In the *source spray* scheme, the source node starts with L bundle copies. Each time the source node encounters a new node, it hands one of the L copies and reduces its number of copies left by one. In the *binary spray* scheme, the source node also starts with L bundle copies. But, whenever a node with $L > 1$ copies encounters a new node, it hands half of the copies that it stores in its buffer.

For both spraying schemes, when a node carries only 1 bundle copyleft, it only forwards it to the final destination. This is the "wait phase." Figs. 1.9 and 1.10 illustrate, respectively, Source Spray and Wait, and Binary Spray and Wait for routing protocol operation. It is important to notice that each bundle has a header field indicating the "number of copies" it represents. As expected, the (actual) bundle is not replicated within a node's buffer.

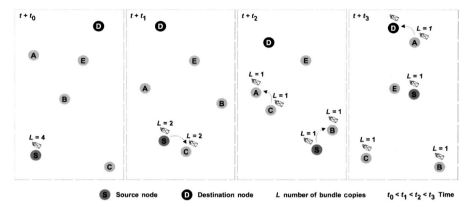

Fig. 1.10 Illustration of the Binary Spray and Wait routing protocol operation.

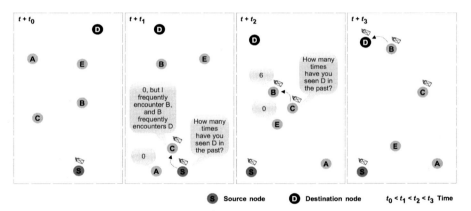

Fig. 1.11 Example of using the historic of node encounters and transitivity information, as a basis for PRoPHET routing decisions.

PRoPHET (Lindgren et al., 2004, 2011) considers that network nodes move in a nonrandom pattern and apply "probabilistic routing." This protocol uses the concepts of the history of node encounters and the transitivity property illustrated in Fig. 1.11, which can be defined as follows. The history of node encounters defines $P_{(a,b)}$ as the probability that two nodes, a and b, meet each other. It is calculated as shown in Eq. (1.2), where P_{init} is an initialization constant:

$$P_{(a, b)} = P_{(a, b)\text{old}} + \left(1 - P_{(a, b)\text{old}}\right) \times P_{\text{init}}, \quad P_{\text{init}} \in [0, 1] \tag{1.2}$$

This probability, called delivery predictability, is renewed with each successive contact between the same nodes. It may also decay over time if these nodes don't meet.

Thus, between two meetings the predictability value ages. The aging equation is shown in Eq. (1.3), where γ is the aging constant and k is the elapsed time since the last aging:

$$P_{(a,b)} = P_{(a,b)\text{old}} \times \gamma^{K}, \quad \gamma \in [0, 1) \tag{1.3}$$

The transitivity property $P_{(a,c)}$ of the delivery predictability is based on the observation that if node A frequently encounters node B, and node B frequently encounters node C, then node B probably would be a good node to forward bundles that are destined to node C and vice versa. Transitivity is calculated as shown in Eq. (1.4), where β is scaling constant that decides the impact of the transitivity on the delivery predictability:

$$P_{(a,c)} = P_{(a,c)\text{old}} + \left(1 - P_{(a,c)\text{old}}\right) \times P_{(a,b)} \times P_{(b,c)} \times \beta, \quad \beta \in [0, 1] \tag{1.4}$$

The delivery predictability metric is calculated in all network nodes for each known destination. It is updated each time a node is encountered and it is used to decide whether or not to forward bundles at communication opportunities. Thus, when a contact opportunity occurs, the involved nodes exchange the delivery predictability information stored on them. Both nodes use this information to update their estimated delivery predictability information. Then, based on this information and the bundle's destination, a bundle is forwarded to the other node if the delivery predictability of the destination of that bundle is higher at the other node.

MaxProp (Burgess et al., 2006) is a generic routing protocol for vehicular DTNs. It performs routing by prioritizing the scheduling of bundles forwarded at contact opportunities, and also the scheduling of bundles to be dropped upon buffer overflow. To calculate these priorities the protocol considers the historical data of path probabilities to nodes, which is determined as follows. Each node i holds a vector f^i that represents its likelihood to meet every other node available in the network. The likelihood of node i meeting node j is initially set according to Eq. (1.5), where n represents the total number of nodes that exist in the network. Each time i meet j the f_j^i entry of the vector is increased by 1. Then, the vector of likelihood is renormalized using incremental averaging according to Eq. (1.6). As a result, nodes that are seen infrequently score lower values over time.

The access to the information contained in the vector of likelihood from the other nodes allows determining which bundles should be sent and deleted from the buffer, based on path calculation using subsequent likelihoods:

$$f_j^i = \frac{1}{n-1} \tag{1.5}$$

$$\sum_{j=1}^{n-1} f_j^i = 1 \tag{1.6}$$

This protocol also includes three principal complementary mechanisms, namely, head starts for new bundles, lists of previous intermediaries, and system-wide acknowledgments. To guarantee that all bundles have a chance of being propagated in the network, a "head start" is given to new bundles. This means that priority is given to the transmission of these bundles. Lists of previous intermediaries are maintained to prevent bundles from being sent to the same node again. System-wide acknowledgments are propagated through the network to notify nodes to eliminate redundant copies of the bundles that have already been delivered to their destination.

MaxProp has been implemented on a real bus-based DTN network called UMass DieselNet (UMass Diverse Outdoor Mobile Environment (DOME) Project, 2006). This protocol has been shown to perform well in a wide variety of DTN environments.

1.5   Conclusion

In this chapter, a general introduction to the delay and disruption tolerant networks topic was addressed. A DTN is an overlay network that aims to enable communications in disruptive network conditions due to environmental or operational factors. DTNs present a big research potential and social impact contributing to enable services and applications in a variety of environments including disaster-recovery/rescue operations, vehicular communications, military battlefields, habitat monitoring, deep-space communications, underwater networks, social networks, and noninteractive Internet access connectivity in rural and developing areas.

This chapter overviewed the state of the art in the area of delay- and disruption-tolerant networking, focusing on DTN architecture concepts, application scenarios, and routing protocols. It aimed to stimulate upcoming studies and dissemination of these technologies, serving as a starting point to researchers wishing to conduct studies in this area as well as experts on the topic.

In the upcoming chapters, key topics about DTN will be addressed and studied in detail.

Acknowledgments

This work has been supported by FCT/MCTES through national funds and when applicable co-funded EU funds under the Project UIDB/EEA/50008/2020; by Brazilian National Council for Research and Development (CNPq) via Grant No. 309335/2017-5.

References

Akyildiz, I.F., Akan, Ö.B., Chen, C., Fang, J., Su, W., 2003. Interplanetary internet: state-of-the-art and research challenges. Comput. Netw. 43, 75–112.

Asplund, M., Nadjm-Tehrani, S., Sigholm, J., 2009. Emerging Information Infrastructures: Cooperation in Disasters. Lecture Notes in Computer Science, Critical Information Infrastructure Security, Springer, Berlin/Heidelberg.

Atkins, C., Guo, J., 2008. Creating a viable vehicular network: a VTP convergence layer for DTN. In: The First International Conference on Wireless Access in Vehicular Environments (WAVE 2008), December 8-9, 2008, Dearborn, Michigan, USA.

Balasubramanian, A., Levine, B.N., Venkataramani, A., 2007. DTN routing as a resource allocation problem. In: ACM SIGCOMM 2007, August 27-31, 2007, Kyoto, Japan,
pp. 373–384.

Broadband & Wireless Networking Laboratory (BWN LAB), 2004. Underwater Acoustic Sensor Networks (UW-ASN). Available from: http://www.ece.gatech.edu/research/labs/bwn/
UWASN/. (Accessed January 2013).

Burgess, J., Gallagher, B., Jensen, D., Levine, B., 2006. MaxProp: routing for vehicle-based
disruption-tolerant networks. In: 25th IEEE International Conference on Computer Communications (INFOCOM 2006), April 23-29, 2006, Barcelona, Catalunya, Spain, pp. 1–11.

Burgess, J., Bissias, G.D., Corner, M., Levine, B.N., 2007. Surviving attacks on disruption-
tolerant networks without authentication. In: Eight ACM International Symposium on
Mobile Ad Hoc Networking and Computing (MobiHoc 2007), September 9-14, 2007,
Montreal, Quebec, Canada, pp. 61–70.

Burleigh, S., Hooke, A., Torgerson, L., Fall, K., Cerf, V., Durst, B., Scott, K., Weiss, H., 2003.
Delay-tolerant networking: an approach to interplanetary internet. IEEE Commun. Mag.
41 (6), 128–136.

CCSU, 2007. Affordable Internet Access Project for Underdeveloped Communities. Available
from: http://www.ccsu.edu/technology/aitislab/ghana/index.htm. (Accessed January
2013).

Cerf, V., Burleigh, S., Hooke, A., Torgerson, L., Durst, R., Scott, K., Fall, K., Weiss, H., April
2007. Delay-tolerant networking architecture. RFC 4838. Available from: http://www.rfc-
editor.org/rfc/rfc4838.txt.

Chaintreau, A., Hui, P., Crowcroft, J., Diot, C., Gass, R., Scott, J., 2006. Impact of human mobility on the design of opportunistic forwarding algorithms. In: 25th IEEE International Conference on Computer Communications (INFOCOM 2006), April 23-29, 2006, Barcelona,
Catalunya, Spain, pp. 1–13.

Chen, L.-J., Chen, Y.-C., Sun, T., Sreedevi, P., Chen, K.-T., Yu, C.-H., Chu, H.-H., 2007. Finding self-similarities in opportunistic people networks. In: 26th IEEE International Conference on Computer Communications (IEEE INFOCOM 2007), May 2007, pp. 2286–2290.

DARPA, 2013. Defense Advanced Research Projects Agency (DARPA). Available from: http://
www.darpa.mil/. (Accessed January 2013).

Defense Advanced Research Projects Agency (DARPA), STO: Solicitations—Disruption Tolerant Networking (DTN). Available from: http://www.darpa.mil/sto/solicitations/DTN/.
(Accessed March 2010).

Demmer, M., Ott, J., November 2008. Delay Tolerant Networking TCP Convergence Layer
Protocol. Available from: http://tools.ietf.org/html/draft-irtf-dtnrg-tcp-clayer-02.

Dong, F., Hu, Y., Tong, M., Ran, X., 2009. Supporting emergency service by retasking delay-
tolerant network architecture. In: Fifth International Conference on Mobile Ad-hoc and
Sensor Networks (MSN 2009), December 14-16, 2009, Fujian, China.

Doria, A., Uden, M., Pandey, D.P., 2002. Providing connectivity to the Saami Nomadic Community. In: 2nd International Conference on Open Collaborative Design for Sustainable
Innovation, December 2002, Bangalore, India.

DTNRG, 2002. Delay Tolerant Networking Research Group. Available from: http://www.
dtnrg.org/wiki. (Accessed January 2013).

Erramilli, V., Chaintreau, A., Crovella, M., Diot, C., 2007. Diversity of forwarding paths in
pocket switched networks. In: 7th ACM SIGCOMM Internet Measurement Conference
2007, October 24-26, 2007, San Diego, California, USA, pp. 161–174.

Fall, K., 2003. A delay-tolerant network architecture for challenged internets. In: ACM
SIGCOMM 2003, August 25-29, 2003, Karlsruhe, Germany, pp. 27–34.

Fall, K., Farrell, S., 2008. DTN: an architectural retrospective. IEEE J. Sel. Areas Commun. 26 (5), 828–836.

Fall, K., Hong, W., Madden, S., 2003. Custody Transfer for Reliable Delivery in Delay Tolerant Networks. Intel Research, Berkeley, California.

Fall, K., Burleigh, S., Doria, A., Ott, J., Young, D., 2009. The DTN URI Scheme. March 28. Available from: http://tools.ietf.org/html/draft-irtf-dtnrg-dtn-uri-scheme-00.

Farrell, S., Symington, S.F., Weiss, H., Lovell, P., March 2009. Delay-Tolerant Networking Security Overview. Available from: http://tools.ietf.org/html/draft-irtf-dtnrg-sec-overview-06.

Fiore, M., Leonardi, A., Matera, A., Casetti, C., Chiasserini, C.-F., Palazzo, S., 2006. Information delivery with geographical forwarding in vehicular wireless networks. In: NEWCOM-ACoRN Joint Workshop, September 20-22, 2006, Vienna, Austria.

Franck, L., Gil-Castineira, F., 2007. Using delay tolerant networks for Car2Car communications. In: IEEE International Symposium on Industrial Electronics 2007 (ISIE 2007), 4-7 June 2007, Vigo, Spain, pp. 2573–2578.

Glance, N., Snowdon, D., Meunier, J.-L., 2001. Pollen: using people as a communication medium. Comput. Netw. 35, 429–442.

Guo, S., Falaki, M.H., Oliver, E.A., Rahman, S.U., Seth, A., Zaharia, M.A., Keshav, S., 2007. Very low-cost internet access using KioskNet. ACM SIGCOMM Comput. Commun. Rev. 37, 95–100.

Haas, Z.J., Small, T., 2006. Evaluating the capacity of resource-constrained DTNs. In: International Wireless Communications & Mobile Computing Conference (IWCMC 2006)—International Workshop on Delay Tolerant Mobile Networks (DTMN), July 3-6, 2006, Vancouver, Canada, pp. 545–550.

Haggle, 2006. Haggle—An innovative Paradigm for Autonomic Opportunistic Communication. Available from: http://www.haggleproject.org/. (Accessed January 2013).

Hofmann-Wellenhof, B., Lichtenegger, H., Collins, J., 2001. Global Positioning System: Theory and Practice. Springer Wien New York, Austria.

Hui, P., Chaintreau, A., Scott, J., Gass, R., Crowcroft, J., Diot, C., 2005. Pocket switched networks and human mobility in conference environments. In: ACM SIGCOMM 2005—Workshop on Delay Tolerant Networking and Related Networks (WDTN-05), August 22-26, 2005, Philadelphia, Pennsylvania, USA, pp. 244–251.

Hui, P., Crowcroft, J., Yoneki, E., 2008. BUBBLE Rap: social-based forwarding in delay tolerant networks. In: 9th ACM International Symposium on Mobile Ad Hoc Networking and Computing (MobiHoc 2008), May 27-30, 2008, Hong Kong.

Internet Society IPN Special Interest Group, 2011. InterPlaNetary Internet Project (IPN). Available from: http://ipnsig.org/home.htm. (Accessed June 2011).

IRTF, 2013. Internet Research Task Force (IRTF). Available from: http://www.irtf.org/. (Accessed January 2013).

Jain, S., Fall, K., Patra, R., 2004. Routing in a delay tolerant network. In: ACM SIGCOMM 2004 Conference on Applications, Technologies, Architectures, and Protocols for Computer Communication, August 30-September 3, 2004, Portland, Oregon, USA, pp. 145–158.

Jain, S., Shah, R., Brunette, W., Borriello, G., Roy, S., 2006. Exploiting mobility for energy efficient data collection in wireless sensor networks. ACM/Kluwer Mob. Netw. Appl. 11, 327–339.

Jones, E.P.C., Ward, P.A.S., 2012. Routing strategies for delay-tolerant networks. ACM Comput. Commun. Rev., 107–114. https://doi.org/10.1145/2387191.2387207.

Jormakka, J., Jormakka, H., Väre, J., 2008. A Lightweight Management System for a Military Ad Hoc Network. Lecture Notes in Computer Science, Information Networking. Towards Ubiquitous Networking and Services, Springer, Berlin/Heidelberg.

Juang, P., Oki, H., Wang, Y., Martonosi, M., Peh, L.S., Rubenstein, D., 2002. Energy-efficient computing for wildlife tracking: design tradeoffs and early experiences with ZebraNet. ACM SIGOPS Oper. Syst. Rev. 36, 96–107.

Katz, I., 2007. A delay-tolerant networking framework for mobile underwater acoustic networks. In: Fifteenth International Symposium on Unmanned Untethered Submersible Technology (UUST'07), August 2007, Durham, NH, USA.

Leontiadis, I., Mascolo, C., 2007. GeOpps: geographical opportunistic routing for vehicular networks. In: IEEE International Symposium on a World of Wireless, Mobile and Multimedia Networks 2007 (WoWMoM 2007), 18-21 June 2007, Espoo, Finland, pp. 1–6.

Lindgren, A., Doria, A., Schelén, O., 2004. Probabilistic routing in intermittently connected networks. In: The First International Workshop on Service Assurance With Partial and Intermittent Resources (SAPIR 2004), August 1-6, 2004, Fortaleza, Brazil, pp. 239–254.

Lindgren, A., Doria, A., Davies, E., Grasic, S., 2011. Probabilistic Routing Protocol for Intermittently Connected Networks. April 3. Available from: http://tools.ietf.org/html/draft-irtf-dtnrg-prophet-09.

Lochert, C., Hartenstein, H., Tian, J., Füssle, H., Hermann, D., Mauve, M., 2003. A routing strategy for vehicular ad hoc networks in city environments. In: 2003 IEEE Intelligent Vehicles Symposium, June 9-11, 2003, Ohio, USA, pp. 156–161.

MITRE, 1997. Applying Systems Engineering and Advanced Technology to Critical National Problems. Available from: http://www.mitre.org/. (Accessed January 2013).

NASA Jet Propulsion Laboratory, 2013a. Jet Propulsion Laboratory. Available from: http://www.jpl.nasa.gov/. (Accessed January 2013).

NASA Jet Propulsion Laboratory, 2013b. Volcano Sensorweb. Available from: http://sensorwebs.jpl.nasa.gov/. (Accessed January 2013).

Nelson, S.C., 2008. Encounter-Based Routing in Disaster Recovery Networks (Degree of Master of Science in Computer Science M.S. thesis). Graduate College of the University of Illinois at Urbana-Champaign.

Nelson, S.C., Bakht, M., Kravets, R., 2009. Encounter-based routing in DTNs. In: 28th IEEE Conference on Computer Communications (INFOCOM 2009), April 19-25, 2009, Rio de Janeiro, Brazil.

Anon., 2008. Networking for Communications Challenged Communities: Architecture, Test Beds and Innovative Alliances. Available from: http://www.n4c.eu/. (Accessed January 2013).

Olajide, T., Washington, A.N., 2008. Epidemic modeling of military networks using group and entity mobility models. In: 5th International Conference on Information Technology: New Generations (ITNG 2008), April 7-9, 2008, Las Vegas, Nevada, USA, pp. 1303–1304.

Partan, J., Kurose, J., Levine, B.N., 2006. A survey of practical issues in underwater networks. In: 1st ACM International Workshop on Underwater Networks (WUWNet 2006), in Conjunction With ACM MobiCom 2006, September 25, 2006, Los Angeles, California, USA, pp. 17–24.

Pentland, A., Fletcher, R., Hasson, A., 2004. DakNet: rethinking connectivity in developing nations. IEEE Comput. 37, 78–83.

Pereira, P.R., Casaca, A., Rodrigues, J.J.P.C., Soares, V.N.G.J., Triay, J., Cervelló-Pastor, C., 2012. From delay-tolerant networks to vehicular delay-tolerant networks. IEEE Commun. Surv. Tutorials 14, 1166–1182.

Pitkänen, M., Keränen, A., Ott, J., 2008. Message fragmentation in opportunistic DTNs. In: Second International IEEE WoWMoM Workshop on Autonomic and Opportunistic Communications (AOC 2008), 23 June 2008, Newport Beach, California, USA, pp. 1–7.

Ramanathan, R., Basu, P., Krishnan, R., 2007. Towards a formalism for routing in challenged networks. In: Second ACM MobiCom Workshop on Challenged Networks (CHANTS 2007), September 14, 2007, Montreal, Quebec, Canada, pp. 3–10.

Schwingenschlogl, C., Kosch, T., 2002. Geocast enhancements of AODV for vehicular networks. ACM SIGMOBILE Mob. Comput. Commun. Rev. 6, 96–97.

Scott, K., 2005. Disruption tolerant networking proxies for on-the-move tactical networks. In: IEEE Military Communication Conference (MILCOM 2005), October 17-20, 2005 Atlantic City, New Jersey, USA, pp. 3226–3231.

Scott, K., Burleigh, S., November 2007. Bundle protocol specification. RFC 5050. Available from: http://www.rfc-editor.org/rfc/rfc5050.txt.

Seth, A., Kroeker, D., Zaharia, M., Guo, S., Keshav, S., 2006. Low-cost communication for rural internet kiosks using mechanical backhaul. In: 12th ACM International Conference on Mobile Computing and Networking (MobiCom 2006), September 24-29, 2006, Los Angeles, CA, USA, pp. 334–345.

Shah, R.C., Roy, S., Jain, S., Brunette, W., 2003. Data MULEs: modeling a three-tier architecture for sparse sensor networks. In: First IEEE International Workshop on Sensor Network Protocols and Applications (SNPA 2003), May 11, 2003, Anchorage, AK, USA, pp. 30–41.

Shao, Y., Liu, C., Wu, J., 2009. Delay-tolerant networks in VANETs. In: Olariu, S., Weigle, M. C. (Eds.), Vehicular Networks: From Theory to Practice. Chapman & Hall/CRC Computer & Information Science Series.

Shen, J., Moh, S., Chung, I., 2008. Routing protocols in delay tolerant networks: a comparative survey. In: The 23rd International Technical Conference on Circuits/Systems, Computers and Communications (ITC-CSCC 2008), July 6-9, 2008, Shimonoseki City, Yamaguchi-Pref., Japan, pp. 1577–1580.

Sherwood, R., Chien, S., 2007. Sensor webs for science: new directions for the future. In: AIAA Infotech@Aerospace 2007, May 7-10, 2007, Rohnert Park, CA.

SiPLABoratory, University of Algarve, 2008. Project UAN. Available from: http://www.siplab.fct.ualg.pt/proj/uan.shtml. (Accessed January 2013).

Small, T., Haas, Z.J., 2003. The shared wireless infostation model—a new ad hoc networking paradigm (or where there is a whale, there is a way). In: 4th ACM International Symposium on Mobile Ad Hoc Networking and Computing (MobiHoc 2003), June 1-3, 2003, Annapolis, MD, USA.

Soares, V.N.G.J., Farahmand, F., Rodrigues, J.J.P.C., 2009. A layered architecture for vehicular delay-tolerant networks. In: Fourteenth IEEE Symposium on Computers and Communications (ISCC'09), July 5-8, 2009, Sousse, Tunisia, pp. 122–127.

Socolofsky, T., Kale, C., January 1991. A TCP/IP tutorial. RFC 1180. Available from: http://tools.ietf.org/html/rfc1180.

Spyropoulos, T., Psounis, K., Raghavendra, C.S., 2004. Single-copy routing in intermittently connected mobile networks. In: First IEEE Communications Society Conference on Sensor and Ad Hoc Communications and Networks (IEEE SECON 2004), October 4-7, 2004, Santa Clara, CA, USA, pp. 235–244.

Spyropoulos, T., Psounis, K., Raghavendra, C.S., 2005. Spray and wait: an efficient routing scheme for intermittently connected mobile networks. In: ACM SIGCOMM 2005—Workshop on Delay Tolerant Networking and Related Networks (WDTN-05), August 22-26, 2005, Philadelphia, PA, USA, pp. 252–259.

Spyropoulos, T., Psounis, K., Raghavendra, C., 2008a. Efficient routing in intermittently connected mobile networks: the single-copy case. IEEE/ACM Trans. Networking 16, 63–76.

Spyropoulos, T., Psounis, K., Raghavendra, C.S., 2008b. Efficient routing in intermittently connected mobile networks: the multiple-copy case. IEEE/ACM Trans. Networking 16, 77–90.

Symington, S., Farrell, S., Weiss, H., Lovell, P., 2011. Bundle Security Protocol Specification. March 11. Available from: http://tools.ietf.org/html/draft-irtf-dtnrg-bundle-security-19.

Tetherless Computing Lab—University of Waterloo, The KioskNet Project. Available from: http://blizzard.cs.uwaterloo.ca/tetherless/index.php/KioskNet. (Accessed June 2011).

Trinity College Dublin, 2006. Sensor Networking With Delay Tolerance (SeNDT). Available from: http://down.dsg.cs.tcd.ie/sendt/. (Accessed January 2013).

UAN, 2007. Underwater Acoustic Network. Available from: http://www.ua-net.eu/. (Accessed January 2013).

UMass Diverse Outdoor Mobile Environment (DOME) Project, 2006. DieselNet. Available from: http://prisms.cs.umass.edu/dome/umassdieselnet. (Accessed January 2013).

UMass Diverse Outdoor Mobile Environment (DOME) Project, 2007. UMass TurtleNet. Available from: http://prisms.cs.umass.edu/dome/turtlenet. (Accessed January 2013).

United Villages, Inc, 2005. First Mile Solutions. Available from: http://www.firstmilesolutions.com/. (Accessed January 2013).

Vahdat, A., Becker, D., 2000. Epidemic routing for partially-connected ad hoc networks. Handbook of Systemic Autoimmune Diseases. Duke University. http://issg.cs.duke.edu/epidemic/epidemic.pdf.

Vallina-Rodriguez, N., Hui, P., Crowcroft, J., 2009. Has anyone seen my goose?—social network services in developing regions. In: IEEE International Conference on Social Computing (SocialCom-09)—Workshop on Social Mobile Web (SMW'09), August 29th-31st, 2009, Vancouver, Canada.

Wizzy Digital Courier, 2003. Wizzy Digital Courier—Leveraging Locality. Available from: http://www.wizzy.org.za/. (Accessed June 2011).

Zhang, Z., 2006. Routing in intermittently connected mobile ad hoc networks and delay tolerant networks: overview and challenges. IEEE Commun. Surv. Tutorials 8, 24–37.

Zhang, P., Sadler, C.M., Lyon, S.A., Martonosi, M., 2004. Hardware design experiences in ZebraNet. In: 2nd International Conference on Embedded Networked Sensor Systems, 2004, pp. 227–238.

Zhang, X., Neglia, G., Kurose, J., Towsley, D., 2007. Performance modeling of epidemic routing. Comput. Netw. 51, 2867–2891.

Zhao, W., Ammar, M.H., 2003. Message ferrying: proactive routing in highly-partitioned wireless ad hoc networks. In: The Ninth IEEE Workshop on Future Trends of Distributed Computing Systems, May 28-30, 2003, San Juan, Puerto Rico, pp. 308–314.

Zhao, W., Ammar, M., Zegura, E., 2004. A message ferrying approach for data delivery in sparse mobile ad hoc networks. In: Press, A. (Ed.), The Fifth ACM International Symposium on Mobile Ad Hoc Networking and Computing (MobiHoc 2004), May 24-26, 2004, Roppongi Hills, Tokyo, Japan, pp. 187–198.

Part One

Types of delay-tolerant networks (DTNs)

Delay-tolerant networks (DTNs) for satellite communications

C. Caini
University of Bologna, Bologna, Italy

2.1 Introduction

Delay/disruption-tolerant networking (DTN) has always been associated with space communications. It originated from a generalization of requirements identified for interplanetary networking (IPN), to encompass all "challenged networks." These networks are basically all networks wherein the ordinary Transmission Control Protocol/Internet Protocol (TCP/IP) architecture fails to provide satisfactory performance due to the presence of one or more of the following impairments: long delays, disruption, intermittent links, network partitioning, etc. (Fall and Farrell, 2008; McMahon and Farrell, 2009; DTNWG, n.d.; Warthmann, 2015). Before going on, it is necessary to dispel the myth that DTN is just for mobile ad hoc networks, where "contacts," i.e., period of connectivity between DTN nodes, are always "opportunistic," i.e., DTN is for all challenged networks, which are either deterministic or random, ranging from interplanetary Internet to wireless sensor networks.

Among the challenged networks there are also the satellite networks, i.e., the networks that include one or more satellite links. Low Earth orbit (LEO) satellite networks were immediately recognized as a perfect candidate for DTN application because of the satellite link intermittency, with many references to them in the RFC4838 (Cerf et al., 2007), which describes the DTN architecture based on the Bundle Protocol (BP), defined in RFC5050 (Scott and Burleigh, 2007). LEO networks have in common with deep space networks the same kind of connectivity, which can be defined as intermittent scheduled. In other words, in deep space and LEO satellite networks, the communication opportunities, or "contacts," are known in advance, i.e., are fully deterministic since they are related to the orbital characteristics of planets and space assets. This kind of connectivity is addressed by specific DTN solutions, such as "scheduled contacts," i.e., transport protocol connections that start and stop at the beginning and at the end of contacts. In these kinds of networks, routing must be designed to cope with scheduled contacts and not with opportunistic connectivity, as in other challenged networks; therefore, specific routing algorithms, such as the contact graph routing (CGR) designed by NASA, are necessary (Araniti et al., 2015; Burleigh et al., 2016; Bezirgiannidis et al., 2016). By contrast, geosynchronous (GEO) satellite networks cannot be immediately classified as challenged networks because they offer continuous connectivity, at least for fixed terminals. However, they are "challenged networks" too, not because of the satellite link intermittency, but

Advances in Delay-tolerant Networks (DTNs). https://doi.org/10.1016/B978-0-08-102793-6.00002-3

because of its long propagation delay. The round trip time (RTT) of a GEO satellite connection is in the order of 600 ms, including some processing time, compared with the few milliseconds of local connections and with about 150 ms of intercontinental connections for terrestrial wired links. As TCP congestion control is ACK-based, the longer the RTT, the worse the performance, with a severe penalization of GEO satellite connections (Caini and Firrincieli, 2004). The usual solution is based on the use of performance-enhancing proxies (PEPs) to isolate the satellite link from the rest of the network, as shown in RFC 3135 (Border et al., 2001). As GEO networks require specific solutions outside the ordinary TCP/IP architecture, they are fully entitled to be considered "challenged networks," although GEO satellite contacts are generally continuous. DTN can, therefore, be successfully applied to them (Caini et al., 2008b, 2011; Caini and Firrincieli, 2011) and will be recalled here as well.

In this chapter, the potential advantages of DTN when applied to satellite communications are evaluated for both GEO and LEO constellations. After a brief introduction on DTN, the chapter discusses the case of GEO constellations first, by highlighting an important analogy between DTN and TCP-splitting PEP architectures. Then, the performance of DTN BP architecture is compared with that of TCP-splitting PEPs and TCP end-to-end. The analysis, performed through a GNU/Linux testbed, considers first continuous links (fixed satellite terminals) and then disrupted links (mobile terminals). In both cases, a variety of scenarios are considered. The chapter continues by considering LEO constellations. The analysis is focused on Earth observation satellites, but conclusions are more general. First, the case of an LEO satellite connected to its control center by means of multiple ground stations is considered. Then the analysis is extended to include also a GEO satellite relay, which is a more innovative but also a more challenging solution for the DTN routing, because of the presence of parallel paths (via ground stations or via GEO relay). As in the GEO case, the analysis is supported by results achieved through a testbed running real implementations of both BP and CGR. Conclusions are finally drawn at the end of the chapter.

2.2 DTN architecture

2.2.1 Bundle protocol

The aim of the BP DTN architecture (Fall and Farrell, 2008; McMahon and Farrell, 2009; DTNWG, n.d.; Cerf et al., 2007) is to support communication in challenging environments. To this end, it relies on the new "Bundle layer" placed between Application and lower layers (usually, Transport) and on its related protocol, the BP (Scott and Burleigh, 2007). The BP interfaces toward lower layers are called "convergence layer adapters" (CLAs). Various CLAs have been defined, including those for TCP (Demmer et al., 2014), UDP (Kruse et al., 2014), and the Licklider Transmission Protocol (LTP) (Burleigh, 2013); the last is primarily designed to cope with the long delays of space networks, which prevent the use of TCP, and is specified in RFC5325 and 5326 (Burleigh et al., 2008; Ramadas et al., 2008). Note that it is not necessary to install the BP on all network nodes, but just on end-points and some

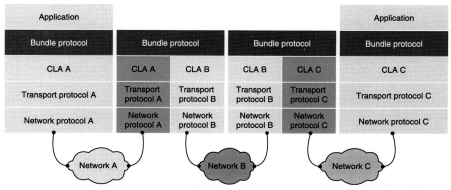

Fig. 2.1 DTN bundle protocol architecture and protocol stack.

selected intermediate nodes, generally located at the borders of homogeneous parts of a heterogeneous network (Fig. 2.1). Nodes with BP become DTN nodes, the others being irrelevant for the DTN architecture. The BP (Scott and Burleigh, 2007) is in charge of the transmission of "bundles," i.e. (usually large) packets of data, such as an image file or a part of it, between DTN nodes.

2.2.1.1   DTN as an overlay

By installing the BP on end-points and some intermediate nodes, the original end-to-end path is divided into multiple DTN hops. The end-to-end Transport semantics is redefined, being now confined inside each DTN hop. This makes it possible to use on each DTN hop the transport protocol (or more generally, the protocol stack) best suited to it. This possibility to match protocols to channel characteristics is the first great advantage offered by the BP DTN architecture. In particular, this enables the use of specialized TCP variants, such as Hybla (Caini and Firrincieli, 2004), on GEO satellite links, or specialized protocols, such as LTP, on LEO satellite and deep space hops. The challenges to be tackled are long propagation delay in GEO, very long propagation delay in deep space, low and/or highly asymmetrical bandwidth in both LEO and deep space, and potential random losses caused by noise on radio links in all these environments.

2.2.1.2   Store-and-forward and custody option

The second advantage of the BP DTN architecture consists of its resilience to disruption and network partitioning. This stems from the possibility of storing bundles in DTN nodes and the store-and-forward policy used for bundle transmission. Bundles, which can be much larger than ordinary IP packets, are first entirely received before being forwarded to the next "proximate" DTN node. If the link to the next node is not immediately available because of link intermittency, they are stored in local databases until the link is available again, even for relatively long periods (e.g., 1 day).

To increase reliability, the bundle source can solicit other nodes on the path to take "custody" of a bundle by setting a bit in the bundle header (Fall et al., 2003). If one of the following nodes accepts the request, it becomes the "custodian," i.e., the node in charge of reliable delivery to another custodian or the destination. The new custodian must notify the old one of the custody acceptance, in a sort of relay race. Once notified, the old custodian is left free to cancel the bundle, which is advantageous when there are memory constraints or security issues. By contrast, if the current custodian does not receive a "custody acceptance" or a "bundle delivered" notification, it has to retransmit the bundle after a timeout expires. The custody option offers a high degree of robustness against disruption thanks to its retransmission feature, and also against hardware and software failures (including reboots), as bundles are not canceled from persistent memory (e.g., a hard disk) until custody has not been successfully passed to another node (or bundle lifetimes expire). The custody option also offers a way to enforce congestion control on intermediate nodes, a challenging problem in DTN networks that is only rarely addressed (Bisio et al., 2009).

2.2.1.3   Fragmentation

Bundle fragmentation is one of the most characteristic features of DTN. It can be either "proactive" to match bundle dimension with limited contact volumes (the maximum amount of data that can be transferred during a contact, given by the product of the contact length and the link bandwidth), known a priori, or "reactive," to avoid retransmitting already acknowledged data in the presence of unpredictable link disruption. Concerning reactive fragmentation, it must be noted that the possibility of restarting a download from an intermediate point is quite common in many file transfer programs. The novelty here is that this feature, like many others, is now offered to upper layers, i.e., to DTN applications, as a "service" by the BP layer itself. In other words, in contrast to TCP applications, DTN applications do not need to include any code to implement it.

2.2.1.4   Intermittent links

DTN applications no longer need to worry about link intermittency and disruption. The introduction of the Bundle layer decouples the decision of sending some data, such as a file, from its actual transmission. The former is in charge of the DTN application, the latter of BP. Cerf et al. (2007) in RFC4838 define several kinds of "contacts" between DTN nodes: persistent (or "always on"), "on-demand," and intermittent. Intermittent contacts are further classified into scheduled, predicted, and opportunistic, depending on the grade of predictability. Scheduled contacts are deterministic, as their availability is known in advance, while predicted and opportunistic contacts are both random. An advantage of scheduled links is that they can be opened and closed by BP at precisely the beginning and end of known contacts, thus improving the link usage efficiency, especially when links are relatively short. This is beneficial in LEO satellites and deep-space communications, both characterized by scheduled intermittent connectivity (with short contacts in LEOs and with scarce, and therefore, precious, bandwidth resources in deep space).

2.2.2   Bundle protocol security

Security is one of the most important aspects of DTN for a variety of reasons. In particular, in many challenged networks the usual security solutions based on keys exchange with remote security servers cannot be applied because of long delays and network partitioning. Moreover, in space applications, it is very important to discard fake data sent by an intruder before they are transmitted on space DTN hops, and not just at reception, as usual, to preserve the scarce bandwidth of space links for authorized traffic only. These and other requirements have led to the definition of a security extension of BP, called Bundle Security Protocol (BSP) in RFC6257 (Symington et al., 2011). Some peculiar BSP aspects defined in this document, however, proved soon hardly compatible with BP fragmentation and other features. Therefore, the security extension has been redesigned and at present, this new version is in an advanced phase of standardization, under the name of Bundle Protocol Security (BPSec) (Birrane and McKeever, 2019).

A key element of BPSec is that different "security services" can be applied to different bundle blocks (the payload block or other extension blocks) because blocks in a bundle represent different kinds of information with different security requirements. Therefore, the two types of security blocks envisaged by BPSec, namely, the Block Integrity Block (BIB) and the Block Confidentiality Block (BCB), do not apply to the entire bundle but only to specific blocks, called "security targets" in the BPSec context. As a result, multiple instances of both BIB and BCB can be added to the same bundle, applying to different security targets, thus allowing for a high level of flexibility.

It is also worth noting that not all DTN nodes are required to implement BPSec, which increases flexibility too. Those that implement BPSec are called "security-aware" nodes and are the only nodes authorized to deal with security blocks. To allow security-unaware nodes to process bundles with security extensions (e.g., for routing), security blocks added to the original bundles must have processing flags appropriately set to prevent bundle discarding.

2.2.3   Bundle protocol implementations

At present, the most important BP implementations are DTN2 and Interplanetary Overlay Network (ION) (Burleigh, 2007), both available as open-source code from SourceForge (DTN2, n.d.; ION, n.d.). Although both of them are compliant with DTN RFCs (Cerf et al., 2007; Scott and Burleigh, 2007), DTN2 is more general while ION is more focused on space environments, as detailed in the following section. Among the other implementations, we cite IBR-DTN, developed at the Technical Braunschweig University, which is available for a wide variety of operating systems, including Android (IBR-DTN, n.d.).

2.2.3.1 DTN2: The bundle protocol reference implementation

DTN2 is the "reference" BP implementation and is still widely adopted, although no more officially updated from ver 2.9.0, which dates back to 2012. It has been designed as an experimental platform to test the BP features on a real implementation, which could also be used for real-world deployment. It supports many convergence layer adapters, such as TCP and UDP. In addition to the BP, the DTN2 package contains some DTN basic applications (dtnping, dtnsend, etc.) and also the DTNperf tool (version 2) for DTN performance evaluation used in many of our experiments. DTN2 implements almost all the features described in RFC4838, except scheduled links, which by contrast are implemented in ION. DTN2 offers a choice of different DTN routing policies, for both static and dynamic routing. Quite interesting, NASA MSFC (Marshall Space Flight Center) provides bug fixes and extensions to the official DTN2 version by releasing patches, downloadable from the DTN2 website (DTN2, n.d.), under the branch "/support/patches." This because the patched version of DTN2 is used on Earth stations involved in DTN experiments with the International Space Station (ISS) and is thus necessary to maintain full compatibility with the ION implementation, used on board of ISS.

2.2.3.2 ION: The bundle protocol implementation by NASA JPL

ION is the BP implementation developed by NASA Jet Propulsion Laboratory (JPL) with contributions from Ohio and other Universities. ION includes several convergence-layer adapters, such as TCP, UDP, and LTP (all interoperable with the patched version of DTN2). The support of LTP is essential in deep space environments because LTP has been designed for very high delay links where TCP cannot operate. The ION version includes some recent advancements, designed to improve performance in the presence of high losses (Alessi et al., 2018). ION supports scheduled links, which are of great importance in all space networks. This support was in the past limited to LTP CL, but from 2018, with the release of ION 3.6.0, it has been extended to all CLs. ION supports proactive but not reactive bundle fragmentation, although is compatible with both. The ION package also includes an implementation of CGR, the DTN routing algorithm by NASA-JPL, conceived to cope with scheduled intermittent connectivity, now renamed SABR (Schedule-Aware Bundle Routing) in the context of the on-going CCSDS (Consultative Committee for Space Data Systems) standardization. The most recent versions of ION contain the SABR version, specified in CCSDS 734 (2018). Although there are no routing protocols in ION specifically designed for random opportunistic connectivity, some opportunistic features have been recently added to CGR itself (Burleigh et al., 2016). The ION package also includes DTNperf_3 (Caini et al., 2013), used in experiments shown in the following section, now compatible with DTN2, ION, and IBR-DTN. As mentioned, the ION implementation of BP is used on board of the ISS for DTN experiments with Earth nodes where DTN2 is used instead, to prove the compatibility of the two BP implementations.

2.3 GEO constellations

2.3.1 Advantages and challenges of GEO constellations

GEO satellites have the great advantage of appearing to be fixed in the sky to an observer on Earth. Their orbit is circular and located in an equatorial plane, with a distance from the Earth's surface of 35,786 km. This high altitude results in a very wide coverage area; a GEO constellation of three satellites, equally spaced at 120 degrees, is enough to provide almost complete coverage of the Earth. However, because the elevation angle of a GEO decreases from the equator to the pole, where it becomes negative, polar regions are necessarily excluded. Moreover, to enable the frequency reuse, modern satellites are multispots, i.e., they have multiple antennae on board covering different areas, which become like cells of terrestrial mobile phone service, only much larger. Note that the borders of these spots on Earth are fixed for GEO, while for LEOs they follow the satellite motion.

Apart from the impossibility of covering polar regions, GEOs have two drawbacks, a high path loss and a long propagation delay, as a consequence of the long distance from Earth. At upper layers, like Transport, this delay results in an RTT of about 600 ms (including some processing time), which is much higher than the usual RTTs on terrestrial connections. In the case of TCP, the reliable transport protocol used on the Internet, this high RTT has a severe impact on performance. The reasons are well-known in the literature, and the interested reader can find an exhaustive explanation, concerning satellite communication (Caini and Firrincieli, 2004; Caini et al., 2008b, 2009a,b). Herein, the TCP transmission rate is ruled by two dynamic windows, the congestion window, or cwnd, related to congestion control, and the advertised window, for flux control. Assuming the advertised window is large enough (to this end it is mandatory to enable the window scale option and to opportunely enlarge the Rx buffers), the Tx rate is determined by the cwnd, being given by cwnd/RTT (apart from scale factors). To achieve the same rate, therefore, a long RTT connection requires a cwnd (and Tx buffers) much higher than a short RTT connection. Ironically, as the cwnd growth is "ACK-based," i.e., triggered by the arrival of packet acknowledgments, or ACKs (Allman et al., 2009), the higher the RTT, the slower the cwnd increase rate, and not faster, as would be necessary. The combination of a larger cwnd required and a slower cwnd increase rate has dramatic effects on performance. A more general explanation is the following: TCP cwnd dynamic is based on a feedback control loop; therefore, it is not strange, but very reasonable, that its performance degrades when the feedback loop time (the RTT) increases, as in all feedback control systems. The performance penalization due to the long RTT is exacerbated when a long RTT connection has to compete with a short RTT connection for the bandwidth of a bottleneck. TCP can share the available bandwidth equally among competing connections only if their RTT is comparable. Otherwise, the connection with the highest RTT is penalized (it starves). This well-known property is called in the literature the RTT-unfairness of TCP.

A second important impairment results from the possible presence of losses due to errors caused by noise on the satellite link. TCP cannot distinguish the origin of packet losses, i.e. whether they are due to congestion or to errors. In fact, because of the presence of parity-check codes, packets with even one bit flipped are discarded by lower layers. As error losses are negligible in terrestrial links, TCP "conservatively" assumes that all losses are due to congestion, and in response to them it halves the cwnd, i.e., the transmission rate, which is harmful if the loss is due to an error (Casetti et al., 2002). This behavior results in performance worsening in the presence of a bit error rate that is not negligible. Moreover, long RTTs amplify the effects of spurious cwnd halving caused by error losses, because it takes more to reopen the cwnd, with a further penalization of GEO satellite connections.

2.3.2 Possible countermeasures against GEO satellite impairments: Specialized protocols design and deployment challenges

The peculiar characteristics of GEO satellite links demand the use of specialized variants of TCP, such as Hybla (Caini and Firrincieli, 2004), a TCP variant designed to counteract long RTTs typical of GEO satellite links, partially included in the Linux kernel. Hybla is based on the following four elements: Hybla enhanced congestion control laws, the SACK option (Mathis and Mahdavi, 1996), packet pacing, and Hoe's initial bandwidth estimation. Hybla congestion control is among the few variants included in the standard Linux kernel, so it can be easily selected using a "sysctl" command and the SACK option is enabled by default in GNU (GNU is not Unix)/Linux; conversely, packet pacing and Hoe's initial bandwidth estimation require the installation of the MultiTCP patch, freely downloadable from MultiTCP (n.d.). On GEO satellite channels Hybla largely outperforms any other TCP variant included in the Linux kernel (Caini et al., 2009a,b), by reducing to a minimum the negative impact of RTT, as will be shown later. Note, however, that for higher RTTs, such as those that can be found on the moon or interplanetary communications, more radical solutions, like the use of LTP instead of TCP, are necessary.

Once a TCP variant, or another transport protocol able to cope with satellite channels, has been found, it is essential to stress that the problem is only theoretically solved. It is not, because in a general open network, like the Internet, the use of a TCP variant (or a transport protocol) specialized for satellite links requires the adoption of different architecture to be deployed, which is a fact many researchers often neglect. Let us explain why. Transport is the first end-to-end layer of the ISO/OSI (Open Systems Interconnection/International Organization for Standardization) architecture, which means that a transport connection goes from source to destination. In most satellite communications, we have a TCP server on terrestrial Internet, serving both terrestrial and satellite clients. As satellite clients are a niche, and as a server can only use one TCP variant unless the technique described in Caini et al. (2008a) is adopted, it is unlikely that a TCP variant specialized to satellites will be adopted on a general server; it is much more plausible that a TCP variant specialized to high-speed networks will be adopted instead. In conclusion, the deployment of protocols specialized to satellite links demands the implementation of a different architecture, where the satellite link is isolated from the rest of the network at the Transport layer, which, in turn, implies the replacement of the end-to-end transport connections with two or more consecutive connections. This can be accomplished in two ways: using TCP-splitting performance-enhancing proxies (PEPs) (Border et al., 2001), which is the usual solution adopted by most satellite service providers, or using the deployment of the DTN architecture in Caini et al. (2008b) and Caini et al. (2011).

2.3.3   DTN as an extension of TCP-splitting PEPs

PEPs are applications running on intermediate nodes to improve performance on network paths where native performance suffers due to characteristics of a link (e.g., a satellite link). Although there are many types of PEPs, here we focus on TCP-splitting PEPs, which are the most common solution (for an exhaustive treatment of PEPs see Border et al. (2001) and ETSI (2009)). A PEP implementation can be either "distributed," with two PEPs at the edges of the satellite link (Fig. 2.2A), or, less frequently, "integrated," with just one PEP at the satellite gateway (Fig. 2.2B). Each TCP-splitting PEP splits the end-to-end connection into two parts. Therefore, in distributed PEPs we have three TCP connections: from the sender to the gateway PEP; between the two PEPs; and from the satellite terminal PEP to the satellite receiver. The aim is twofold: first, to isolate the challenges of the satellite link, i.e., long RTTs and possible high Packet Error Rate (PER), from the rest of the network [as a result, the first and the last connections can use ordinary TCP, such as NewReno (Henderson et al., 2012)]; second, to have the possibility to take advantage of a TCP version (or another transport protocol) specialized for the satellite link. The aim of integrated PEPs is the same, but we have just two connections: the first from the satellite sender to the integrated PEP (ordinary TCP), the other on the satellite link (specialized TCP). The lack of PEP on the user terminal is a significant deployment advantage (terminal PEP vendors may disagree on that), considering that for one satellite gateway we may have hundreds or thousands of terminals. However, to keep an integrated PEP transparent to end-users, the choice of transport protocols on the satellite link is limited to enhanced versions of TCP, and more specifically to TCP variants that are compatible with standard TCP receivers.

In brief, TCP-splitting PEPs split the end-to-end path into three or more hops. The same result, which is the key to improve performance, can be achieved using the DTN architecture. For example, the DTN equivalent of the integrated PEP in Fig. 2.2B is shown in Fig. 2.2C. Therefore, we can state that DTN architecture includes the concept of TCP-splitting PEPs, while being, at the same time, much more general (Caini et al., 2011).

If commonalities are many, there are, however, also some important differences. First, the DTN solution is not as transparent as PEP, because BP must be installed on end-nodes; however, it is much more "elegant." While TCP-splitting is based on a sort of trick (fake ACKs are sent back by the first PEP on the path, not by destination, which is effective but neither elegant nor robust), in DTN the same result is achieved thanks to the different enhanced architecture. Second, but quite important in

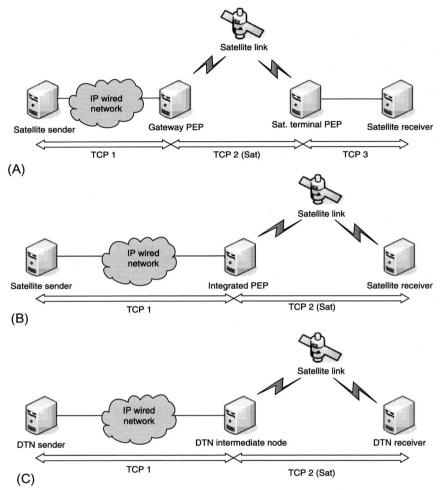

Fig. 2.2 PEP and DTN architecture comparison: (A) distributed PEP; (B) integrated PEP; and (C) DTN equivalent of integrated PEP.

practice, TCP-splitting is incompatible with Internet Protocol security (IPsec) (ETSI, 2009; Cruickshank et al., 2009), while DTN can take advantage of the many features offered by the security extensions of BP (Birrane and McKeever, 2019).

2.3.4   Performance with fixed terminals: Continuous channels

The performance of DTN in contrast to PEPs and end-to-end TCP was first studied (Caini et al., 2008b) concerning satellites with continuous links, usual for fixed terminals, and later extended by making use of more recent DTN2 BP implementations in Caini et al. (2011) and Caini et al. (2009c).

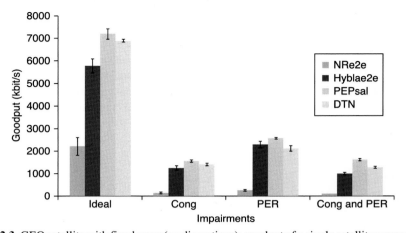

Fig. 2.3 GEO satellite with fixed users (no disruptions); goodput of a single satellite connection (RTT 600 ms); averaged values, 90% confidence intervals.
Reproduced with permission from Caini, C., Cruickshank, H., Farrell, S., Marchese, M., 2011. Delay- and disruption-tolerant networking (DTN): an alternative solution for future satellite networking applications. Proc. IEEE 99(11), 1980–1997. © IEEE.

Results obtained using a GNU/Linux testbed, are summarized here in Fig. 2.3. Four techniques are compared: two TCP variants used end-to-end, NewReno and Hybla, and two architectures enabling the use of TCP specialized to satellite environments, PEPsal and DTN. More precisely, PEPsal (n.d.) is an integrated TCP-splitting PEP (Caini et al., 2007), released as free software for GNU/Linux, while DTN is its equivalent DTN architecture (Fig. 2.2B and C). In both PEPsal and DTN cases, Hybla is assumed to be used on the satellite link, for a direct comparison with Hybla end-to-end. The four techniques are evaluated in four scenarios: ideal, congestion, PER, and congestion plus PER. Bars show the performance in terms of goodput on a 180 s satellite data transfer.

2.3.4.1 Ideal channel

Although in this case, the only impairment is the long RTT of the GEO satellite connection (600 ms, of which 575 ms is due to the satellite link and 25 ms to the terrestrial leg; see Fig. 2.2), NewReno is unable to exploit the satellite bandwidth (10 Mbit/s). End-to-end Hybla performs better (5.8 Mbit/s), but the best performance is achieved by PEPsal. DTN reaches the same performance as PEPsal, due to the similarity between the two architectures.

2.3.4.2 Congestion

The second environment highlights the severity of the RTT-unfairness problem. Here a satellite connection (RTT = 600 ms) competes with five terrestrial connections (RTT = 25 ms) for the bandwidth of a 10 Mbit/s bottleneck. The maximum fair share, i.e., 1/6 of the available bandwidth, should be the ideal target (1.6 Mbit/s) for all connections. While this target is reached by competing for terrestrial connections (data are not shown in the figure), the performance achieved on the satellite connection by New Reno is poor (only 50 kbit/s). In contrast, the other three techniques substantially remedy the RTT-unfairness problem, with performance near to the 1.6 Mbit/s target. Note, however, that the good results stem from different reasons: in fact, for end-to-end Hybla the good performance should be mainly ascribed to its improved congestion control, while for both PEPsal and DTN it is due to the isolation of the satellite link from the wired network (Caini et al., 2007).

2.3.4.3   PER

This is as in the ideal case, but with huge losses (PER $= 1\%$) on the satellite link. Although the impairment is different, results are qualitatively the same as in the congestion case. Note, however, that, even for the best-performing techniques, the goodput is always far from the available bandwidth (10 Mbit/s), because of the very high PER. In contrast to the congestion case, here the better performance of PEPsal and DTN is mainly due to the use of a TCP variant optimized to the satellite link, and not to the isolation of the satellite channel from the wired network.

2.3.4.4   PER and congestion

This environment suffers from congestion and PER simultaneously. Results are close to the congestion-only case, which shows that congestion is the dominant impairment.

Results of Fig. 2.3 prove the essential equivalence of DTN and PEP performance in the presence of continuous links, i.e., for fixed terminals, which is a fundamental result. It shows that DTN can provide the same goodput performance as the most effective and widely adopted solutions even in the absence of disruptions and/or network partitioning. The challenge here is the long RTT of the GEO satellite, which is enough to justify the adoption of DTN.

2.3.5   Performance with mobile terminals: Disruptive channels

Terminal mobility causes frequent disruptions on the satellite link, due to obstacles (buildings, tunnels, etc.) that impair satellite visibility. TCP (and PEPs based on TCP) offers a certain amount of robustness against disruption using the retransmission timeout (RTO) mechanism (Paxson et al., 2011). In brief, in the absence of feedback from the server, a TCP client retransmits the last unacknowledged segment many times, at increasing intervals, until either positive feedback is eventually achieved or either a maximum number of retries or a maximum time (the choice is left by RFC to the implementation) is reached, in which case the TCP connection aborts. The mechanism is intrinsically complex and cannot be fully described here; see Paxson et al. (2011) and Caini et al. (2010) for further details. Let us just point out that the maximum tolerable disruption length is not fixed but can be set either directly or indirectly by deciding the maximum number of trials (the Linux default is 15 retries, roughly corresponding to 20 min for a GEO satellite connection). Because it is highly unlikely that these default settings will be changed on a general server just to increase the disruption resilience of a few satellite mobile terminals, in practice we can reasonably consider 20 min as the maximum tolerable disruption length for TCP. In comparison, DTN BP offers superior robustness. Let us focus on the DTN2 BP implementation with TCP as CL, for a brief insight. Here "short" disruptions, i.e., disruption shorter than 30 s (default), are directly tackled by TCP with the usual RTO mechanism; in contrast, "long" disruptions, i.e., disruptions longer than 30 s, trigger the BP to close the disrupted TCP connection and make a series of attempts to open a new TCP connection. After each failed attempt the BP inserts an increasing interval with a ceiling of 10 min. The retries go on for 24 h (default), which is much longer than the usual 20 min TCP resilience (Caini et al., 2010). In GEO satellite communications with mobile terminals, most disruptions are likely shorter than 20 min; therefore, they can be tackled either by TCP or by DTN plus TCP CL. Which is the best is uncertain and needs further investigation.

To this end, in Caini et al. (2011) and Caini et al. (2009c) disruption was introduced in the previous DTN comparative evaluation, by considering a satellite terminal on board a train, connected to the Internet via GEO. Disruptions are caused by tunnels (39% of the line, the Bologna-Florence "Direttissima") and are of various durations. Assuming a constant speed of 120 km/h, we have 30 "short" disruptions and only three "long" disruptions (namely, 214, 92, and 553 s). Short disruptions are directly counteracted by TCP retransmission, so their impact is null in comparative evaluations. In contrast, longer disruptions differentiate the performance of DTN (with TCP CLA) and TCP alone. Notably, the forced TCP closure caused by a long disruption triggers the BP reactive bundle fragmentation, which prevents the retransmissions of already acknowledged parts.

The tests consider a persistent file transfer to a satellite terminal on board a train moving from Bologna to Florence, with goodput averaged over the total trip time (2865 s). Results are shown in Fig. 2.4 and should be compared with those presented in Fig. 2.3 (all the settings are the same). From the comparison, the following remarks can be made. First, performance is reduced in all cases (note that the ordinate scale is different), which is expected due to the satellite channel unavailability (39% of the time) and other penalization factors, such as the "restart-delay" (i.e., the time elapsed before TCP restarts transmission after the channel is available again at the physical layer) and the time needed by TCP to reach again a steady-state (Caini et al., 2010).

The second remark is that, although the qualitative behavior is the same as without disruptions, in this case, DTN and not PEPsal, is the best-performing technique. The difference is small but real. Finally, it should be stressed that, if we had considered a 60 km/h train speed, the longest disruption would have become greater than 1200 s, which is approximately the "maximum tolerable disruption length" with Linux TCP defaults. In this case, all techniques but DTN would have aborted the data transfer.

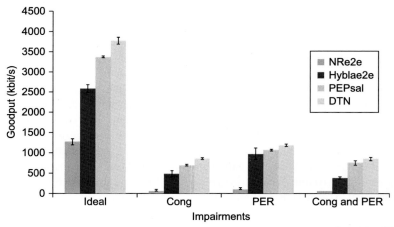

Fig. 2.4 GEO satellite with mobile users (disruptions due to railway tunnels); goodput of a single satellite connection (RTT = 600 ms); averaged values, 90% confidence intervals. Reproduced with permission from Caini, C., Cruickshank, H., Farrell, S., Marchese, M., 2011. Delay- and disruption-tolerant networking (DTN): an alternative solution for future satellite networking applications. Proc. IEEE 99(11), 1980–1997. © IEEE.

Summing up, with mobile terminals DTN may become advantageous also in terms of goodput, the actual gain depending on the duration and frequency of disruptions. For very long disruptions (>20 min), DTN is the only choice.

2.4   LEO constellations

2.4.1   *Advantages and challenges of LEO constellations*

LEO satellites have a reduced distance from Earth (from 160 to 2000 km), which makes them better suited than GEOs to Earth observation. Moreover, the reduced distance results in a lower path loss, which helps in closing the link budget, and a shorter propagation delay, which solves the RTT-penalization of GEO TCP connections. On the other hand, for a fixed observer on Earth they move in the sky (with a typical orbital period of around 90–100 min). Note that, while this is an advantage for Earth observation (as the Earth rotates on its polar axis, all the Earth can be explored by one LEO satellite on a polar or quasipolar orbit), for a continuous global coverage (e.g., for providing ubiquitous phone service and Internet access) large constellations with tens of satellites are required. In the case of either one LEO satellite or an incomplete constellation, satellite links are scheduled intermittently, as in deep space communications, and represent an interesting study case for DTN applicability (Cerf et al., 2007; Scott and Burleigh, 2007). Some pioneering experiments with a real observation satellite have already been successfully carried out (Ivancic et al., 2010). Here, we will focus on the case of one satellite for Earth observation, but most remarks are more general and could apply also to incomplete constellations.

2.4.1.1 Link intermittency and scheduled contacts

Due to orbital motion, a single LEO satellite is in line of sight to a ground station for only brief periods separated by long intervals. During these transmission opportunities or "contact windows," the maximum amount of data that can be transmitted, called "contact volume," is given by the product of the nominal Tx speed of the link and the contact length. As contacts are both short and few, they need to be efficiently exploited by minimizing the time wasted when the channel is available but not used at upper layers (e.g., at Transport). This is the first challenge posed by LEOs at upper layers, and DTN scheduled contacts seem particularly suitable to tackle it.

2.4.1.2 Delay between data creation and availability: Multiple ground stations or GEO relays

The Earth rotation is an advantage for Earth observation, but it also implies that an LEO satellite on a polar orbit does not pass over the same ground station every orbital period but at longer intervals (from 100 min to 12 h). This, of course, causes a delay between data creation (e.g., a photo taken by a camera on board the satellite) and actual data available on Earth (e.g., at LEO control center), which represents the second challenge posed by LEOs, being critical for many applications (e.g., disaster monitoring, military observation). A partial remedy is the use of multiple ground stations, while a more radical solution is the use of GEO satellites as relays instead of ground stations on Earth, as suggested in Greda et al. (2010), Katona (2012), and Johnston et al. (2012). LEO-GEO connectivity is not continuous, as might appear at first glance, but scheduled intermittent, at least for LEOs on polar orbits, because of the GEOs' inability to cover polar regions. The two solutions, via ground stations and via a GEO relay, are more complementary than alternative and could be advantageously used in concert (Apollonio et al., 2013).

2.4.1.3 Routing with scheduled intermittent links

Although the introduction of multiple ground stations or a GEO relay reduces the average delay between data creation and availability, on the other hand, it poses a new routing problem, which can be considered as a third challenge, because data can reach the destination through multiple alternative paths, all characterized by scheduled intermittent connectivity. Crucially, these intermittent links require solutions specifically designed for DTNs. It must be stressed once again that this intermittency is not random, as in mobile ad hoc networks, but deterministic, because it is related to the motion of the satellite and the Earth. Therefore, BP routing algorithms proposed for opportunistic and unpredictable links cannot be applied. Conversely, the use of the CGR algorithm (Araniti et al., 2015) is very appealing, as suggested by CGR's authors themselves and as investigated in some studies (Caini and Firrincieli, 2012) and (Apollonio et al., 2013) and (Bezirgiannidis et al., 2016). CGR is part of ION (Burleigh, 2007), the DTN package developed by NASA JPL used in all the experiments described in the following section.

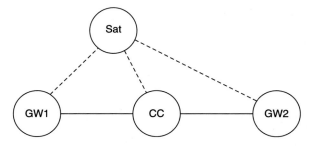

Fig. 2.5 LEO satellites for Earth observation: multiple ground stations. *Dotted lines* denote scheduled satellite links; *continuous lines* denote continuous wired links. LTP on all links.

2.4.2 LEO satellites for Earth observation: Multiple ground stations

The scenario considered in Caini and Firrincieli (2012) applies to Earth observation satellites and is freely inspired by the COSMO-SkyMed Italian constellation (COSMO, n.d.). A series of image files have to be transferred from the satellite to a control center for inspection. As shown in Fig. 2.5, data can be downloaded either directly at control center premises or via two additional gateway stations (GW1 and GW2). Satellite links are scheduled intermittently, while terrestrial links between GW1, GW2, and the control center are continuous (see Fig. 2.5). LTP is used in the Transport layer on satellite links because of its ability to copy with scheduled contacts, while TCP is used on terrestrial links.

2.4.2.1 Experiment description

Contact durations of satellite links, Tx rates, and contact volumes are specified in the "contact plan" presented in Table 2.1. The Tx rate is 1 Mbit/s in the downlink and 64 kbit/s in the uplink. Although the bandwidth asymmetry is large, it does not affect performance, as the uplink is used only by LTP ACKs and BP status reports, both of which are very short. In the experiment, we assume the transfer of 125 MB (divided into 500 bundles of 250 kB), which requires three consecutive satellite contacts

Table 2.1 Earth observation: multiple ground stations scenario (contact plan).

Link	Contact #	Start-stop time (min)	Link speed (Mbit/s)	Contact volume (MB)
Sat–>Earth Gw1	1	60–68	1	60
Sat –>Control C	1	428–436	1	60
Sat–>Earth Gw2	1	796–804	1	60
GW1_CC	Dummy (TCP cont.)	1–840	10	
GW2_CC	Dummy (TCP cont.)	1–840	10	

(namely, to GW1, to the control center, and GW2). On the satellite link, we have an RTT of 130 ms and a negligible packet loss; the BP custody option is always on. Concerning routing, for each bundle the task of CGR is to select the best route from the satellite to the control center, bearing in mind both the temporal sequence of satellite contacts and their limited contact volumes.

2.4.2.2 Analysis of results

Results, i.e., time series of "sent" signals and "delivered" status reports collected on the satellite, are given in Fig. 2.6. The sent series indicates when bundles are transferred from the application to the Bundle layer on the satellite, where they await the next contacts (Table 2.1). The delivered series indicates when bundles are delivered to the application on the destination node, i.e., the control center. Results prove that CGR can exploit all contacts in a very effective way: 210 bundles are delivered through GW1 (first contact), another 223 directly (second contact), and the remaining 67 through GW2 (third contact). Link utilization efficiency (data transferred normalized to the contact volume) is high (88% and 93% in the first two contacts; the third figure cannot be computed as the last contact is only partially used). These excellent results prove the effectiveness of both ION scheduled links and LTP.

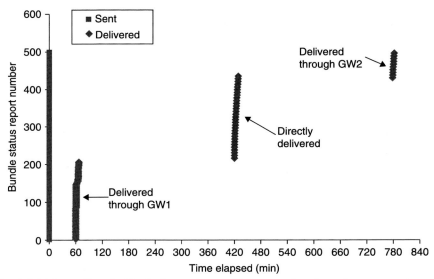

Fig. 2.6 LEO satellites for Earth observation: multiple ground stations. Status report logs. Reproduced with permission from Caini, C., Firrincieli, R., 2012. Application of contact graph routing to LEO satellite DTN communications. In: Proc. of IEEE ICC 2012, Ottawa, Canada, June 2012, pp. 3340–3344. © IEEE.

2.4.3 LEO satellites for Earth observation: Ground stations and GEO relays

In Apollonio et al. (2013) a GEO relay connects an LEO sat to its control center, as suggested in Johnston et al. (2012) regarding the Inmarsat SB-Sat system. However, in contrast to the cited paper, the possibility to use a ground station is preserved. The aim is to show that GEO relays are complementary rather than an alternative to ground stations and that the enabling technology for their combined use is the DTN BP architecture with CGR.

The DTN layout consists of five nodes: an LEO satellite, a ground station used by the LEO, an LEO Control Centre, a GEO control center, and a final user (see Fig. 2.7). Note that, as the GEO relay is transparent to BP traffic (it is not a DTN node), it is not present in the figure. As usual, dotted and continuous lines denote intermittent satellite links and terrestrial wired links, respectively; the LTP CLA is used here only on satellite links, and TCP CLA on the terrestrial ones, as before. In the experiments, we consider the transfer of both telemetry data and image files, with different priorities. The source for both is the LEO satellite, while the destination is the LEO control center for telemetry and the final user for images.

2.4.3.1 Experiment description

Contact durations of satellite links, link speeds, and resulting contact volumes are specified in the "contact plan" presented in Table 2.2. First, let us emphasize the dual characteristics of the two satellite hops, i.e., the LEO_sat-GEO_CC and LEO_sat-GS. In the former, the time availability is relatively high, roughly 70% of every orbital period (100 min), with two 15 min disruptions in correspondence to the poles. By contrast, the Tx rate is relatively low, having assumed a fixed rate of 128 kbit/s, which is the maximum rate for a dedicated channel in accordance with Johnston et al. (2012). The LEO_sat-GS hop has opposite characteristics, with a contact window of only 10 min and a Tx rate of 10 Mbit/s (thanks to the low path loss). For the sake of simplicity, only one orbital period, in which we have one contact with the ground station, is assumed. Note that the contact with the ground station is nested in the larger second

Fig. 2.7 LEO satellites for Earth observation: ground stations and GEO relays. *Dotted lines* denote scheduled satellite links (LTP); *continuous lines* denote continuous wired links (TCP).

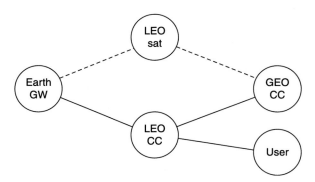

LEO_sat-GEO_CC contact, thus adding complexity to the routing problem, as two paths to the destination are active in parallel. Moreover, for the sake of simplicity, in the experiments contact windows (and contact volumes) are scaled down by a factor of 60 (i.e. we have seconds instead of minutes, see contact windows at the bottom of Fig. 2.8).

Table 2.2 Earth observation: ground station and GEO relay (contact plan).

Link	Contact #	Start-stop time (min)	Link speed	Contact volume (MB)
LEO_sat-GS	1	45–55	10 Mbit/s	750
LEO_sat-GEO_CC	1	1–20	128 kbit/s	19.2
	2	35–70	128 kbit/s	33.6
	3	85–100	128 kbit/s	14.4
GS-LEO_CC	Dummy (TCP cont.)	1–200	10 Mbit/s	
GEO_CC-LEO_CC	Dummy (TCP cont.)	1–200	10 Mbit/s	

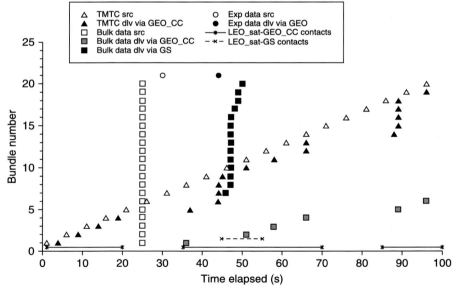

Fig. 2.8 Overall bundle traffic. Bundle telemetry (from LEO_sat to LEO_CC, always via GEO_CC) and bundle data (from LEO_sat to User, either via GEO_CC or GS). Reproduced with permission from Apollonio, P., Caini, C., Lülf, M., 2013. DTN LEO satellite communications through ground stations and GEO relays. In: Dhaou, R., et al. (Eds.), Personal Satellite Services PSATS 2013, LNICST, vol. 123. Springer-Verlag, Heidelberg, pp. 1–12. ©Springer.

For telemetry traffic, a new bundle of 5 kB is generated every 5 s, as shown in Fig. 2.8 (TMTC src series). These bundles have high priority and should be delivered to the LEO control center as soon as possible, either through the GEO relay or through the ground station, the faster the better. For image traffic, a burst of 20 low priority bundles of 100 kB is generated during the first polar disruption, at 25 s (Bulk data-src series). Their destination is the User node, which, however, can be reached only through LEO_CC, as before. After 5 s, these bundles are followed by a high-priority bundle always of 100 kB, representing a high-value target (Expedited src).

2.4.3.2 Analysis of results

The experiment has been carried out on a Linux-based virtual testbed (one virtual machine running ION for each DTN node), created using Virtualbricks virtual testbed manager (Apollonio et al., 2014). The results referring to the overall bundle traffic (time series of bundle sourced and delivered) are given in Fig. 2.8. The evident complexity of the picture reflects the complexity of the experiments and, as an exhaustive explanation would be out of the scope here, we will limit ourselves to a summary. The interested reader is referred to Apollonio et al. (2013). Let us start with telemetry traffic, which has the highest priority. All TMTC bundles are delivered to the destination via the GEO relay path, which is the fastest. Bulk bundles generated during the first polar disruption are delivered in part via the GEO relay and in part via the ground station. More precisely, they are forwarded by CGR to the GEO relay until the contact volume of the second LEO_sat-GEO_CC contact is exhausted (bulk bundles 1–6), then to the ground station (bulk bundles 7–20). This decision is challenged by the later generation of higher-priority bundles (both TMTC and the expedited data), which causes bulk bundles 5 and 6 to be transmitted only during the third LEO_sat-GEO_CC contact.

Although suboptimal for bulk bundles 3, 4, 5, and 6, which could have been delivered first via the ground station, results are satisfactory. They show that the GEO relay and the ground station links can be successfully pooled thanks to CGR (no need to reserve a path for a specific kind of traffic) and that BP priorities are generally respected. All high-priority bundles are routed via GEO-relay, not as a result of a fixed assignment, but because this route offers the fastest delivery in the scenario considered; conversely, bulk bundles, which have the minimum priority, are not prevented from using the GEO-relay; in fact, they can use this dynamically, i.e. when there is enough residual capacity, and otherwise are routed to the ground station.

2.5 Conclusion

In this chapter, the potential advantages of DTN when applied to satellite communications have been evaluated for both GEO and LEO constellations.

For GEO, an important analogy between DTN and PEP architectures has been highlighted. In particular, the DTN architecture has been presented as an extension of TCP-splitting PEPs, which represent the usual solution to overcome the RTT-penalization suffered by GEO satellite TCP connections. To solve the problem, both architectures split the original satellite connections into multiple connections, to isolate the satellite link from the rest of the network. The difference is that, while TCP-splitting PEPs rely on a sort of trick, to disguise the sender and induce him to transmit at a higher rate, the DTN architecture is a much more elegant solution, as it relies on the redefinition of the protocol architecture with the introduction of the new Bundle layer between Transport and Application. As well as isolating the satellite link and allowing in it the use of transport protocol specialized to satellite environments, as PEPs, the bundle protocol offers a series of new services, including a superior resilience to medium and long disruptions and much more advanced security features.

For LEO constellations, the main challenge is no longer the long RTT, but the intermittency of the satellite link. Even in this case, the application of DTN is useful, thanks to scheduled links and CGR. In particular, in this chapter the problem of intermittent connectivity from LEO sat to gateway stations has been studied. Then, the analysis has been extended including a GEO relay. Results show that DTN BP and CGR are not only able to cope with intermittent links and DTN routing (first experiments) but also to enforce QoS (Quality of Service) based on bundle priority and to dynamically pool parallel paths (second experiments), which greatly improves bandwidth exploitation, as any fixed assignment is avoided. In conclusion, in this chapter, the reasons for a successful DTN application to both GEO and LEO satellites have been thoroughly examined. Results achieved with the two major BP implementations (DTN2 and ION) prove that the DTN architecture, although still experimental in some aspects, is mature enough for real deployment.

Acknowledgments

The author would like to thank all those who contributed to the DTN research reported here: in particular, Rosario Firrincieli, for his precious contribution in the development of Hybla and the realization of most experiments, Piero Cornice and Marco Livini for developing the DTNperf_2 tool, Michele Rodolfi for the fully redesigned DTNperf_3, Daniele Lacamera for Hybla and PEPsal implementations in GNU/Linux, Nicola Alessi for implementing LTP advancements in ION and Marco Giusti for maintaining Virtualbricks. A special thanks to all other students who cannot be explicitly mentioned here, but whose contribution to the research was fundamental.

References

Alessi, N., Burleigh, S., Caini, C., De Cola, T., 2018. Design and performance evaluation of ltp enhancements for lossy space channel. Int. J. Satell. Commun. Netw. (March), 1–12.

Allman, M., Paxson, V., Blanton, E., September 2009. TCP congestion control. IETF RFC 5681,.

Apollonio, P., Caini, C., Lülf, M., 2013. DTN LEO satellite communications through ground stations and GEO relays. In: Dhaou, R., et al. (Eds.), Personal Satellite Services PSATS. LNICST, vol. 123. Springer-Verlag, Heidelberg, pp. 1–12.

Apollonio, P., Caini, C., Giusti, M., Lacamera, D., July 2014. Virtualbricks for DTN satellite communications research and education. In: Proc. of PSATS 2014, Genoa, Italy, pp. 1–14.

Araniti, G., Bezirgiannidis, N., Birrane, E., Bisio, I., Burleigh, S., Caini, C., Feldmann, M., Marchese, M., Segui, J., Suzuki, K., 2015. Contact graph routing in DTN space networks: overview, enhancements and performance. IEEE Commun. Mag. 53 (3), 38–46.

Bezirgiannidis, N., Caini, C., Tsaoussidis, V., 2016. Analysis of contact graph routing enhancements for DTN in space. Int. J. Satell. Commun. Netw. 34 (Sept./Oct.), 695–709.

Birrane, E., McKeever, K., April 2019. Bundle protocol security specification. Internet Draft, Work-In-Progress, Available from: https://datatracker.ietf.org/doc/draft-ietf-dtn-bpsec.

Bisio, I., Cello, M., de Cola, T., Marchese, M., 2009. Combined congestion control and link selection strategies for delay tolerant interplanetary networks. In: Proc. of IEEE Globecom 2009, Honolulu, Hawaii, November–December.

Border, J., Kojo, M., Griner, J., Montenegro, G., Shelby, Z., June 2001. Performance enhancing proxies intended to mitigate link-related degradations. IETF RFC 3135,.

Burleigh, S., 2007. Interplanetary overlay network (ION) an implementation of the DTN bundle protocol. In: Proc. of 4th IEEE Consumer Communications and Networking Conference, pp. 222–226.

Burleigh, S., October 2013. Delay tolerant networking LTP convergence layer (LTPCL) adapter. Internet-Draft, Work-in-Progress. Available from: https://tools.ietf.org/html/draft-burleigh-dtnrg-ltpcl-05.

Burleigh, S., Ramadas, M., Farrell, S., September 2008. Licklider transmission protocol—motivation. Internet RFC 5325,.

Burleigh, S., Caini, C., Messina, J.J., Rodolfi, M., September 2016. Toward a unified routing framework for delay-tolerant networking. In: Proc. of IEEE WiSEE 2016, Aachen, Germany, pp. 82–86.

Caini, C., Firrincieli, R., 2004. TCP hybla: a TCP enhancement for heterogeneous networks. Int. J. Satell. Commun. Netw. 22, 547–566.

Caini, C., Firrincieli, R., November 2011. DTN and satellite communications. In: Vasilakos, A., Zhang, Y., Spyropoulos, T. (Eds.), Delay Tolerant Networks: Protocols and Applications. CRC Press, New York, pp. 283–318.

Caini, C., Firrincieli, R., June 2012. Application of contact graph routing to LEO satellite DTN communications. In: Proc. of IEEE ICC 2012, Ottawa, Canada, pp. 3340–3344.

Caini, C., Firrincieli, R., Lacamera, D., 2007. PEPsal: a performance enhancing proxy for TCP satellite connections. IEEE Aerosp. Electron. Syst. Mag. 22 (8), b-9–b-16.

Caini, C., Firrincieli, R., Lacamera, D., 2008a. The TCP "adaptive-selection" concept. IEEE Syst. J. 2 (1), 83–89.

Caini, C., Cornice, P., Firrincieli, R., Lacamera, D., 2008b. A DTN approach to satellite communications. IEEE J. Sel. Areas Commun. 26 (5), 820–827. Special issue on Delay and Disruption Tolerant Wireless Communication.

Caini, C., Firrincieli, R., Lacamera, D., 2009a. Comparative performance evaluation of TCP variants on satellite environments. In: Proc. of IEEE ICC 2009, Dresden, Germany, June, pp. 1–5.

Caini, C., Firrincieli, R., Lacamera, D., De Cola, T., Marchese, M., Marcondes, C., Sanadidi, M. Y., Gerla, M., 2009b. Analysis of TCP live experiments on a real GEO satellite testbed. Perform. Eval. 66 (6), 287–300.

Caini, C., Cornice, P., Firrincieli, R., Lacamera, D., Livini, M., 2009c. TCP, PEP and DTN performance on disruptive satellite channels. In: Proc. of IEEE IWSSC'09, Siena, Italy, pp. 371–375.

Caini, C., Firrincieli, R., Livini, M., 2010. DTN bundle layer over TCP: retransmission algorithms in the presence of channel disruptions. J. Commun. 5 (2), 106–116.

Caini, C., Cruickshank, H., Farrell, S., Marchese, M., 2011. Delay- and disruption-tolerant networking (DTN): an alternative solution for future satellite networking applications. Proc. IEEE 99 (11), 1980–1997.

Caini, C., d'Amico, A., Rodolfi, M., September 2013. DTNperf_3 at work: aims and use. In: Proc. of ACM CHANTS 2013, Miami, USA, pp. 53–55.

Casetti, C., Gerla, M., Mascolo, S., Sanadidi, M.Y., Wang, R., 2002. TCP Westwood: end-to-end congestion control for wired/wireless networks. Wirel. Netw. 8, 467–479.

CCSDS 734.3-R-1, July 2018. Schedule-aware bundle routing. In: CCSDS Red Book. Issue 1.

Cerf, V., Hooke, A., Torgerson, L., Durst, R., Scott, K., Fall, K., Weiss, H., April 2007. Delay-tolerant networking architecture. IETF RFC 4838,.

COSMO-SkyMed, Website http://www.cosmo-skymed.it/en/index.html.

Cruickshank, H., Mort, R., Berioli, M., 2009. Broadband satellite multimedia (BSM) security architecture and interworking with performance enhancing proxies. In: Sithamparanathan, K., Marchese, M. (Eds.), Personal Satellite Services PSATS 2009. LNICST, vol. 15. Springer-Verlag, Heidelberg, pp. 132–142.

Demmer, M., Ott, J., Perreault, S., June 2014. Delay tolerant networking TCP convegence layer protocol. IETF RFC 7242,.

DTN2 reference code, Available from: http://sourceforge.net/projects/dtn/.

DTNWG, Website https://datatracker.ietf.org/wg/dtn/charter/.

ETSI, September 2009. Technical report on performance enhancing proxies (PEPs) for the European ETSI Broadband Satellite Multimedia (BSM) working group. ETSI Report TR 102 676 http://portal.etsi.org.

Fall, K., Farrell, S., 2008. DTN: an architectural retrospective. IEEE J. Sel. Areas Commun. 26 (5), 828–836.

Fall, K., Hong, W., Madden, S., July 2003. Custody transfer for reliable delivery in delay tolerant networks. Technical Report IRB-TR-03-030, Intel Research, Berkeley, pp. 1–6. Available from: http://www.intel-research.net/Publications/Berkeley/081220030852_157.pdf.

Greda, A., Knupfer, B., Knogl, J.S., Heckler, M.V.T., Bischl, H., Dreher, A., April 2010. A multibeam antenna for data relays for the German communications satellite Heinrich-Hertz. In: Proc. of EuCAP 2010 Conf, pp. 1–4.

Henderson, T., Floyd, S., Gurtov, A., Nishida, Y., 2012. The NewReno modification to TCP's fast recovery algorithm. IETF RFC 6582,.

IBR-DTN, Website http://www.ibr.cs.tu-bs.de/projects/ibr-dtn/.

ION Code, Available from: http://sourceforge.net/projects/ion-dtn.

Ivancic, W., Eddy, W.M., Stewart, D., Wood, L., Northam, J., Jackson, C., 2010. Experience with delay-tolerant networking from orbit. Int. J. Satell. Commun. Netw. 28 (5–6), 335–351.

Johnston, B., Haslam, M., Trachtman, E., Goldsmith, R., Walden, H., McGaugh, P., September 2012. SB-SAT-persistent data communication LEO spacecraft via the Inmarsat-4 GEO constellation. In: Proc. of ASMS 2012 Conf., Baiona, Spain, pp. 21–28.

Katona, Z., September 2012. GEO data relay for low earth orbit satellites. In: Proc. of ASMS 2012 Conf., Baiona, Spain, pp. 81–88.

Kruse, H., Jero, S., Ostermann, S., March 2014. Datagram convergence layers for the delay- and disruption-tolerant networking (DTN) bundle protocol and licklider transmission protocol (LTP). Internet RFC: 7122,.

Mathis, M., Mahdavi, J., October 1996. TCP selective acknowledgment options. IETF RFC 2018,.

McMahon, A., Farrell, S., 2009. Delay- and disruption-tolerant networking. IEEE Internet Comput. 13 (6), 82–87.

MultiTCP Package, Available from: http://sourceforge.net/projects/multitcp.

Paxson, V., Allman, M., Chu, J., Sargent, M., June 2011. Computing TCP's retransmission timer. IETF RFC 6298,.

PEPsal Code, Available from: http://sourceforge.net/projects/pepsal/.

Ramadas, M., Burleigh, S., Farrell, S., September 2008. Licklider transmission protocol—specification. Internet RFC 5326,.

Scott, K., Burleigh, S., November 2007. Bundle protocol specification. IETF RFC 5050,.

Symington, S., Farrell, S., Weiss, H., Lovell, P., May 2011. Bundle security protocol specification. Internet RFC 6257,.

Warthmann, F., September 2015. Delay-Tolerant Networks (DTNs), A Tutorial, Version 3.2. Available from: http://ipnsig.org/wp-content/uploads/2015/09/DTN_Tutorial_v3.2.pdf.

Delay-tolerant networks (DTNs) for deep-space communications [star]

N.L. Clarke, B.V. Ghita, and S.M. Furnell
Plymouth University, Plymouth, United Kingdom

3.1 Introduction

This chapter examines a particular application area for delay-tolerant networks (DTNs) technology, looking at the requirements associated with acquiring data collected by satellites in a deep-space context.

The main discussion begins in Section 3.2, which sets the scene for the topic of space communications and the related data needs. Notably, as an application area, it represents a context with extreme demands for delay tolerance, thanks to the distances and potential for interference and interruption that are involved. On this basis, Section 3.3 then examines the specific networking requirements associated with space data communications, and considers the resulting applicability of DTNs as a solution.

Having established the applicability of DTNs in the space data context, Section 3.4 moves on to consider how such a solution could be realized in practice. In doing so, particular attention is given to how additional requirements can be served in terms of securing the communications, which represents a likely requirement to safeguard the space data, as well as an additional challenge from the DTN perspective when looking at the applicability of existing security protocols.

3.2 Data communications in deep space

The DTN concept genetically describes unreliable environments, whereby the communication with endpoints is likely to be impaired by external factors. Among the existing environments that may characterize such communication, deep space communication is likely to represent by far the most extreme scenario, given the distances involved as well as the levels of noise and node distribution. To begin with, transmission distances for data communication are currently spanning from Earth-orbiting satellites to the Mars exploration mission, with no intermediate nodes for data store-and-forwarding. The delays involved emanate from the start of the usage of traditional reliable end-to-end protocols, such as the transmission control protocol (DARPA, 1981), which require maximum delays of no more than several seconds in order to operate. The level of space and radiation noise adds another obstacle to

---

[star] This chapter is a reprint of the chapter originally published in the first edition of Advances in Delay-Tolerant Networks (DTNs): Architecture and Enhanced Performance.

Advances in Delay-tolerant Networks (DTNs). https://doi.org/10.1016/B978-0-08-102793-6.00003-5

reliable communication, as does the intermittent connectivity when there is no line of sight between the communicating entities or when the interference duration is beyond temporary.

Traditionally, space communication has been performed using bespoke communication and protocols, implemented specifically for each individual space mission. During the past decade, with the increase in the number of space missions as well as the diversity and volume of data they transferred, the internet engineering task force (IETF) put forward a request for comments on the DTN architecture, which (after being tested in a number of space conditions) became the standard architecture to be used by the consultative committee for space data systems (CCSDS), the body that governs the direction and strategy of space communication (Cerf et al., 2007).

Beyond the protocols and transmission conditions involved, the critical aspect of space communication is the data, more particularly its evolution over recent years. In the early days, transmissions of data ranged from regular independent readings (such as radiation or temperature) to low-speed communication streams, potentially including small-sized data objects. However, with improvements in data collection and sensor technology came the requirement to send more data, including both high-definition imaging and streaming. While the streaming data, due to its real-time nature, has more flexibility in terms of the tolerable interference and packet drop rates, in the case of imaging the data transmission requires a guarantee of integrity and end-to-end delivery. This, in turn, matches very well with the resilient character of store-and-forward DTN functionality, ensuring that data bundles are not lost.

DTN is, by design, a very good choice for the transport of space data, and there are a number of contributing factors to its success, which may vary from environment to environment. To begin with, the architecture relies upon redundancy and meshing to deliver data, as the lack of continuous connectivity would be compensated by node redundancy. However, the current environment has limited cooperation between the nodes, particularly due to separation between different missions, and therefore it does not fully support this feature. Even when cooperation exists, the scarce resources available within each node serve to significantly reduce the storage capacity available for relaying data bundles. Although this is likely to be a temporary issue, as hardware is upgraded in space station instruments, a better separation between the operational plane and the bundle store-and-forward component, ensuring the integrity of the node and not jeopardizing its functionality, would lead to an even more reliable infrastructure.

3.3   Networking requirements for deep-space data

The operational characteristics and the data transmission parameters define each space mission. To begin with, their distance from Earth is directly proportional to the data rate at which they operate, as well as their on-board storage capacity. Further, the storage capabilities, in conjunction with the processing power onboard, serve to define the capacity of such a node to store and forward data from other space missions located further from the Earth's surface.

Based on these characteristics, the first category of missions is the Earth observation satellites, located on an orbit at hundreds of kilometers from the surface, typically collecting Earth imaging, either visible or within the infrared (heat map) spectrum. Such nodes would be the ideal candidates for handling store-and-forward DTN bundles, assuming cooperation between missions. A typical example of such an observation satellite is the Proba series of satellites. The Proba-1 satellite, owned by the European Space Agency (ESA), provides a typical example of a geostationary satellite (Teston et al., 2004). Orbiting at an altitude of 570–680 km, the satellite provides Earth-imaging and monitoring services. Its on-board compact high resolution imaging spectrometer (CHRIS) provides 15×15 km Earth spectral imaging with a resolution of 20 m, while the high resolution camera (HRC) generates 4×4 km Earth imaging with a resolution of 8 m. The satellite is controlled by ESA from their ground station in Redu, Belgium, and has a transmission rate of 1 Mbps downlink and 4 kbps uplink. In spite of its relatively fast downlink, CHRIS spectral images provide an interesting operational challenge, typical for space missions: using its onboard 1.2 Gbit memory, it can store several sets of images, which are on average 131 Mbits each. The satellite transmits the data to Redu (up to four downloads per day) and to Kiruna, Sweden (four to five additional downloads per day). However, should the satellite have the capacity and flexibility to communicate the acquired data with other ground stations or orbiting satellites with faster downlink speeds, it would have the ability to transmit a larger amount of data to Earth and, therefore, provide a more efficient exploitation of its capabilities. Further, the onboard memory could also be used for store-and-forward by other missions, especially to improve direct communication with ground stations. Summarizing, a complete DTN infrastructure would be able to utilize the satellite as a relay node for other sources of data, as well as using other satellites to relay the data collected by the satellite to Earth in order to improve its usage.

Taking another example, the group of polar operational environmental satellites (POES) jointly run by the National Oceanic and Atmospheric Administration (NOAA) and ESA (National Environmental Satellite, Data, and Information Services (NESDIS), 2013) provide an even more comprehensive scenario for DTN usage. POES has a high number of ground stations that collect data from its satellites, due to the fact that access ranges from free to subscription-only and the required hardware can be relatively low-cost. The network currently includes six satellites (NOAA 14-19 and METOP-A) with a polar orbit at 800-900 km, collecting data with the onboard advanced very high resolution radiometer (AVHRR/3). To take METOP as a typical (and more recent) example, the data rates range from bi-directional 4 kb/s downlink/2 kb/s uplink for telemetry, tracking, and command (TT&C) to imagery and data transmission between 1 and 3.5 Mb/s broadcast and targeted 70 Mb/s. Given the relatively high number of participating nodes, coupled with the ability to transmit to a large number of ground stations, POES would represent an excellent vehicle for DTN infrastructure.

The examples provided above are a few orders of magnitude below the current typical speed rates for the Internet, which, even at access level, reach 20–50 Mbps. At the other end of the communication spectrum, the satellites due to be deployed in the near future include data rates that are comparable with high-speed access Internet speeds. A very good example of this is provided by the Sentinel satellite network, which is due to be deployed under the ESA Copernicus initiative (previously known as the Global Monitoring for Environment and Security project). The two satellites which form the Sentinel 1 mission, running at an altitude of 693 km, will include a TT&C rate of up to 2 Mb/s downlink/64 kb/s uplink and will deliver data to stations on earth at 520 Mb/s, delivering 2.4 TB of data daily to the participating ground stations. Further, the satellites will also include onboard storage of 1.4 Tb, which would provide an excellent relay environment for low-speed distant space missions (Attema et al., 2007).

Communication between missions in general, and more specifically intercommunication between space exploration objects, would be part of standard functionality in a (ground-based) Internet scenario, with each node providing routing or message-relaying functionality for the other nodes. However, this is a rather unlikely scenario given the current operational separation between missions. In essence, each mission was designed independently, providing for communication needs only between the measurement instruments onboard the space object and the ground station, with measurement data and telemetry information sent by the space object and commands sent by the ground station. Traditionally, the exchange of data has been implemented using the CCSDS data communication reference model (CCSDS, 2007). This provides for IP-based connectivity, but includes as its typical network protocol either the Space Communication Network Specification-Network Protocol (SCPS-NP) (CCSDS, 1999) or the Space Packet Protocol (CCSDS, 2003). While SPP does include some basic routing functionality, SCPS-NP routing protocols are only at the conceptual stage (Hong and Hu, 2010). These shortcomings have recently been formally acknowledged by CCSDS, who have been working toward formally including DTN as the de-facto option for space intercommunication (CCSDS, 2010).

Although representing a proposition for the future rather than a current operational reality, DTN would allow the individual missions to interconnect and exchange data. Beyond the direct advantages of increasing reliability, wide deployment of DTN would allow a scaled-down redesign of the transmission power capabilities of space objects, in favor of increased data rates due to the meshed infrastructure. At a basic level, communication would include one-hop communication scenarios, involving an orbiter object that would relay the transmission between a remote spacecraft and the ground station. Wider deployment and interoperability between missions would further expand this to enable multiple-hop communication, involving either groups of orbiters or intermediate spacecraft acting as redundant or alternative relays for data bundles.

3.4   Implementing a deep-space DTN solution

Having considered the requirements and challenges relating to space data communication, it is also relevant to examine how a related DTN approach could be realized in practice. As such, this section considers the technology architecture that could be used to support a deep-space DTN solution, including the application requirements to receive and share the data with stakeholders, and significant supporting issues around security and protection of the data.

The implementation of a deep-space DTN requires careful consideration of the overall requirements, the stakeholders involved, and the subsequent challenges that arise. In addition to the networking requirements of the DTN, previously identified, operational aspects must also be studied, such as providing efficient and effective access to metadata—to allow a user to know what data exists to be accessed, to provide a framework to allow authentication, authorization, and accountability of transactions, and to ensure the confidentiality and integrity of the data itself is maintained.

Unfortunately, when considering many of the mainstream and well-accepted protocols for communication security, such as transport layer security (TLS), their challenge–response nature renders them infeasible for use within a DTN—as a response from one node to another cannot be guaranteed within the required time frame. However, when considering options to allow endusers to have quick and immediate access to relevant information and yet allow for the provision of large volumes of DTN-based datasets, two distinct approaches can be taken. Similarly to other network implementations, such as Universal Mobile Telecommunication Systems (UMTS—3G mobile networks) and IP Multimedia Subsystem (IMS), upon which many 4G networks are designed, signaling, or management-related data can be communicated independently of the actual content data. This separation of data responsibilities allows:

- Management-based information operating on the traditional Internet-based network. Operating at the application layer, this will enable users to be given access to all metadata related to the datasets, permit authentication, authorization, and accountability policies to be enforced, and ensure confidentiality and integrity of all management-based communications using standardized protocols such as TLS.
- Space-data information to be sent via the DTN. The actual data itself, due to its size and configuration, will be communicated via the space-data DTN. It will, therefore, be subject to the normal unpredictable operating environment of a DTN.

This separation allows a vastly simplified architecture and removes the necessity to redesign (and, importantly, validate) network protocols, with the application enabling a large number of small delay-tolerant signaling-based communications to be sent securely. The result of capitalizing upon the pre-existing Internet infrastructure permits the use of standardized web-based components for both the delivery (via a three-tier web application) and consumption of metadata (via an Internet browser).

While secure delivery of all management-based information can be achieved utilizing pre-existing solutions, an issue does still arises with respect to securing the space data itself. The widely adopted Bundle Protocol allows secure delivery of data; however, an open issue exists with respect to the distribution and management of cryptographic keys (Symington et al., 2011; Scott and Burleigh, 2007; Farrell, 2007; Farrell et al., 2009; Burgin and Hennessy, 2012). While, from an application-layer perspective, it would be possible to incorporate a key management server to enable key distribution between high-level DTN nodes (via the Internet-based connectivity), this would (with large DTN networks) result in a significantly complex and cumbersome system. Current research is examining how best to achieve this more generally, with Menesidou and Katos (2012) looking at the implementation of a one-pass protocol, and the utilization of computationally and communication-heavy long-term keys from which short-term session keys can be derived. However, such work has to be validated and the issue remains unresolved.

3.4.1 A space-data DTN architecture

The composition and architecture of a space-data DTN can take various forms, but for the purposes of illustration and discussion, we can consider one possible model in more detail, in order to examine how the required functionality can be achieved (Clarke et al., 2012). A key advantage of this model is the focus upon the reuse of proven, standardized, and accepted protocols.

Fig. 3.1 illustrates the principal interactions of the key components within a space-data DTN. In contrast to typical DTN implementations, this architecture relies upon access to normal network communications (as defined by a standard Internet access) in addition to the DTN. This capability permits the use of standard security mechanisms to protect key services—mechanisms whose operation could not be relied upon in a DTN where delay and disruption are present.

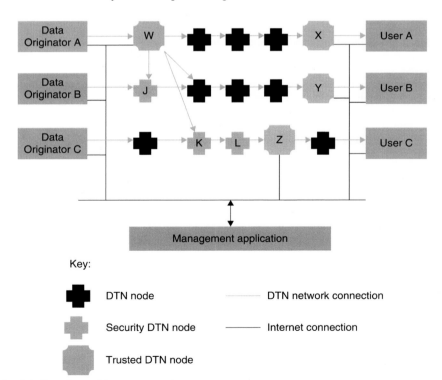

Fig. 3.1 Security architecture overview.

The security architecture is comprised of four key components:

- Management Application (MA)—a web application that enables endusers to obtain space data. The application provides authentication, authorization, and accountability services.
- Data Originator (DO)—the original source of space data that is placed within the DTN. These components are assumed to be trusted.
- End Users (EU)—the final destination of space data.
- Trusted DTN nodes (TDTN) —a subset of the DTN nodes that are able to deliver space data datasets.

Note that for simplicity and ease of understanding some DTN network connectivity between nodes is omitted from the figure. However, node "W" provides an indication of the interconnectivity of nodes that would actually exist within the DTN. The figure more generally presents three different types of network connectivity for illustration. This again is not a definitive set, but rather an example of the interactions between the principal components. Data Originators A, B, and C are all storing their datasets within the DTN network—at the Bundle layer within both the Security and Trusted DTN nodes. Complete datasets are stored at the Trusted DTN nodes. Users A, B, and C are also downloading datasets from the DTN network from the Trusted DTN nodes. In all three examples, data is sent within the DTN to untrusted DTN nodes with security being maintained between Security and Trusted DTN nodes (as specified by the Bundle Protocol security). The Management Application provides the mechanism for TDTN Nodes and Users to communicate and request datasets. Fig. 3.2 illustrates the network interactions that are sent when downloading space data from the DTN.

A user requests a dataset by logging into the Management Application and clicking upon the available datasets. A one-time URL (with sufficiently long freshness) is generated by the Management Application and sent to the most appropriate (frequently this would be geographically nearest) Trusted DTN node that is currently storing

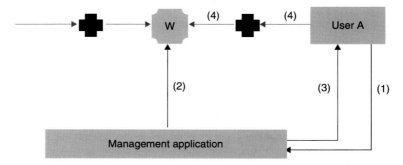

(1) User A selects the dataset they wish to download from the MA website.
(2) MA generates a unique one-time URL and provides this information to the nearest trusted DTN node (W) that is storing the requested dataset.
(3) MA also sends this one-time URL to user A.
(4) User A utilises the one-time URL to request the dataset from W.

Fig. 3.2 Data request process.

the dataset. The same URL is then sent to the user so that they can directly request the data across the DTN. All communication sent across the Internet-based network is secured. It should be noted that the process relies upon a number of assumptions (which would typically hold true for the space-data scenarios that currently exist):

- A process exists for datasets to be distributed from Data Originators onto the DTN.
- A process exists for the management application to be knowledgeable of where the datasets are distributed throughout the DTN.
- The management application, users, and trusted DTN nodes can communicate via a normal Internet-type connection.

In reality, the communication path indicated by (4) in Fig. 3.2 could be any combination of untrusted DTN nodes, security DTN nodes, and trusted DTN nodes. Indeed, for some data requests, the user might find themselves a single hop from a Trusted DTN node with the necessary datasets, whereas on other occasions the datasets might need to traverse large segments of the network.

3.4.2   Trusted DTN nodes

The term "Trusted DTN" is created in order to differentiate it from the concept of a Security DTN, which is already defined by the Bundle Protocol and provides the communication security between Security DTN nodes at the bundle layer (Symington et al., 2011). Trusted DTN nodes are still Security DTN nodes, but also include additional functionality:

- Operating above the bundle layer, they provide the functionality to store (and subsequently forward) complete datasets, rather than simply bundles (as defined by the bundle protocol).
- Trusted DTN nodes have standard Internet-based communication capabilities with the management application—that is, all management signaling information between the MA and TDTN conforms to standard Internet-based traffic conditions and is not subject to the delay and disruption that could affect a DTN network connection.

The creation of TDTN nodes, while adding an additional administrative layer to the architecture, serves to resolve three issues: providing efficient cryptographic support of datasets, enabling effective key management, and delivery of datasets. Unfortunately, providing datasets at the bundle layer would significantly increase the administrative overhead of managing and distributing the space data. Furthermore, cryptographic support at the bundle layer would significantly increase the complexity of key distribution and management. Through managing the space data as complete datasets rather than at the bundle layer, it is possible to minimize the administrative overhead of tracking and encrypting bundles. With TDTNs also connected to the Internet, an efficient and effective mechanism exists for instructing the node to forward data through the network.

The TDTN node incorporates functionality to process and store whole datasets. As illustrated in Fig. 3.3, the architecture consists of an agent and database system. What the TDTN stores, for how long, and who can access the data are all determined by the Management Application. Although it is not included within the diagram, it is

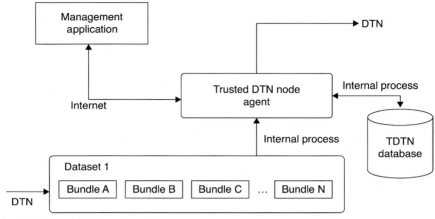

Fig. 3.3 Trusted DTN node architecture.

assumed that the bundle layer security functionality is providing the necessary data confidentiality and integrity services.

The TDTN, therefore, includes all the functionality of a standard DTN node, plus the enhanced functionality of a Security DTN node, and an additional layer (as illustrated) that provides higher-layer access that deals with complete datasets of data rather than merely bundles.

3.4.3 Data Originators

The Data Originator node includes an identical set of functionality and security mechanisms to the Trusted DTN, with the additional functionality of allowing data originator owners to interface with the agent, rather than operating completely autonomously as the Trusted DTN nodes do (as illustrated in Fig. 3.4). This

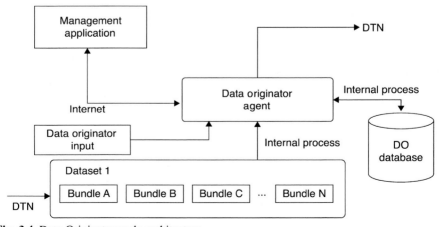

Fig. 3.4 Data Originator node architecture.

functionality is required to provide the capacity of owners to upload space data into the database for distribution via the DTN.

The Data Originator Agent continuously monitors the DO database for new space datasets. Upon identification, the agent informs the MA that such data exists—initially by automatically completing the database name and owner fields. These are derived from the file name and the owner of the DO node. It is then the responsibility of the DO owner to log in to the MA and complete the necessary metadata information regarding the dataset. Functionality within the MA will also allow automatic completion of this information for recurring or scheduled events. The DO owner will also configure the access rights for this dataset.

3.4.4   Security considerations

The suggested architecture offers a number of advantages from a security perspective, enabling services for confidentiality, integrity, authentication, authorization, and accountability. The centralized management application is key to the provision of the latter three, and it is worth briefly considering how these are addressed.

Authentication to the Management Application is required by two aspects of the architecture:

• People—end users, data originators and administrators
• Trusted components—data originators and trusted DTN nodes

Authentication of the trusted components can be managed by the previously stated communications security and the use of mutual authentication within the TLS Protocol. Authorization policies will principally control and maintain what datasets individual users are able to access. All interactions performed on the MA are subject to accountability policies. The type and time frame of the records are subject to administrative settings, but it is suggested that all interactions, from registration to password logins and data requests to data accesses, should be logged.

The MA itself can take the form of a three-tier web application: web front end, back-end system, and database. The database will store all information regarding the space data system. It is therefore a critical system and a potential single point of failure. To mitigate against such issues, the MA can take the same procedural steps as modern e-commerce systems, as they exhibit the same potential threats and issues. The system can be mirrored and regularly backed up to protect availability.

The database will also include appropriate encryption and access control policies so as to mitigate against a variety of web-based attacks, including detection for distributed denial of service (DDoS) attacks, cross-side scripting, buffer overflows, etc., providing for confidentiality and integrity services of the metadata and account information.

Finally, the servers that will operate the web application will be hardened against attack through the provision of regular patching, installation, and configuration of security countermeasures and monitoring via the system administrator.

3.5 Summary

This chapter has highlighted the clear role that DTNs can play within the context of space communications, with the domain posing particular challenges that delay-tolerant approaches are well placed to address. However, while the DTN itself addresses the difficulties associated with distance and potential interruption, it also introduces considerations in terms of how to deliver other requirements (such as data security) on top of the infrastructure.

Given the requirement to ensure secure delivery of all data, the architecture presented here offers an approach that provides protection against a wide range of DTN-based and non-DTN-based threats to which the system is vulnerable. Key to this architecture is the use of both DTN and non-DTN networks that permit the use of a combination of both DTN-specific security protocols and well-established (and thus accepted) security protocols typically found within secure Internet-based services. While the creation of the Trusted DTN nodes introduces an additional layer of complexity, they represent an efficient solution to the secure delivery of space data.

This chapter has highlighted the particular considerations and issues that arise when considering the application of DTNs to the space environment. While the architecture presented goes some way in showing how to develop a usable, flexible, and secure approach, it does not successfully address the issues of securing the DTN-based data. Further research needs to examine and develop reliable and lightweight protocols that do not rely upon the traditional challenge-response approaches.

References

Attema, E., Bargellini, P., Edwards, P., Levrini, G., Lokas, S., et al., August 2007. Sentinel 1: The radar mission for GMES operational land and sea services. ESA Bulletin 131.

Burgin, K., Hennessy, A., 2012. Suite B Ciphersuites for the Bundle Security Protocol. internet-draft http://tools.ietf.org/html/draft-irtf-dtnrg-bundle-security-19. (Accessed 7 July 2014).

Cerf, V., Burleigh, S., Hooke, A., Torgerson, L., Durst, R., et al., April 2007. Delay-Tolerant Networking Architecture. Request for Comments 4838 (RFC4838).

Clarke, N., Katos, V., Menesidou, S., Ghita, B., Furnell, S., 2012. D2.3 Security Mechanisms. Report from EU FP7 Space-Data Routers for Exploiting Space Data. http://www. spacedatarouters.eu/deliverables/. (Accessed 7 July 2014).

Consultative Committee for Space Data Systems (CCSDS), May 1999. Space Communications Protocol Specification (SCPS)—Network Protocol (SCPS-NP). Informational Report, CCSDS713.0-B-1.

Consultative Committee for Space Data Systems (CCSDS), September 2003. Space Communications Protocol Specification (SCPS)—Network Protocol (SCPS-NP). Informational Report, CCSDS133.0-B-1.

Consultative Committee for Space Data Systems (CCSDS), December 2007. Overview of Space Communications Protocols. Informational Report, CCSDS 130.0-G-2.

Consultative Committee for Space Data Systems (CCSDS), August 2010. Rationale, Scenarios, and Requirements For DTN in Space. Informational Report, CCSDS 734.0-G-1.

DARPA, September 1981. Transmission Control Protocol. Request for Comments 793 (RFC793).

Farrell, S., 2007. DTN Key Management Requirements. Work in progress as an internet-draft http://tools.ietf.org/html/draft-farrell-dtnrg-km-00. (Accessed 7 July 2014).

Farrell, A., Symington, S.F., Weiss, H., Lovell, P., 2009. Delay-Tolerant Networking Security Overview. internet-draft http://tools.ietf.org/html/draft-irtf-dtnrg-sec-overview-06. (Accessed 7 July 2014).

Hong, J., Hu, K., 2010. Research on SCPS-NP routing protocol. In: 2nd International Conference on Information Engineering and Computer Science (ICIECS), 25–26 December 2010.

Menesidou, S., Katos, V., 2012. Authenticated key exchange (AKE) in delay tolerant networks. In: Proc. of the 27th IFIP International Information Security and Privacy Conference, Springer IFIP AICT, Greece.

National Environmental Satellite, Data, and Information Services (NESDIS), 2013. Polar Operational Environmental Satellite—Polar Orbiting Satellites. http://www.oso.noaa.gov/poes/index.htm. (Accessed 1 February 2013).

Scott, K., Burleigh, S., 2007. Bundle Protocol Specification. Request for Comments, RFC5050,.

Symington, S., Farrell, S., Weiss, H., Lovell, P., 2011. Bundle Security Protocol Specification. Request for Comments, RFC 6257, http://datatracker.ietf.org/doc/rfc6257. (Accessed 7 July 2014).

Teston, F., Vuilleumier, P., Hardy, D., Bernaerts, D., October 2004. The PROBA-1 micro satellite. In: Shen, S.S., Lewis, P.E. (Eds.), Imaging Spectrometry X. Proceedings of the SPIE, vol. 5546, pp. 132–140.

Vehicular delay-tolerant networks

Vasco N.G.J. Soares[a] and Joel J. P. C. Rodrigues[b,c]
[a]Instituto de Telecomunicações, Polytechnic Institute of Castelo Branco, Castelo Branco, Portugal, [b]Federal University of Piauí (UFPI), Teresina - PI, Brazil, [c]Instituto de Telecomunicações, Portugal

4.1 Introduction

Vehicular networks have been defined in the literature as spontaneous self-organized networks, where vehicles equipped with short to medium range wireless communication, persistent storage, and processing capabilities cooperate to enable communication with other vehicles or roadside infrastructure equipment (Toor et al., 2008; Jakubiak and Koucheryavy, 2008; Yousefi et al., 2006). In these networks, nodes can be located in the line of sight or out of the radio range if a multihop network is built among several nodes.

Although research in vehicular networks dates back to the 80s (Gillan, 1989), the field exploded around the year 2000 with the development of adequate and affordable wireless communication technologies (Li and Wang, 2007). Recently, these networks are attracting much attention from governments, industries, and academic research communities all over the world. Many automobile manufacturers are already developing prototypes of vehicles equipped with sensing, computing, and wireless communication devices (CAR 2 CAR, 2008; NOW, 2007; GM-CMU Collaborative Lab, 2003). Hence, it is envisioned that vehicular networks are very likely to be deployed in a near future.

One of the main reasons, if not the most important, for the growing interest in these networks is the wide range of envisioned applications that can have a direct impact on everyday life (Khaleda et al., 2009; Toor et al., 2008), from time-critical applications to delay-tolerant applications. These networks are regarded as a key technology for improving road safety, optimizing the traffic flow, and road capacity. They can also be used as monitoring networks for sensor data collection. Several commercial and entertainment applications have been envisioned. Vehicular networks can also be employed to provide connectivity to remote rural communities and regions, or to assist in communication between rescue teams and other emergency services in catastrophe hit areas lacking a conventional communication infrastructure.

However, to harness the advantages of vehicular networks, several technical challenges need to be overcome before these networks can be widely deployed. Some of these challenges are common to other wireless networks, while others are caused by the unique properties of vehicular networks. Various authors (Jakubiak and Koucheryavy, 2008; Yousefi et al., 2006; Cruces, 2008; Franck and Gil-Castineira,

Advances in Delay-tolerant Networks (DTNs). https://doi.org/10.1016/B978-0-08-102793-6.00004-7

2007; Schoch et al., 2008), have observed that difficult communication problems arise due to highly dynamic network topology and short contact durations caused by the mobility and speed of vehicles. Limited transmission range, radio obstacles due to physical factors (e.g., buildings, tunnels, terrain, and vegetation), and interferences (e.g., high congestion channels caused by a high density of nodes), can cause disruption, intermittent connectivity, and significant loss rates. Furthermore, vehicular networks may have large node density variations due to location and time (e.g., dense in a traffic jam, sparse in suburban traffic, and extremely sparse in rural areas).

These factors make vehicular networks susceptible to frequent fragmentation and often lack continuous end-to-end connectivity, which renders data dissemination and routing a challenging task. Delay-tolerant networking concepts may provide a generic solution to this problem that will be addressed in this chapter.

This chapter is organized as follows. Section 4.2 presents a selection of the numerous applications envisioned for vehicular networks. The next Section 4.3 analyzes traditional routing and data dissemination strategies for vehicular networks and discusses the introduction of delay-tolerant networking concepts to vehicular communications. Then, in Section 4.4 a layered architecture for vehicular delay-tolerant networks (VDTNs) is presented as well as an overview of the research done with this architecture. Finally, Section 4.5 concludes the chapter presenting a summary of the review.

4.2   Vehicular networks applications

Communication among vehicles and between vehicles and roadside infrastructure using wireless technology has a large potential for enabling a plethora of applications and services, ranging from time-critical applications to delay-tolerant applications. These applications can be classified into the following main six categories: road safety, traffic optimization, commercial, entertainment, rural connectivity, and disaster scenario connectivity illustrated in Fig. 4.1. The applications presented in the following section, compiled from several sources, represent a subset of an ever-growing list.

Vehicular networks are regarded as a key technology in improving road safety (IEEE ITSS, 2009; European Standard, 2010; RITA, 2005). Vehicular communication can assist drivers to prevent an accident from happening, or if an accident occurs to prevent car pile-up. Examples of a diverse range of applications to increase road safety include cooperative collision avoidance (Tatchikou et al., 2005), collision warning (Yang et al., 2004; Elbatt et al., 2006), blind-spot warning, vision enhancement, emergency break warning, work zone warning, road hazard notification (e.g., icy road,

Fig. 4.1 Delay tolerance of vehicular network applications.

fog), emergency video streaming (Park et al., 2006), approaching emergency vehicle warning, speed limit notification, curve speed notification, and cooperative driving.

Vehicular communication can be of great use in enhancing the efficiency of transportation systems. The goal is to improve the traffic flow and road capacity, through the use of applications and services, such as traffic condition monitoring (Nadeem et al., 2004; Fahmy and Ranasinghe, 2008), platooning (i.e., vehicle following), cooperative notification systems (Buchenscheit et al., 2009), vehicle tracking, lane-changing assistance, freeway management (Anda et al., 2005), road congestion prevention (Fahmy and Ranasinghe, 2008), cooperative driving (Hubaux et al., 2004), and toll collection. Vehicular networks can also collect and relay data gathered by a wireless sensor network (Akyldiz et al., 2002; Khemapech et al., 2005) such as weather conditions (e.g., temperature, humidity, rainfall, wind), pollution measurements (e.g., smoke, visibility, noise) (Lahde et al., 2007), road surface conditions, and construction zones (Lee et al., 2006a).

Vehicular networks have other applications beyond road safety and traffic optimization. Examples of some promising commercial applications include dissemination of commercial advertisements (Lee et al., 2007) (e.g., hotels, restaurants, and gas stations), marketing data, travel (e.g., estimated bus arrival time), tourist and leisure information, and parking space availability (Lu et al., 2009). Entertainment applications can provide value-added services to users. For example, the passengers in a vehicle may access the Internet (Chen and Chan, 2009) and do cooperative downloads (Fiore and Barcelo-Ordinas, 2009; Nandan et al., 2005), or play games (Palazzi et al., 2007), chat (Smaldone et al., 2008), and share multimedia content through peer to peer (P2P) systems (e.g., music, videos) (Lee et al., 2006b) with passengers in other vehicles.

Vehicular networks are also particularly important in remote regions and rural areas that lack a fixed communication infrastructure. They can enable several nonreal-time applications, such as file transfer, electronic mail (email), cached Web access, and health monitoring (telemedicine) (N4C, 2008; Wizzy Digital Courier, 2003; Pentland et al., 2004; Doria et al., 2002; Seth et al., 2006; Farahmand et al., 2008). Finally, catastrophe hit areas lacking a conventional communication infrastructure can benefit from the deployment of a vehicular network to provide support for communication between rescue teams and assist communication between the rescue teams and other emergency services (Asplund et al., 2009).

4.3 Vehicular communications

Vehicular ad hoc networks (VANETs) (Füssle et al., 2005; Huang et al., 2009; Jakubiak and Koucheryavy, 2008; Khaleda et al., 2009; Toor et al., 2008; Torrent-Moreno et al., 2006; Yousefi et al., 2006) have been proposed as a specific type of mobile ad hoc network (MANET) (Corson and Macker, 1999; IETF MANET Working Group, 1997) where mobile nodes are vehicles like cars, trucks, buses, and motorcycles. Vehicles do not move randomly, but rather follow the road infrastructure, within the constraints of traffic flow and traffic regulations. They move at high speed, and their behavior is influenced by road signs, traffic lights, and other vehicles. The network density changes very dynamically, depending on location, time of day, or recent events (e.g., accidents).

In a VANET, communication between nodes can be classified as vehicle-to-vehicle (V2V), vehicle-to-roadside (V2R), or vehicle-to-infrastructure (V2I). Roadside units (RSUs) are static nodes deployed along the road, which are used to improve connectivity and service provision. RSUs can be connected to a core network and the Internet. These concepts are illustrated in Fig. 4.2. Several approaches and architectures have been considered to implement vehicular communications (Wu et al., 2004). Examples include a pure V2V ad hoc network, a wired backbone with wireless last-hop, or a hybrid architecture combining the previous two.

MANET routing protocols aim at establishing end-to-end connections among network nodes and ensure end-to-end semantics of existing transports and applications (Zhou, 2003; Feeney, 1999; Meghanathan, 2009; Liu and Kaiser, 2005; Abolhasan et al., 2004; Ott et al., 2006). However, several authors state that these protocols do not perform well in VANETs, mainly due to their difficulty in dealing with rapid topology changes and frequent fragmentation (Chennikara-Varghese et al., 2006; Li and Wang, 2007; Lee et al., 2010). Therefore, traditional MANET routing protocols must be modified to suit VANETs characteristics or new protocols must be designed. This has been a subject of interest for several years and has resulted in a large number of routing protocol proposals for VANETs. Theoretical background and surveys of these protocols are also presented (Lin et al., 2010; Li and Wang, 2007; Chennikara-Varghese et al., 2006; Broustis and Faloutsos, 2006; Lee et al., 2010).

Different VANET applications have different requirements. Thus a single routing protocol may not be capable of efficiently handling all the inherent characteristics of the multiplicity of previously mentioned applications, as they, for instance, may use unicast, broadcast, or multicast transmission facilities. Based on this observation, the effort has been conducted to develop routing protocols designed for particular

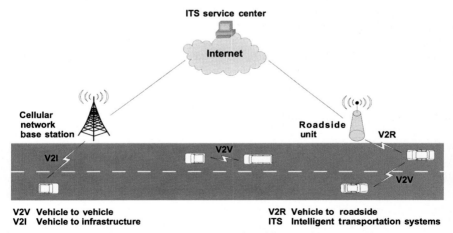

Fig. 4.2 Illustration of communication scenarios in vehicular ad hoc networks.

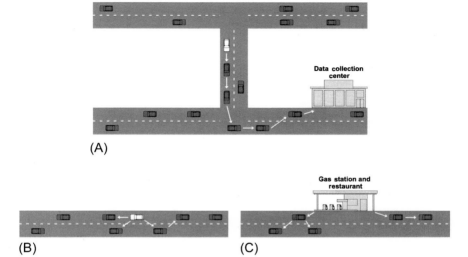

Fig. 4.3 Illustration of (A) unicast, (B) multicast/geocast, and (C) broadcast routing schemes.

applications. This observation was used to classify VANET routing protocols into three categories: unicast, multicast/geocast, and broadcast (Lin et al., 2010). Unicast routing constructs a source-to-destination path. Multicast routing is used to deliver data from one source to many interested recipients. Geocast routing is used to deliver data to a predefined geographic region. Finally, broadcast routing is used to deliver data to all nodes in the network. Fig. 4.3 illustrates these routing principles.

Unicast routing protocols can be classified into two categories: topology-based and position-based (Lee et al., 2010). Fig. 4.4 shows an illustration of these routing strategies. Topology-based routing protocols use network information about links to make routing decisions for packets. This type of routing protocols can be further subdivided into reactive and proactive protocols. Reactive routing protocols determine routes on a demand or need basis. Proactive routing protocols propagate topology information periodically and find routes continuously between any two nodes in the network, regardless of whether they are needed or not.

Position-based routing protocols, also called geographic routing protocols, do not exchange link-state information, and do not maintain established routes. They make forwarding decisions based on the geographic location of the destination node and the location of neighboring nodes. Hence, it is required that nodes have location capabilities, which can be provided by global positioning system (Hofmann-Wellenhof et al., 2001) devices or location services (Flury and Wattenhofer, 2006; Li et al., 2000; Yu et al., 2004; Capkun et al., 2001).

Most routing protocol research studies for VANETs are generally focused on scenarios with high node densities such as highways and city roads (Zhang and Wolff, 2008, 2010). However, different conditions occur in urban and rural roads/environments due to moderately or low node densities, little or no fixed roadside

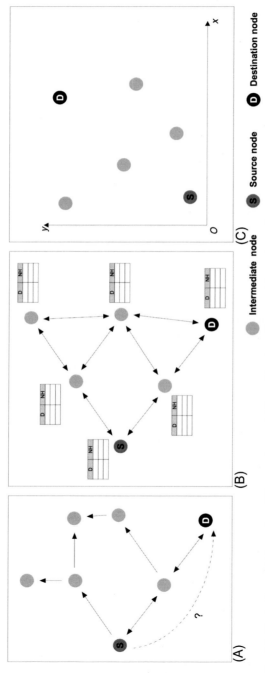

Fig. 4.4 Illustration of (A) reactive, (B) proactive, and (C) position-based/geographic routing schemes.

infrastructure available, and terrain effects. These conditions lead to long periods where V2V or V2I communications are infrequent, interrupted, or simply not possible. It has also been observed that the uneven nature of vehicle traffic and market penetration of the technology may be responsible for network fragmentation (Wisitpongphan et al., 2007). Other authors have made similar observations and concluded that vehicular networks frequently form partitions and thus prevent end-to-end communication strategies (Little and Agarwal, 2005; Jakubiak and Koucheryavy, 2008; Abuelela and Olariu, 2007; Yousefi et al., 2006).

The vehicular network performance in sparse scenarios is thus an important problem that needs careful study. Since routing protocols designed for fully connected networks are not suitable, it is necessary to adopt routing strategies that still work properly even for frequently disconnected scenarios. To address these challenges, researchers have been employing the nontraditional store-carry-and-forward (SCF) routing paradigm, which was proposed for delay tolerant networks (DTNs) (Cerf et al., 2007), in VANETs (Shao et al., 2009; Cruces, 2008; Franck and Gil-Castineira, 2007; Morillo-Pozo et al., 2008; Burgess et al., 2006; Leontiadis and Mascolo, 2007; Cabrera et al., 2009; Pereira et al., 2012).

According to the SCF paradigm, a communication opportunity between two nodes occurs when they come within the transmission range of one another, and an opportunistic communication path between a source and a destination node consists of one or more communication links over different time instants. This means that nodes are allowed to buffer messages and carry them along for long periods. A source node can directly deliver a message if it comes in contact with the destination node, otherwise, it will forward/replicate the message to other intermediate nodes in contact (according to an SCF routing scheme), hoping that the message will eventually be delivered to the destination node yet at the cost of an increased routing delay.

The SCF communication paradigm removes the need for an end-to-end path from a source to a destination node and, therefore, is attractive to a set of nonreal-time (i.e., delay-tolerant) vehicular applications. Vehicular networks following this approach are usually referred in the literature to as "DTN-enabled VANETs" (Franck and Gil-Castineira, 2007), "delay-tolerant VANETs" (Shao et al., 2009), "delay-tolerant vehicular networks" (Morillo-Pozo et al., 2008), "vehicle-based disruption-tolerant networks" (Burgess et al., 2006), or "VDTNs" (Soares et al., 2009a). There are many other examples of real DTN-based networks that follow this approach, such as, interplanetary networks, underwater networks, wildlife tracking networks, military networks, wireless sensor networks, people networks, networks for developing/rural communities, and disaster recovery networks.

In situations where the number of vehicles that can act as communication nodes is insufficiently low, direct contacts between vehicles can be so infrequent that even the SCF paradigm is insufficient, by itself, to accomplish data delivery. It is interesting to note, however, that although vehicles may not come in direct contact with each other, they may pass at the same location, at different times, one after the other. This has motivated the introduction of a special type of nodes, called stationary relay nodes (Farahmand et al., 2008) or throw boxes (Zhao et al., 2006), as roadside infrastructure elements that are strategically placed to increase contact opportunities between vehicles. These fixed battery-powered relays are installed at road intersections and have radio, storage, and processing capabilities, allowing passing-by vehicles to collect and leave data on them. Several studies have shown the importance of these nodes to improve the delivery rate as well as to reduce the delivery delay in VDTNs, and have analyzed the placement problem to improve network performance (Rodrigues et al., 2011; Farahmand et al., 2011; Banerjee et al., 2008; Zhao et al., 2006). A hardware and software architecture for energy efficiency that aims to maximize performance and simultaneously meet the energy constraints of these nodes has also been proposed (Banerjee et al., 2007).

Fig. 4.5 illustrates the concepts of SCF and stationary relay nodes in a VDTN. Vehicle A forwards its data to vehicle B as contact opportunity occurs at time $t + t_0$. Then,

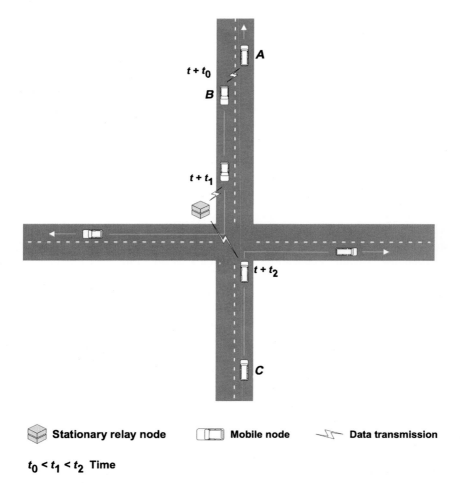

Stationary relay node   **Mobile node**   **Data transmission**

$t_0 < t_1 < t_2$ **Time**

Fig. 4.5 Illustration of store-carry-and-forward data forwarding among vehicles and a stationary relay node deployed at a road intersection.

vehicle B stores the data in its buffer and follows its route. Later, vehicle B finds a stationary relay node at time $t + t_1$ and forwards data to it. Following a different route, vehicle C passes along the stationary relay node at a later time $t + t_2$, collecting the data left there by vehicle B. Thus, the stationary relay node creates an additional contact opportunity that would not exist before, since vehicles B and C would not meet each other.

4.4   Vehicular delay-tolerant networks

The DTN architecture offers a promising solution for improving the performance of routing and data dissemination in sparse or partitioned opportunistic vehicular networks that have received increasing attention. Nevertheless, this architecture departs from the Internet model that is defined by the end-to-end use of the Internet protocol (IP) rather than bundling (Burleigh et al., 2003). Besides, it can be observed that IP itself provides an asynchronous packet delivery mechanism and IP packet delivery can be delayed if applications tolerate large delays and asynchronous communications (Ochiai et al., 2009).

These observations motivated the proposal of a network architecture for vehicular communications in sparse and partitioned network environments called VDTN-layered architecture (Soares et al., 2009a). VDTN architecture follows the SCF paradigm proposed for DTNs, however, contrary to DTN architecture that introduces the overlay bundle layer over the transport layer to allow the interconnection of highly heterogeneous networks, VDTN architecture places the bundle layer below the network layer, introducing an IP over VDTN approach. A comparison between DTN and VDTN layers is illustrated in Fig. 4.6.

Another distinctive characteristic of VDTN architecture is the proposal of out-of-band signaling with separation of the control and data planes. As may be observed in Fig. 4.6, the VDTN bundle layer is divided into two layers: the bundle signaling

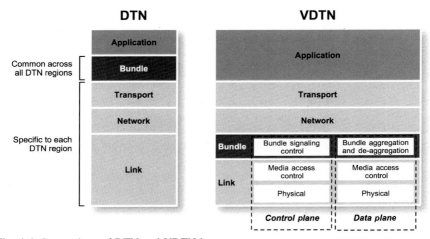

Fig. 4.6 Comparison of DTN and VDTN layers.

control layer (BSC) and the bundle aggregation and de-aggregation layer (BAD). BSC layer performs the control plane functions including, among others, signaling, routing, node localization, resources reservation (at the data plane), and other network protocols that are used to set up, maintain, and terminate data plane connections. BAD performs the data plane functions. The data plane is responsible for the transport of incoming IP packets, which are aggregated into data bundles, from a source node to single or multiple destination nodes. Hence, the functions executed at this plane include, among others, buffer management (queuing), scheduling, traffic classification/differentiation, data aggregation/de-aggregation, and forwarding.

The main idea is to assemble IP packets into variable-length data bundles, and route them asynchronously through the network, using the data plane. The data plane connection is set up using out-of-band signaling information previously transmitted through a separate control plane connection. This means that the control plane can exchange signaling information through a separate, dedicated, low-power, low bandwidth, and long-range link, which is always active to allow node discovery. On the contrary, the data plane may use a high-power, high bandwidth, and short-range link to exchange data bundles, which is active only during the estimated contact duration time and if there are data bundles to be exchanged between the network nodes. The use of out-of-band signaling procedures ensures the optimization of the data plane resources (e.g., storage and bandwidth) and also allows saving power, which is very important for power-limited network nodes such as stationary relay nodes (Soares et al., 2010c).

The concept of out-of-band signaling and data-plane link activation and de-activation is illustrated in Fig. 4.7. At the time $t + t_0$, a mobile node (i.e., vehicle) and a stationary relay node discover each other and initiate signaling messages

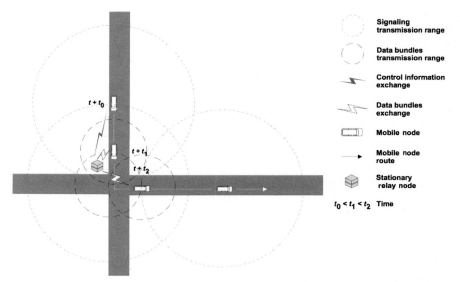

Fig. 4.7 Illustration of control information and data bundles exchange between VDTN network nodes.

exchange through the control plane link connection. Using this information, the data plane connection is configured and activated on both nodes at the time $t+t_1$. Then, the data bundles are exchanged until the time $t+t_2$. After this time, the data plane link connection is deactivated because the nodes are no longer in the data plane link range of each other.

IP over VDTN architecture supports a class of vehicular network applications characterized by delay tolerant, asynchronous data traffic. Such applications can even tolerate some data loss. Several research works have been published in the last few years that investigate the use of VDTN network architecture. A survey of the most relevant studies is presented next.

Given the particularities of VDTN network architecture, a simulation tool called VDTNsim was created to support performance studies related to the development, experimentation, and performance evaluation of protocols, algorithms, services, and applications for these networks (Soares et al., 2010b). A laboratory testbed prototype for VDTNs called VDTN@Lab (Dias et al., 2011b) was developed to contribute to the validation of simulation models, as well as a real-world testbed (Paula et al., 2011) that demonstrates and validates the technical concepts of the VDTN architecture in a real environment. Figs. 4.8 and 4.9 show pictures of these testbeds.

VDTN bundle layer also defines the "bundle" as its protocol data unit. However, in the case of VDTN architecture, a bundle consists of an aggregate of IP packets with common attributes (e.g., destination address and quality of service). The aggregation of IP packets to form the payload of data bundles is expected to result in fewer packets processing and routing decisions, which can be translated to less complexity, faster routing, lower network cost, and energy savings. This topic was studied in detail (Isento et al., 2011), wherein several algorithms for the assembly process were evaluated.

The separation of the control and data plane with out-of-band signaling was explored to introduce the function of node localization at the control plane (Soares et al., 2010c). This functionality can be used to predict the contact durations and

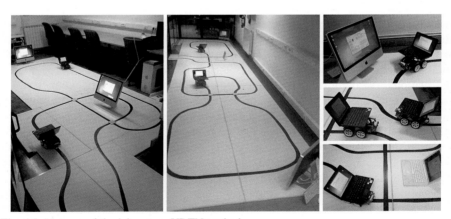

Fig. 4.8 Pictures of the laboratory VDTN testbed.

Fig. 4.9 Pictures of the real-world VDTN testbed.

the maximum number of bytes that can be transmitted during contact opportunities, thus preventing incomplete bundle transmissions and the waste of data link capacity. It was also shown that knowing the contact duration, makes it possible to determine the period during which the data plane link should be activated, and thus extending the battery life of power-limited nodes, such as stationary relay nodes.

VDTN architecture also considers the use of fragmentation and reassembly to ensure that contact volumes are fully utilized, thus avoiding the retransmission of partially transmitted bundles. In this sense, proactive and reactive DTN fragmentation schemes have been evaluated in VDTNs (Dias et al., 2011a).

Concerning routing in VDTNs, a geographic routing protocol called GeoSpray was proposed Soares et al., 2011b. GeoSpray is based on the following design principles: (i) supporting an opportunistic networking paradigm and the delivery of data bundles based on the SCF paradigm; (ii) using geographical location information provided by positioning devices to assist in routing decisions; (iii) employing a multiple-copy routing scheme, with a strict upper bound on the number of copies per bundle, combined with a forwarding routing strategy, to improve the timely delivery of bundles across multihop routes; (iv) clearing bundles that have already been delivered to the destinations; (v) optimizing the resources used in the network, including storage, bandwidth, and energy; and (vi) maximizing the bundle delivery probability, minimizing the bundle delivery delay, and overhead. Results of the performance assessment studies on this protocol have shown that it significantly improves the delivery probability, reduces the delivery delay, and is efficient in terms of storage and bandwidth resources utilization when compared to well-known routing protocols for DTN-based networks Soares et al., 2011b, 2012.

Routing mechanisms rely on the assumption that network nodes are willing to cooperate with the SCF services. However, fully cooperative behaviors cannot be taken for granted. For example, some network nodes might not be willing to unconditionally accept bundles sent by other nodes, to save buffer resources for their data. The same applies to schedule bundle forwarding. Some nodes might schedule their bundles for transmission first, instead of bundles stored in the buffer received from other nodes. The effects of node cooperation in the VDTN network performance were also investigated (Soares and Rodrigues, 2011). The study focused on evaluating the impact of node cooperation at the data plane level. From the analysis of the results, it was made clear that noncooperative behavior severely affects the delivery probability and the delivery delay. This study paves the way for further research on mechanisms for incentivizing cooperation between nodes in VDTNs.

To achieve multihop data delivery, the SCF paradigm can be combined with replication-based routing protocols. However, since nodes in a VDTN are resource-constrained (e.g., bandwidth and storage capacity), a key challenge is to provide scheduling and dropping policies that contribute to performance improvement of the network. This topic has been studied (Soares et al., 2010a) and the results obtained have shown the importance of using a combination of a scheduling and a dropping policy that gives preferential treatment to less replicated bundles. It was demonstrated that such a combination could greatly improve the delivery probability and the delivery delay when compared with the traditional head-drop, first-in-first-out scheduling.

The "best-effort" SCF service is not adequate in a scenario where several nonrealtime VDTN applications, with different performance requirements, compete for scarce network resources like bandwidth and storage. This motivated the study of algorithms, mechanisms, and protocols for the support of traffic differentiation in these networks (Soares et al., 2011a). A priority class of service model was considered and it was shown how buffer management strategies, dropping, and scheduling policies, can provide strict priority-based services or custom allocation of network resources. The observed results confirmed the importance of supporting traffic differentiation on VDTNs and motivate further research on this topic.

Another topic studied was the introduction of content storage and retrieval mechanisms to VDTNs (Silva et al., 2011). The motivation for this comes from the observation that if application protocols in DTN-based networks have a request-response or publish-subscribe paradigm and its contents are relevant for various users, then such mechanisms would allow resource retrieval with caching and distributed resource storage in conjunction with retrieval. This would minimize end-to-end interactions and increase application performance (Ott and Pitkänen, 2007). Results have shown that these mechanisms improve VDTN network performance in terms of delivery probability and latency (Silva et al., 2010).

Various studies have also been conducted to evaluate the use of VDTNs to enable nonreal-time data exchange in rural connectivity scenarios. The potential of stationary relay nodes in improving the network performance was further investigated (Rodrigues et al., 2011). The research indicated that even in such occasional opportunistic connectivity scenarios these nodes effectively increase the number of contact opportunities and thus contribute to significantly improve the delivery probability and the delivery delay. Also of interest was the potential impact of mobile node density and mobile node movement models on the number of contact opportunities, the delivery probability, and the delivery delay (Soares et al., 2009c). Another relevant study about the influence of storage capacity constraints on the delivery probability was presented (Soares et al., 2009b). It was concluded that routing protocols with different replication strategies react in a different way to the increase of the buffer size in specific network nodes.

4.5   Conclusion

This chapter has focused on the problems and challenges of vehicular networking. Vehicular communication paradigms were overviewed and compared, discussing their influence on routing and data dissemination strategies. A particular interest was given to the application of delay-tolerant networking concepts as a research approach to solve the unique communication problems that vehicular networks exhibit.

A VDTN architecture was presented. VDTN seeks novel and effective solutions for communicating in vehicular environments where continuous end-to-end connectivity cannot be assumed. This network architecture adopts a SCF paradigm combined with an IP over the VDTN approach and out-of-band signaling with control and data plane separation. Finally, an overview of research conducted in VDTN architecture was also provided and open research issues were highlighted.

This chapter aimed at stimulating research and contributes to further advances in the design and study of VDTNs. It is meant to serve as a good starting point for interested researchers who want to get quickly into this field that has been receiving increasing attention in recent years.

Acknowledgments

This work has been supported by FCT/MCTES through national funds and when applicable co-funded EU funds under the UIDB/50008/2020; by Brazilian National Council for Research and Development (CNPq) via Grant No. 309335/2017-5.

References

Abolhasan, M., Wysocki, T., Dutkiewicz, E., 2004. A review of routing protocols for mobile ad hoc networks. Ad Hoc Netw. 2, 1–22.
Abuelela, M., Olariu, S., 2007. Traffic-adaptive packet relaying in VANET. In: The Fourth ACM International Workshop on Vehicular Ad Hoc Networks (VANET 2007), in Conjunction With ACM MobiCom 2007, September 10, 2007, Montréal, QC, Canada, pp. 77–78.
Akyldiz, I.F., Su, W., Sankarasubramaniam, Y., Cayirci, E., 2002. Wireless sensor networks: a survey. Comput. Netw. 38, 393–422.
Anda, J., Lebrun, J., Ghosal, D., Chuah, C.-N., Zhang, M., 2005. VGrid: vehicular adhoc networking and computing grid for intelligent traffic control. In: IEEE 61st Semiannual Vehicular Technology Conference (VTC 2005), May 30-June 1, 2005, Stockholm, Sweden, pp. 2905–2909.
Asplund, M., Nadjm-Tehrani, S., Sigholm, J., 2009. Emerging Information Infrastructures: Cooperation in Disasters. Lecture Notes in Computer Science, Critical Information Infrastructure Security, Springer, Berlin/Heidelberg.

Banerjee, N., Corner, M.D., Levine, B.N., 2007. An energy-efficient architecture for DTN throwboxes. In: 26th IEEE International Conference on Computer Communications (INFOCOM 2007), May 6-12, 2007, Anchorage, Alaska, USA, pp. 776–784.

Banerjee, N., Corner, M.D., Towsley, D., Levine, B.N., 2008. Relays, base stations, and meshes: enhancing mobile networks with infrastructure. In: 14th ACM International Conference on Mobile Computing and Networking (MobiCom 2008), September 14-19, 2008, San Francisco, California, USA, pp. 81–91.

Broustis, I., Faloutsos, M., 2006. Routing in Vehicular Networks: Feasibility, Security and Modeling Issues. Department of Computer Science and Engineering, University of California, Riverside.

Buchenscheit, A., Schaub, F., Kargl, F., Weber, M., 2009. A VANET-based emergency vehicle warning system. In: First IEEE Vehicular Networking Conference (IEEE VNC 2009), October 28-30, 2009, Tokyo, Japan, pp. 1–8.

Burgess, J., Gallagher, B., Jensen, D., Levine, B., 2006. MaxProp: routing for vehicle-based disruption-tolerant networks. In: 25th IEEE International Conference on Computer Communications (INFOCOM 2006), April 23-29, 2006, Barcelona, Catalunya, Spain, pp. 1–11.

Burleigh, S., Hooke, A., Torgerson, L., Fall, K., Cerf, V., Durst, B., Scott, K., Weiss, H., 2003. Delay-tolerant networking: an approach to interplanetary internet. IEEE Commun. Mag. 41, 128–136.

Cabrera, V., Ros, F.J., Ruiz, P.M., 2009. Simulation-based study of common issues in VANET routing protocols. In: IEEE 69th Vehicular Technology Conference (VTC2009-Spring), April 26-29, 2009, Barcelona, Spain.

Capkun, S., Hamdi, M., Hubaux, J.-P., 2001. GPS-free positioning in mobile ad-hoc networks. In: 34th Annual Hawaii International Conference on System Sciences (HICSS-34), January 3-6, 2001, Maui, Hawaii.

CAR 2 CAR, 2008. CAR 2 CAR Communication Consortium. Available from: http://www.car-to-car.org/. (Accessed January 2013).

Cerf, V., Burleigh, S., Hooke, A., Torgerson, L., Durst, R., Scott, K., Fall, K., Weiss, H., April 2007. Delay-tolerant networking architecture. RFC 4838. Available from: http://www.rfc-editor.org/rfc/rfc4838.txt.

Chen, B.B., Chan, M.C., 2009. MobTorrent: a framework for mobile internet access from vehicles. In: 28th IEEE Conference on Computer Communications (IEEE INFOCOM 2009), April 19-25, 2009, Rio de Janeiro, Brazil, pp. 1404–1412.

Chennikara-Varghese, J., Chen, W., Altintas, O., Cai, S., 2006. Survey of routing protocols for inter-vehicle communications. In: 3rd Annual International Conference on Mobile and Ubiquitous Systems: Networks and Services (MOBIQUITOUS 2006) Workshops— Second International Workshop on Vehicle-to-Vehicle Communications (V2VCOM 2006), July 17-21, 2006, San Jose, California, USA.

Corson, S., Macker, J., January 1999. Mobile ad hoc networking (MANET): routing protocol performance issues and evaluation considerations. RFC 2501. Available from: http://www.ietf.org/rfc/rfc2501.txt.

Cruces, O.T., 2008. Applying Delay Tolerant Protocols to VANETs (Master thesis). Universitat Politècnica de Catalunya.

Dias, J.A., Isento, J.N., Rodrigues, J.J.P.C., Pereira, P.R., Lloret, J., 2011a. Performance implications of fragmentation mechanisms on vehicular delay-tolerant networks. In: 11th International Conference on ITS Telecommunications (ITST 2011), August 23-25, 2011, St. Petersburg, Russia, pp. 436–441.

Dias, J.A., Isento, J.N., Soares, V.N.G.J., Rodrigues, J.J.P.C., 2011b. Impact of scheduling and dropping policies on the performance of vehicular delay-tolerant networks. In: 2011 IEEE

International Conference on Communications (IEEE ICC 2011)—Communication Software, Services and Multimedia Applications Symposium (ICC'11 CSMA), June 5-9, 2011, Kyoto, Japan, pp. 1–5.

Doria, A., Uden, M., Pandey, D.P., 2002. Providing connectivity to the saami nomadic community. In: 2nd International Conference on Open Collaborative Design for Sustainable Innovation, December 2002, Bangalore, India.

Elbatt, T., Goel, S.K., Holland, G., Krishnan, H., Parikh, J., 2006. Cooperative collision warning using dedicated short range wireless communications. In: Third ACM International Workshop on Vehicular Ad Hoc Networks (VANET 2006), in Conjunction With ACM MobiCom 2006, September 29, 2006, Los Angeles, California, USA, pp. 1–9.

European Standard, March 2010. Intelligent Transport Systems (ITS); Communications Architecture. Draft ETSI EN 302 665 V1.0.0. Available from: http://www.etsi.org/deliver/etsi_en/302600_302699/302665/01.00.00_20/en_302665v010000c.pdf.

Fahmy, M.F., Ranasinghe, D.N., 2008. Discovering automobile congestion and volume using VANET's. In: 8th International Conference on ITS Telecommunications (ITST 2008), October 22-24, 2008, Thailand.

Farahmand, F., Patel, A.N., Jue, J.P., Soares, V.G., Rodrigues, J.J., 2008. Vehicular wireless burst switching network: enhancing rural connectivity. In: 3rd IEEE Workshop on Automotive Networking and Applications (Autonet 2008), Co-located With IEEE GLOBECOM 2008, December 4, 2008, New Orleans, LA, USA, pp. 1–7.

Farahmand, F., Cerutti, I., Patel, A.N., Jue, J.P., Rodrigues, J.J.P.C., 2011. Performance of vehicular delay-tolerant networks with relay nodes. Wirel. Commun. Mob. Comput. 11, 929–938.

Feeney, L.M., 1999. A Taxonomy for Routing Protocols in Mobile Ad Hoc Networks. Swedish Institute of Computer Science.

Fiore, M., Barcelo-Ordinas, J.M., 2009. Cooperative download in urban vehicular networks. In: Sixth IEEE International Conference on Mobile Ad-hoc and Sensor Systems (IEEE MASS 2009), October 12-15, 2009, University of Macau, Macau SAR, P.R.C, pp. 20–29.

Flury, R., Wattenhofer, R., 2006. MLS: an efficient location service for mobile ad hoc networks. In: Seventh ACM International Symposium on Mobile Ad Hoc Networking and Computing (MobiHoc 2006), May 22-25, 2006, Florence, Italy, pp. 226–237.

Franck, L., Gil-Castineira, F., 2007. Using delay tolerant networks for Car2Car communications. In: IEEE International Symposium on Industrial Electronics 2007 (ISIE 2007), 4-7 June 2007, Vigo, Spain, pp. 2573–2578.

Füssle, H., Torrent-Moreno, M., Transier, M., Festag, A., Hartenstein, H., 2005. Thoughts on a protocol architecture for vehicular ad-hoc networks. In: 2nd International Workshop on Intelligent Transportation (WIT 2005), March 15-16, 2005, Hamburg, Germany.

Gillan, W.J., 1989. PROMETHEUS and DRIVE: their implications for traffic managers. In: Vehicle Navigation and Information Systems Conference, September 11-13, 1989, Toronto, Ont., Canada, pp. 237–243.

GM-CMU Collaborative Lab, 2003. General Motors Collaborative Research Lab @ Carnegie Mellon University. Available from: http://gm.web.cmu.edu/. (Accessed January 2013).

Hofmann-Wellenhof, B., Lichtenegger, H., Collins, J., 2001. Global Positioning System: Theory and Practice. Springer Wien New York, Austria.

Huang, C.-M., Chen, J.-L., Chang, Y.-C., 2009. Telematics Communication Technologies and Vehicular Networks: Wireless Architectures and Applications. Information Science Publishing, Germany.

Hubaux, J.-P., Čapkun, S., Luo, J., 2004. The security and privacy of smart vehicles. IEEE Secur. Priv. Mag. 2, 49–55.

IEEE ITSS, 2009. IEEE Intelligent Transportation Systems Society. Available from: http://ewh. ieee.org/tc/its/. (Accessed January 2013).

IETF MANET Working Group, 1997. Mobile Ad-hoc Networks. Available from: http:// datatracker.ietf.org/wg/manet/charter/. (Accessed January 2013).

Isento, J.N., Dias, J.A., Rodrigues, J.J.P.C., Chen, M., Lin, K., 2011. Performance assessment of aggregation and de-aggregation algorithms for vehicular delay-tolerant networks. In: 8th IEEE International Conference on Mobile Ad-hoc and Sensor Systems (IEEE MASS 2011), October 17-22, 2011, Valencia, Spain, pp. 158–160.

Jakubiak, J., Koucheryavy, Y., 2008. State of the art and research challenges for VANETs. In: Fifth IEEE Consumer Communications & Networking Conference (CCNC 2008)—2nd IEEE Workshop on Broadband Wireless Access, January 10-12, 2008, Las Vegas, Nevada, USA, pp. 912–916.

Khaleda, Y., Tsukadaa, M., Santab, J., Choia, J., Ernst, T., 2009. A usage oriented analysis of vehicular networks: from technologies to applications. J. Commun. 4, 357–368.

Khemapech, I., Duncan, I., Miller, A., 2005. A survey of wireless sensor networks technology. In: 6th Annual Postgraduate Symposium on the Convergence of Telecommunications, Networking and Broadcasting, June 27-28, 2005, Liverpool, UK.

Lahde, S., Doering, M., Pöttner, W.-B., Lammert, G., Wolf, L., 2007. A practical analysis of communication characteristics for mobile and distributed pollution measurements on the road. Wirel Commun Mob Comput. 7, 1209–1218.

Lee, U., Magistretti, E., Zhou, B., Gerla, M., Bellavista, P., Corradi, A., 2006a. MobEyes: smart mobs for urban monitoring with a vehicular sensor network. IEEE Wirel. Commun. 13, 52–57.

Lee, U., Park, J.-S., Yeh, J., Pau, G., Gerla, M., 2006b. CodeTorrent: content distribution using network coding in VANET. In: 1st International Workshop on Decentralized Resource Sharing in Mobile Computing and Networking (ACM MobiShare 2006), in Conjunction With ACM MobiCom 2006, September 25, 2006, Los Angeles, CA, USA.

Lee, S.-B., Pan, G., Park, J.-S., Gerla, M., Lu, S., 2007. Secure incentives for commercial ad dissemination in vehicular networks. In: 8th ACM International Symposium on Mobile Ad Hoc Networking and Computing (MobiHoc 2007), September 9-14, 2007, Montreal, Quebec, Canada, pp. 150–159.

Lee, K.C., Lee, U., Gerla, M., 2010. Survey of routing protocols in vehicular ad hoc networks. In: Watfa, M. (Ed.), Advances in Vehicular Ad-hoc Networks: Developments and Challenges. Information Science Reference (IGI Global).

Leontiadis, I., Mascolo, C., 2007. GeOpps: geographical opportunistic routing for vehicular networks. In: IEEE International Symposium on a World of Wireless, Mobile and Multimedia Networks 2007 (WoWMoM 2007), 18-21 June 2007, Espoo, Finland, pp. 1–6.

Li, F., Wang, Y., 2007. Routing in vehicular ad hoc networks: a survey. IEEE Veh. Technol. Mag. 2, 12–22.

Li, J., Jannotti, J., Couto, D.S.J.D., Karger, D.R., Morris, R., 2000. A scalable location service for geographic ad hoc routing. In: 6th Annual International Conference on Mobile Computing and Networking (MobiCom 2000), August 6-11, 2000, Boston, Massachusetts, pp. 120–130.

Lin, Y.-W., Chen, Y.-S., Lee, S.-L., 2010. Routing protocols in vehicular ad hoc networks: a survey and future perspectives. J. Inf. Sci. Eng. 26, 913–932.

Little, T.D.C., Agarwal, A., 2005. An information propagation scheme for VANETs. In: 8th International IEEE Conference on Intelligent Transportation Systems, September 13-16, 2005, Vienna, Austria, pp. 155–160.

Liu, C., Kaiser, J.R., 2005. A Survey of Mobile Ad Hoc Network Routing Protocols. University of Magdeburg.

Lu, R., Lin, X., Zhu, H., Shen, X.S., 2009. SPARK: a new VANET-based smart parking scheme for large parking lots. In: 28th IEEE Conference on Computer Communications (INFOCOM 2009), April 19-25, 2009, Rio de Janeiro, Brazil, pp. 1413–1421.

Meghanathan, N., 2009. Survey and taxonomy of unicast routing protocols for mobile ad hoc networks. Int. J. Appl. Graph Theory Wirel Ad hoc Netw. Sens. Netw. 1, 1–21.

Morillo-Pozo, J., Barcelo-Ordinas, J.M., Trullos-Cruces, O., Garcia-Vidal, J., 2008. Applying cooperation for delay tolerant vehicular networks. In: Fourth EuroFGI Workshop on Wireless and Mobility, January 16-18, 2008, Barcelona, Spain, pp. 12–13.

Nadeem, T., Dashtinezhad, S., Liao, C., Iftode, L., 2004. Traffic view: a scalable traffic monitoring system. In: 2004 IEEE International Conference on Mobile Data Management, January 2004, Berkley, GA, pp. 13–26.

Nandan, A., Das, S., Pau, G., Gerla, M., Sanadidi, M.Y., 2005. Co-operative downloading in vehicular ad-hoc wireless networks. In: Second Annual Conference on Wireless On-demand Network Systems and Services (WONS'05), January 19-21, 2005, St. Moritz, Switzerland, pp. 32–41.

Anon., 2008. Networking for Communications Challenged Communities: Architecture, Test Beds and Innovative Alliances. Available from: http://www.n4c.eu/. (Accessed January 2013).

NOW, 2007. NOW Project: Network on Wheels. Available from: http://dsn.tm.kit.edu/english/projects_now-project.php. (Accessed January 2013).

Ochiai, H., Shimotada, K., Esaki, H., 2009. IP over DTN: large-delay asynchronous packet delivery in the internet. In: International Conference on Ultra Modern Telecommunications (ICUMT 2009)—Workshop on the Emergence of Delay-/Disruption-Tolerant Networks (E-DTN 2009), October 14, 2009, St.-Petersburg, Russia.

Ott, J., Pitkänen, M., 2007. DTN-based content storage and retrieval. In: IEEE International Symposium on a World of Wireless, Mobile and Multimedia Networks (WoWMoM 2007)—First IEEE Workshop on Autonomic and Opportunistic Communications (AOC 2007), June 18-21, 2007, Helsinki, Finland.

Ott, J., Kutscher, D., Dwertmann, C., 2006. Integrating DTN and MANET routing. In: ACM SIGCOMM Workshop on Challenged Networks (CHANTS 2006), September 15, 2006, Pisa, Italy, pp. 221–228.

Palazzi, C.E., Roccetti, M., Ferretti, S., Pau, G., Gerla, M., 2007. Online games on wheels: fast game event delivery in vehicular ad-hoc networks. In: Third International Workshop on Vehicle-to-Vehicle Communications 2007 (V2VCOM 2007), in Conjunction With IEEE Intelligent Vehicles Symposium 2007, June 12, 2007, Istanbul, Turkey.

Park, J.-S., Lee, U., Oh, S.Y., Gerla, M., Lun, D.S., 2006. Emergency related video streaming in VANET using network coding. In: The Third ACM International Workshop on Vehicular Ad Hoc Networks (VANET 2006), September 29, 2006, Los Angeles, California, USA, pp. 102–103.

Paula, M.C.G., Isento, J.N., Dias, J.A., Rodrigues, J.P.C., 2011. A real-world VDTN testbed for advanced vehicular services and applications. In: 16th IEEE International Workshop on Computer Aided Modeling Analysis and Design of Communication Links and Networks (IEEE CAMAD 2011), June 10-11, 2011, Kyoto, Japan.

Pentland, A., Fletcher, R., Hasson, A., 2004. DakNet: rethinking connectivity in developing nations. IEEE Comput. 37, 78–83.

Pereira, P.R., Casaca, A., Rodrigues, J.J.P.C., Soares, V.N.G.J., Triay, J., Cervelló-Pastor, C., 2012. From delay-tolerant networks to vehicular delay-tolerant networks. IEEE Commun. Surv. Tutorials 14, 1166–1182.

RITA, 2005. Intelligent Transportation Systems (ITS)—Joint Program Office. Available from: http://www.its.dot.gov/index.htm. (Accessed January 2013).

Rodrigues, J.J.P.C., Soares, V.N.G.J., Farahmand, F., 2011. Stationary relay nodes deployment on vehicular opportunistic networks. In: Denko, M.K. (Ed.), Mobile Opportunistic Networks: Architectures, Protocols and Applications. CRC Press—Taylor & Francis Group (hardcover), USA.

Schoch, E., Kargl, F., Weber, M., Leinmüller, T., 2008. Communication patterns in VANETs. IEEE Commun. Mag. 46, 119–125.

Seth, A., Kroeker, D., Zaharia, M., Guo, S., Keshav, S., 2006. Low-cost communication for rural internet kiosks using mechanical backhaul. In: 12th ACM International Conference on Mobile Computing and Networking (MobiCom 2006), September 24-29, 2006, Los Angeles, CA, USA, pp. 334–345.

Shao, Y., Liu, C., Wu, J., 2009. Delay-tolerant networks in VANETs. In: Olariu, S., Weigle, M. C. (Eds.), Vehicular Networks: From Theory to Practice. Chapman & Hall/CRC Computer & Information Science Series.

Silva, B.M., Soares, V.N.G.J., Proença Jr., M.L., Rodrigues, J.J.P.C., 2010. Impact of content storage and retrieval mechanisms on the performance of vehicular delay-tolerant networks. In: 18th International Conference on Software, Telecommunications and Computer Networks (SoftCOM 2010), September 23-25, 2010, Split—Bol, Croatia, pp. 407–411.

Silva, B.M., Soares, V.N.G.J., Rodrigues, J.J.P.C., 2011. Towards intelligent caching and retrieval mechanisms for upcoming proposals on vehicular delay-tolerant networks. J. Commun. Softw. Syst. 7, 1–8.

Smaldone, S., Han, L., Shankar, P., Iftode, L., 2008. RoadSpeak: enabling voice chat on roadways using vehicular social networks. In: European Conference on Computer Systems (EuroSys 2008)—1st Workshop on Social Network Systems, 31st March-4th April 2008, Glasgow, Scotland.

Soares, V.N.G.J., Farahmand, F., Rodrigues, J.J.P.C., 2011. GeoSpray: a geographic routing protocol for vehicular delay-tolerant networks. In: Information Fusion. vol. 15. Elsevier, pp. 102–113, https://doi.org/10.1016/j.inffus.2011.11.003.

Soares, V.N.G.J., Rodrigues, J.J.P.C., 2011. Chapter 7: Cooperation in DTN-based network architectures. In: Misra, S., Obaidat, M. (Eds.), Cooperative Networking. Wiley, pp. 101–115, https://doi.org/10.1002/9781119973584.ch7.

Soares, V.N.G.J., Farahmand, F., Rodrigues, J.J.P.C., 2009a. A layered architecture for vehicular delay-tolerant networks. In: Fourteenth IEEE Symposium on Computers and Communications (ISCC'09), July 5-8, 2009, Sousse, Tunisia, pp. 122–127.

Soares, V.N.G.J., Farahmand, F., Rodrigues, J.J.P.C., 2009b. Evaluating the impact of storage capacity constraints on vehicular delay-tolerant networks. In: Second International Conference on Communication Theory, Reliability, and Quality of Service (CTRQ 2009), July 20-25, 2009, Colmar, France, pp. 75–80.

Soares, V.N.G.J., Farahmand, F., Rodrigues, J.J.P.C., 2009c. Impact of vehicle movement models on VDTN routing strategies for rural connectivity. Int. J. Mob. Netw. Des. Innov. 3, 103–111.

Soares, V.N.G.J., Farahmand, F., Rodrigues, J.J.P.C., 2010a. Performance analysis of scheduling and dropping policies in vehicular delay-tolerant networks. Int. J. Adv. Internet Technol. 3, 137–145.

Soares, V.N.G.J., Farahmand, F., Rodrigues, J.J.P.C., 2010b. VDTNsim: a simulation tool for vehicular delay-tolerant networks. In: 15th IEEE International Workshop on Computer-Aided Modeling Analysis and Design of Communication Links and Networks (IEEE CAMAD 2010), December 3-4, 2010, Miami, FL, USA, pp. 101–105.

Soares, V.N.G.J., Rodrigues, J.J.P.C., Farahmand, F., Denko, M., 2010c. Exploiting node localization for performance improvement of vehicular delay-tolerant networks. In: 2010 IEEE International Conference on Communications (IEEE ICC 2010)—General Symposium on

Selected Areas in Communications (ICC'10 SAS), May 23-27, 2010, Cape Town, South Africa, pp. 1–5.

Soares, V.N.G.J., Farahmand, F., Rodrigues, J.J.P.C., 2011a. Traffic differentiation support in vehicular delay-tolerant networks. Telecommun. Syst. 48, 151–162.

Soares, V.N.G.J., Rodrigues, J.J.P.C., Farahmand, F., 2012. Performance assessment of a geographic routing protocol for vehicular delay-tolerant networks. In: 2012 IEEE Wireless Communications and Networking Conference (WCNC 2012), April 1-4, 2012, Paris, France, pp. 2526–2531.

Tatchikou, R., Biswas, S., Dion, F., 2005. Cooperative vehicle collision avoidance using inter-vehicle packet forwarding. In: IEEE Global Telecommunications Conference (IEEE GLOBECOM 2005), 28 Nov.-2 Dec. 2005, St. Louis, MO, USA, pp. 2762–2766.

Toor, Y., Muhlethaler, P., Laouiti, A., Fortelle, A.D.L., 2008. Vehicle ad hoc networks: applications and related technical issues. IEEE Commun. Surv. Tutorials 10, 74–88.

Torrent-Moreno, M., Festag, A., Hartenstein, H., 2006. System design for information dissemination in VANETs. In: 3rd International Workshop on Intelligent Transportation (WIT 2006), March 14-15, 2006, Hamburg, Germany, pp. 27–33.

Wisitpongphan, N., Bai, F., Mudalige, P., Sadekar, V., Tonguz, O., 2007. Routing in sparse vehicular ad hoc wireless networks. IEEE J. Sel. Areas Commun. 25, 1538–1556.

Wizzy Digital Courier, 2003. Wizzy Digital Courier—Leveraging locality. Available from: http://www.wizzy.org.za/. (Accessed June 2011).

Wu, H., Fujimoto, R., Guensler, R., Hunter, M., 2004. MDDV: a mobility-centric data dissemination algorithm for vehicular networks. In: First ACM Workshop on Vehicular Ad Hoc Networks (VANET 2004), in Conjunction With ACM MobiCom 2004, October 1 2004 Philadelphia, PA, USA, pp. 47–56.

Yang, X., Liu, J., Zhao, F., Vaidya, N.H., 2004. A vehicle-to-vehicle communication protocol for cooperative collision warning. In: First Annual International Conference on Mobile and Ubiquitous Systems: Networking and Services (MobiQuitous 2004), August 22-26, 2004, Boston, Massachusetts, USA, pp. 114–123.

Yousefi, S., Mousavi, M.S., Fathy, M., 2006. Vehicular ad hoc networks (VANETs): challenges and perspectives. In: 6th International Conference on ITS Telecommunications (ITST 2006), June 21-23, 2006, Chengdu, China, pp. 761–766.

Yu, Y., Lu, G.-H., Zhang, Z.-L., 2004. Enhancing location service scalability with HIGH-GRADE. In: 1st IEEE International Conference on Mobile Ad-hoc and Sensor Systems (IEEE MASS 2004), October 24-27, 2004, Fort Lauderdale, Florida, USA, pp. 164–173.

Zhang, M., Wolff, R.S., 2008. Routing protocols for vehicular ad hoc networks in rural areas. IEEE Commun. Mag. 46, 126–131.

Zhang, M., Wolff, R.S., 2010. A border node based routing protocol for partially connected vehicular ad hoc networks. J. Commun. 5, 130–143.

Zhao, W., Chen, Y., Ammar, M., Corner, M., Levine, B., Zegura, E., 2006. Capacity enhancement using throwboxes in DTNs. In: Third IEEE International Conference on Mobile Ad-hoc and Sensor Systems (MASS 2006), October 9-12, 2006, Vancouver, Canada.

Zhou, H., 2003. A Survey on Routing Protocols in MANETs. Department of Computer Science and Engineering, Michigan State University.

Delay-tolerant networks (DTNs) for underwater communications [*]

R.H. Rahman and M.R. Frater
University of New South Wales, Canberra, ACT, Australia

5.1 Introduction

Almost 70% of our planet is covered by ocean. But less than 10% of the total ocean volume has been investigated because of the harshness of this environment. Underwater networks (UWNs) are therefore gaining popularity among researchers, as well as industry personnel. UWNs consist of distributed autonomous underwater devices and vehicles using sensors/actuators to cooperatively monitor or affect physical or environmental conditions, such as temperature, sound, vibration, pressure, motion, etc., at different locations. Potential underwater applications can be realized by enabling wireless communications among the underwater devices. Some of these applications are listed below (Akyildiz et al., 2005; Pompili and Akyildiz, 2009):

- Ocean sampling network: A network of sensors and autonomous underwater vehicles (AUVs) can perform cooperative 3D sampling of the coastal environment.
- Environmental monitoring: In order to understand and predict the effect of human activities on the marine ecosystems, UWNs can perform pollution and biological monitoring. UWNs can also be used to monitor ocean currents and winds, which in turn can improve weather forecasting.
- Undersea exploration: UWNs can help to discover underwater oilfields and assist in exploration of valuable minerals.
- Disaster prevention: UWNs can be used to measure seismic activities, which can provide early warnings for tsunamis.
- Military tactical surveillance: AUVs and fixed underwater sensors can collaboratively monitor areas for the purpose of surveillance and intrusion detection.

A typical example of a UWN is depicted in Fig. 5.1.

Contrary to terrestrial wireless networks, which operate using radio frequency (RF), UWNs typically use acoustic communication technology. The reason for not using radio waves in underwater communication is that they require extra low frequencies (30–300 Hz) to propagate through the electrically conductive salty water. These extra low frequencies need large antennae and high transmission power (Stojanovic, 1999). On the other hand, optical waves do not suffer from high attenuation but are affected by scattering (Smart, 2005). Moreover, they require highly focused beams for accurate transmission. Hence, UWNs are normally based on

---

[*] This chapter is a reprint of the chapter originally published in the first edition of Advances in Delay-Tolerant Networks (DTNs): Architecture and Enhanced Performance.

Advances in Delay-tolerant Networks (DTNs). https://doi.org/10.1016/B978-0-08-102793-6.00005-9

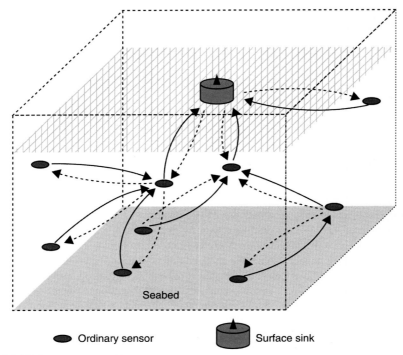

Fig. 5.1 Underwater network.

acoustic wireless communication. Due to the aforementioned reasons, UWNs are also referred to as underwater acoustic sensor networks (UASNs) or underwater acoustic networks (UANs) or underwater wireless sensor networks (UWSNs) or underwater sensor networks (USNs). These names are used interchangeably throughout this chapter.

Although many researchers have been engaged in developing networking solutions for terrestrial wireless networks, the unique characteristics of the underwater acoustic communication channel impose many challenges, such as (Akyildiz et al., 2005):

- Propagation delay is five orders of magnitude higher than in RF terrestrial channels, and variable. This is due to the low speed of sound underwater (approximately 1500 m/s).
- The bandwidth is extremely limited, which causes the data rate to be low.
- The channel is impaired by multipath and fading.
- Limited battery power.
- High bit error rates and temporary loss of connectivity.

Issues such as high bit error rates, fading, limited and distance-dependent bandwidth, and the inter-symbol interference (ISI) caused by multipath, can be addressed at the physical layer by designing appropriate receivers. On the other hand, the higher layers must address characteristics such as long and variable propagation delays and temporary loss of connectivity (Pompili and Akyildiz, 2009). Delay-tolerant networking

(DTN) approaches can be utilized to deal with the aforementioned issues of long propagation delay and temporary loss of connectivity (Fall, 2003). Hence, the focus of this
chapter is on DTN solutions that try to solve the networking problems associated
with UANs.

The remainder of this chapter is organized as follows. Section 5.2 provides a
detailed survey of the state-of-the-art DTN solutions for UASNs. Section 5.3 gives
a contemporary view of underwater DTNs. Section 5.4 discusses some possible future
research directions, and conclusions are drawn in Section 5.5.

5.2 Related work

In recent times, there have been quite a few research works related to underwater
DTN. In this section, we will explore some of these works and analyze their features.

5.2.1 A framework for mobile underwater acoustic networks

Katz (2007) has developed a delay-tolerant framework for mobile UANs. He has
designed an augmented networking model that can adapt itself to the environment.
In addition to the standard TCP/IP protocol suite, the model uses four data classification planes, namely, perception, intuition, management, and adaptation. The perception plane contains the set of data on each TCP/IP layer and it can be used to build a
more accurate view of the network's immediate state. For example, the perception
plane monitoring the data link layer and the network layer can provide information
about the source and destination addresses for all detected packets, as well as loss
and latency measurements between them. The intuition plane tries to deduce contex-
tual information from the perceived data. For example, perception data from the physical layer can be combined with node positions to produce intuition data such as
regions of remarkably strong or weak signal quality. Similarly, node velocities can
be utilized to estimate the duration for which current links will be operational. The
management plane represents a shared database that assists the intuition and adaptation. In order to meet the requirement of a message being sent onto the network, the
adaptation plane reads data from the management plane to make educated decisions
on behalf of a particular layer.

The framework offers a mechanism for balancing a node's mobility with its
requirement for reliability and latency. To achieve this, the applications indicate
the reliability and latency requirements of a message at the time of sending. Then
the individual layers decide whether they can use adaptation to meet these requirements. For example, the framework queries each layer in order to determine whether
the message is deliverable within its specified time bounds. If an affirmative response
is obtained from all the layers, the message is sent through the layers to be wrapped by
the protocol headers so that adaptation may take place. When the message reaches the
physical layer and is sent onto the network, the application is notified of the success.

The limitation of this approach is that it is only a framework: no specific protocol
definition is given for any of the introduced planes. The framework simply gives an

overview of the functionality of the perception, adaptation, intuition and management planes. Therefore, there is no way of knowing whether the framework will actually work and, if it does, how it will perform.

5.2.2   Delay-tolerant data dolphin (DDD)

Acoustic signals propagating in water tend to experience greater losses in the channel than radio signals propagating in air or in a vacuum. For the same link length and data rate, a relatively higher transmission power is usually required for an acoustic signal than would be required for a radio signal. As a result, energy preservation is more critical in UASNs than in traditional wireless sensor networks. Magistretti et al. (2007) propose a delay-tolerant data dolphin (DDD) protocol for underwater delay-tolerant applications. The protocol aims to increase the energy efficiency of the resource-constrained underwater environment. DDD utilizes the mobility of specialized collector nodes (known as *dolphins*) to collect information sensed by the ordinary static sensor nodes. In DDD, each sensor node is only required to transmit its sensed data directly to the nearest dolphin within its communication range. In this way, the protocol avoids energy-consuming multihop communication.

In this scheme, static sensor nodes are deployed on the sea bed in the region of interest. These nodes gather information from the environment and store the sensed data locally after processing. The acoustic modem in each sensor has two components. The first component is used for acoustic communication with a nearby dolphin. The second component is a low-power transceiver which is used to detect the presence of a dolphin in the vicinity and trigger the first component. The detection process is carried out by a special signal transmitted from the dolphins. A number of dolphin nodes are used to collect the stored data packets when they move within the direct communication range of the static sensor nodes. The mobility of the dolphins can be random or controlled based on the network requirements. The dolphin nodes broadcast special beacon signals to announce their presence. The advertisement period depends on three factors: (i) the deployment of the sensor nodes, (ii) the maximum communication range of the sensor nodes, and (iii) the speed of dolphins. Finally, the dolphins deliver the collected data packets as soon as they reach a base station on the surface.

The performance of DDD heavily depends on the quantity of dolphin nodes. If there are not enough dolphin nodes, all the stored data packets may not be collected from the scattered sensor nodes. Since the dolphin nodes move randomly, it is possible that some sensor nodes remain completely unvisited, which will result in the loss of data packets. Increasing the number of dolphin nodes may increase the data delivery possibility, but in that case cost will become a major issue.

5.2.3   Adaptive routing

Sometimes an UASN is required to meet different delivery rates according to different application requirements. Therefore, an intelligent routing technique that can adapt to different application requirements is desirable. In this regard, a novel routing technique called adaptive routing for underwater delay/disruption-tolerant sensor networks is proposed by Guo et al. (2008), where it is assumed that all nodes know their own 3D locations. Nodes make routing decisions based on the characteristics of the data packets and the network conditions. The goal of this protocol is twofold: (i) to satisfy different application requirements, and (ii) to achieve a good tradeoff among delivery ratios, end-to-end delays, and energy consumption. Each packet in transit at a node is assigned a priority that is calculated depending on four factors: (i) packet emergency level, (ii) packet age, (iii) density of the neighbors around the node, and (iv) residual energy of the node. The calculated priority indicates the importance of a packet at a specified node at a given time. Based on the priority level, a forwarding node creates one or more copies of a packet, that is, the higher the priority, the more copies the packet will have. Moreover, routing decisions are also made based on the packet priority. In order to make the protocol flexible, all of the aforementioned four factors except the emergency level are variable. The protocol divides the entire routing spectrum into four states. The priority scale is evenly partitioned into four intervals, which correspond to the four routing states. A node choosing routing state i for a packet can create and forward i copies of the packet. The protocol limits the range of i between 1 and 4 inclusive.

Simulation results show that this strategy can satisfy different application requirements such as delivery ratio, average end-to-end delay, and energy consumption. However, in order to calculate the priority level of each packet, the proposed protocol relies heavily on periodic and frequent communication among neighboring nodes. This may increase the energy consumption of individual nodes, as well as the overall end-to-end latency.

5.2.4 An integrated data retrieval protocol

Chan and Motani (2007) developed an efficient and reliable data retrieval protocol to collect data from individual network regions in a customizable and optimal manner. The protocol aims to overcome the challenges associated with the underwater channel while maximizing energy and bandwidth efficiency. As channel conditions and application requirements vary, the protocol dynamically adapts itself to achieve efficiency. The protocol is able to operate with standard half-duplex modems for practical purposes, which reduces costs and node size as compared with using full-duplex modems.

In this protocol, a retrieval device is sent out into the underwater region of interest to collect data from the sensor nodes. Since these receivers can easily be recharged, they are responsible for maintaining network connectivity and reliability. This is achieved through polling, which controls the communication links and the transmission sequences of the senders in order to avoid contention in transmission. Rather than using an explicit acknowledgment technique, the protocol utilizes Luby transform (LT) codes (Luby, 2002) and forward error correction (FEC) codes to ensure reliability.

Each sender node (sensor nodes situated in field) performs the following protocol operational steps:

1. Perform data sensing while passively listening for receiver's announcement beacons.

2. When a beacon is received, time synchronization is set up. In addition, status packets are transmitted, which contain the size of the actual file to be transmitted.

3. Wait for a broadcast message from the receiver. If the broadcast message contains the ID of the sender, data packets, after being encoded using LT code parameters, are sent out. The LT encoding parameters are obtained from the received broadcast message.

4. Step 3 is repeated until the *finish* signal is received, at which point the node reverts to the passive mode in Step 1.

Each receiver node (data retrieval AUVs) performs the following protocol operational steps:

1. Enter the underwater region of interest. Announce presence to all senders by broadcasting beacon signals. Set up time synchronization.

2. Receive status packets from n senders. Estimate the transmission time for each sender. Select m out of n senders which have similar transmission times.

3. Broadcast messages containing the LT encoding parameters to the m senders.

4. Receive and decode encoded packets from the m senders. Out of the remaining $n - m$ senders, select the next m senders in the same manner.

5. Steps 3 and 4 are repeated until packets from all of the n senders have been successfully decoded. A *finish* signal is broadcast to all senders, indicating the end of the data communication session.

The major limitation of this protocol is that it is very restricted in the applications it can support. Moreover, data communication is always receiver-initiated and within a single hop. In order for the data to be delivered reliably, the receiver node(s) have to come within the communication range of all the sender nodes. Hence, the success of this protocol heavily depends on how accurately and efficiently the receiver nodes can visit all the sender nodes. However, this may have a negative impact on the end-to-end latency. Increasing the number of receiver nodes may increase reliability, but it will also increase the cost of the network.

5.2.5 An underwater convergence layer

Merani et al. (2011) have designed and implemented an open-source underwater convergence layer (UCL) which provides DTN support for the WHOI micro-modem (Freitag et al., 2005). This layer is positioned between the DTN layer and the underlying acoustic communications protocols. The goal of this framework is to allow access to acoustic modems from different vendors in the form of a DTN convergence layer adapter (CL A). The role of the CLA is to manage protocol-specific details associated with acoustic communication, while presenting a consistent interface to the DTN framework. The UCL consists of a number of modules. Their features are briefly described below.

1. The *Platform Access and Abstraction Module* logs performance statistics to files, and configures and performs bi-directional communication with serial port devices and TCP/IP sockets.

2. The *Data Link Module* is in charge of sending and receiving frames, maintaining a list of reachable neighbors and advertising the local node in the network. The node discovery

algorithm states that every node that does not have any data frames to transmit within a certain amount of time must actively advertise itself. This allows other nodes to learn about the presence and reachability of this node.

3. The *NMEA Module* contains C++ classes that are used to parse WHOI micro-modem messages. The goal is to translate native modem messages into more generic structures that can be used by other modules.
4. The *WHOI Micro-Modem Driver Module* provides a high-level interface to control the modem. The module processes the requests to transmit data and notifies other modules when some data/acknowledgments arrive or collisions occur.
5. The job of the *Network Module* is to exchange data in the form of packets.
6. The *External Convergence Layer Module* parses, validates, and generates XML messages that are compliant with the DTN application and informs about the available acoustic links within range.

The UCL framework has been tested using four fixed nodes and three mobile nodes. However, the paper does not demonstrate any concrete results. So it is difficult to ascertain the actual performance of the framework from a broader perspective.

5.2.6 *Data collection with multiple mobile actors*

In Wang et al. (2008), an underwater application model is presented for collecting data from USNs with multiple mobile actors. The proposed scheme works in two ways: the local base station collects sensor data at regular time intervals, and mobile actors collect data from virtual clusters with high temporal resolution. The paper presents three algorithms to accomplish the tasks: (a) area partitioning and actors scattering algorithm, (b) subregion-optimizing algorithm, and (c) virtual cluster formation algorithm. The scheme assumes that it can acquire the approximate position of each sensor node and the actors have self-navigating ability. All sensor nodes are randomly deployed in the same depth of water.

The algorithm for area partitioning and actors scattering (APAS) works as follows:

1. The edge nodes are identified.
2. The coverage rectangle, which should enclose all nodes in the network, is calculated.
3. The rectangle is divided into subregions according to the aspect ratio and total number of nodes in the network.
4. The data collection duration is estimated for each subregion.
5. The subregions are optimized (explained next).
6. After Step 5, the number of subregions is known. One actor node is positioned in each subregion, where it serves as the CH.

The algorithm for subregion optimization works as follows:

1. Each subregion keeps a value of estimated time delay.
2. Each region that cannot fulfill the given constraints is divided into four new subregions. This process is repeated until all subregions fulfill the delay constraints.
3. Neighboring subregions having low-density nodes are merged from left to right and from top to bottom.
4. Subregions having no sensor nodes are deleted.

The algorithm for virtual cluster formation works as follows:

1. Request messages are broadcast by actors which contain the actor-ID.
2. Upon receiving a request message, the sensor node saves the actor-ID and rebroadcasts the message. It also sends back a response using the reverse route.
3. Once a sensor node has saved one actor-ID, it will neither accept any more request messages nor rebroadcast them.
4. The entire process is repeated until all the sensor nodes receive at least one request message.

Wang et al. (2008) have in effect proposed a clustering algorithm for underwater acoustic communications. The major drawback of this technique is that every node needs to know its location information. Furthermore, although the paper outlines an optimized region-partitioning algorithm, it does not say how this is achieved, that is, whether it is achieved in a centralized or distributed manner.

5.2.7   The PASR protocol

The prediction assisted single-copy routing (PASR) proposed by Guo et al. (2010) is designed to operate in different underwater mobility models. PASR utilizes a greedy algorithm which tries to capture the characteristics of network mobility patterns and provide guidance on how to use historical data. Here, the authors propose a trace-based greedy algorithm called aggressive chronological projected graph (ACPG), which captures the mobility properties of the network, and, taking guidance from ACPG, a heuristic prediction-assisted single-copy routing protocol is designed.

The protocol assumes a network having M layers where multiple sensor nodes are placed in each layer. The nodes can move passively with water currents in the horizontal plane while vibrating slightly in the vertical direction. The protocol assumes that one data sink is placed in the middle of the water surface. In addition, the network operates in a slotted manner. Sensors in the lowest layer generate packets to be transmitted to the sink using nodes in the middle layers as relays. A packet received or generated in one slot can be forwarded in future slots. Each node broadcasts a *HELLO* message to its neighbors at the beginning of each slot to announce its presence and exchange necessary information. The ACPG algorithm is used to identify the network mobility properties along with the common properties of near-optimal routes. At each time slot t ($t > 0$), ACPG uses two functions: (i) *edge projection,* during which connections in a time slot are projected on to a graph as edges and (ii) *routes reservation and graph update*, during which routes are discovered and the graph is updated. These two routines operate alternately until all traffic demands are met. Using the aggressive route discovery process, ACPG quickly discovers low-delay routes with reduced complexity. It also summarizes the properties of the routes, which in turn portray the characteristics of the underlying mobility patterns.

Based on the guidance from ACPG, the PASR protocol chooses appropriate historical information to predict future contacts in order to perform single-copy routing. ACPG is able to capture the following properties of routes and node contacts which can be used to deduce mobility pattern: geographic preference, contact periodicity, inter-contact time distribution, and contact probability. For example, (i) if the guidance displays geographic preference, a node can use it to determine whether to forward packets to a neighbor or not, (ii) if mobility shows contact periodicity, PASR can utilize the last contact time to estimate the next contact time with high accuracy, and (iii) if the inter-contact time follows some well-known distribution, then the last contact time can be used to predict whether a node is approaching or departing from another node.

A major drawback of this scheme is its dependency on network structure. The network is assumed to be layered and all communication direction is vertically upwards. This constrains the application domain. In addition, the protocol depends on periodic *HELLO* messages for establishing routes between neighboring layers, leading to high communication overheads. Finally, the network operates in a slotted manner, which means there should be some sort of synchronization among the nodes, and this is very hard to achieve in the slow underwater acoustic environment.

5.2.8 The TCBR protocol

Temporary cluster-based routing (TCBR) (Ayaz et al., 2010) is a routing protocol designed for underwater monitoring missions. In the TCBR architecture, multiple sinks are deployed on the water surface and data packets received at any sink are considered as delivered successfully. Two types of nodes are used: (1) ordinary nodes and (2) courier nodes. Ordinary nodes sense data and try to forward these data packets to a nearby courier node. Courier nodes can sense as well as carry data packets from ordinary nodes to a surface sink. The courier announces itself using *Hello* packets. The *Hello* packets can be forwarded a maximum of three hops. When an ordinary node receives a *Hello* packet from a courier node, it can compute its own distance (in terms of the number of hops) relative to that courier node by observing the hop count in the packet. The protocol, however, does not say what happens to nodes that do not fall within three hops of a courier node. Data packets are forwarded by identifying suitable next hops using *Inquiry Request* packets. Although simulation results show a good packet delivery ratio, the result itself, however, is very questionable because it is based on the IEEE 802.11 (Tanenbaum, 2002) medium access control (MAC) protocol (originally designed for terrestrial wireless networks), whereas earlier literature (Akyildiz et al., 2005) has clearly opposed using this MAC directly in underwater acoustic environments due to its low utility. Furthermore, TCBR's performance has not been compared with any existing routing protocol.

5.2.9 The resilient routing protocol

Pompili et al. (2006) proposed a routing solution based on some form of centralized planning of the network topology. The proposed solution relies on a virtual circuit routing technique, where multihop connections are established a priori between each source and sink, and each packet associated with a particular connection follows the same path. The protocol allows exploitation of powerful optimization tools by the central manager (e.g., the surface station) to achieve optimal performance. The proposed routing solution follows a two-phase approach. In the first phase, the network manager determines optimal node-disjoint primary and backup multihop data paths such that the energy consumption of the nodes is minimized. It is formulated as an integer linear problem (Ahuja et al., 1993). In phase two, an online distributed solution guarantees survivability of the network by locally repairing paths (in the case of disconnections or failures) or by switching the data traffic on the backup paths (in the case of severe failures). If a source node or a relaying node detects that the quality of a link on the primary path has degraded, it can start transmitting duplicate packets on its backup path to the sink to increase reliability. On the other hand, if it finds that a link on the primary path is totally impaired, it will stop transmitting on the primary path and switch the data traffic completely to the backup path. A disadvantage of this protocol is its reliance on centralized coordination.

5.2.10   The DUCS protocol

As underwater batteries are hard to recharge, energy efficiency is a major concern for UASNs. In this regard, Domingo and Prior (2007) have presented an energy-aware distributed underwater clustering scheme (DUCS) for long-term applications. The protocol supports random node mobility and is self-organizing and adaptive in nature.

In DUCS, sensor nodes are organized into multiple clusters, where one node is selected as the CH for each cluster. The non-CH nodes are known as regular nodes. Regular nodes transmit data packets to their respective CHs using single-hop communication. After receiving data packets from all the regular nodes within a cluster, the CH applies postprocessing on the received data (e.g., data aggregation) and transmits them towards the sink using multihop routing with the help of other CHs. It is the responsibility of the CHs to coordinate and maintain intra-cluster and inter-cluster communications. In order to avoid fast draining of the battery of any particular sensor node, the role of the CH is rotated among different nodes within a cluster in a randomized fashion. The operation of DUCS is completed in two rounds, (i) *setup* and (ii) *network operation*:

- In the *setup* round, the network is divided into clusters.
- In the *network operation* round, transfer of data packets takes place.

Simulation results show that DUCS increases the packet delivery ratio and throughput and at the same time reduces the network overhead.

The protocol, however, has a couple of performance issues. First, node movements due to water currents can affect the structure of clusters. This in turn causes the *setup* phase to be repeated many times, which consequently decreases the cluster lifetime. In the same manner, water currents can move two neighboring CH nodes away so that they cannot communicate with each other directly. As a result, during the *network operation* phase, a CH may fail to transmit its collected data towards another CH.

5.2.11   The EDETA protocol

*E*nergy-efficient a*D*aptive hi*E*rarchical and robus*T* *A*rchitecture (EDETA) (Climent et al., 2011, 2012) is a hierarchical routing protocol in which nodes arrange themselves in clusters and one of them performs the role of the CH. In order to send the collected data from the nodes to the sink using multiple hops, the CHs form a tree among themselves. The protocol has two phases: (i) initialization and (ii) normal. During phase (i), clustering is performed and CHs are elected. During phase (ii), nodes send their data periodically to their respective CHs at their scheduled times, and CHs forward their data to their parents until the data reaches the sink. EDETA incurs additional communication overhead because it needs to construct and maintain a cluster-like structure. In addition, the protocol requires scheduling for data transmission, which is difficult to achieve in the delay-prone underwater environment.

5.2.12   The AURP protocol

The AURP (Yoon et al., 2012) protocol utilizes heterogeneous acoustic communication channels and controlled mobility of multiple AUVs to perform routing. Here, AUVs are employed as carrier nodes to improve the network performance in terms of data delivery ratio and energy consumption. AURP attempts to minimize the total number of transmissions by using AUVs as relay nodes. First, ordinary underwater sensor nodes send the data to a gateway (GW) in a multihop manner. Then AUVs collect sensed data from GW nodes and forward the data to the sink. To determine the route from ordinary sensors to the gateways, the GWs periodically flood PHE (pheromone) messages to the network. Upon receiving PHE messages, each ordinary sensor determines the next hop to forward the data based on the length of the path (the number of hops) that those messages have taken, that is, the node selects the next hop so that the path length is minimized. An ordinary sensor also forwards PHE messages received from other ordinary sensors or GWs. The AUVs move along a predefined path. When a GW detects that an AUV is in its proximity and is close enough for data transmission, the GW forwards all the data that it has collected from ordinary sensor nodes to the AUV. The main limitation of AURP is its periodic flooding of PHE messages, which causes large communication overhead.

5.2.13   The utility-based protocol

Spyropoulos et al. (2009) have proposed a utility-based routing protocol for heterogeneous DTNs. Although the protocol is primarily designed for terrestrial DTNs, according to the authors, it can be readily applied to underwater acoustic networks because of its generic nature. The aim here is to find the L best relay nodes in the network (given some optimization criterion) for packet forwarding. The protocol defines mechanisms to identify the "better" relays (nodes having better *utility* value) and avoid using the least useful ones. Utility-based routing uses forwarding tokens to grant a node the right to further forward message copies. In addition, each node x defines a utility function $Ux(y)$ for every other node y in the network. $Ux(y)$ denotes the probability that node x will deliver a message to node y. If a node x, carrying a message for destination d and $c > 1$ forwarding tokens for this message, encounters another node y with no copy of the message, then x creates and forwards a copy of that message to y if and only if one of the following conditions are met:

- $U_y(d) > U_{th}$, where U_{th} is a threshold value
- $U_y(d) > U_x(d)$

x also hands over half of its forwarding tokens to *y*. During each encounter, packet forwarding can be done based on either of the conditions or a combination of both (e.g., use the first condition to ensure a minimum utility and then the second condition among the nodes that qualify). The calculation of *utility* can be based on a number of different parameters. In a real scenario, there are a number of different reasons why a given node might be a "better" relay (might have better *utility*) than another. Some of these reasons are: special "bonds" with the destination, higher availability of resources, higher degree of reliability or trustworthiness, etc. The authors have provided three different ideas based on which the *utility* function may be calculated:

1. The first idea is to choose those nodes as relays which have seen the destination most recently.
2. The second idea is to choose as relays those nodes which have the most mobility. The assumption is that the more mobile the node, the larger the (geographic) area it can cover and the more chance it has of encountering other potential relay nodes or the destination.
3. The final idea is to choose nodes which are more *sociable*. This indicates nodes that visit "hub" locations (e.g., café) more often than others. Visiting these "hub" locations causes these nodes to have more *social links* than average.

The protocol, however, has some limitations, such as its explicit dependency on node mobility. Moreover, the first packet forwarding condition (mentioned above) requires an appropriate threshold value (U_{th}) to be set in every case, which is difficult to achieve. The second packet forwarding condition is easier to implement, yet it does not guarantee that all "bad nodes" will be avoided (e.g., "very low utility" nodes could still give copies to "low utility" nodes).

5.2.14   The AMCTD protocol

The adaptive mobility of courier nodes in threshold-optimized depth-based (AMCTD) (Jafri et al., 2013) protocol employs proficient depth threshold and optimal weight function to achieve longer network lifetime. AMCTD is composed of four phases: (i) initialization, (ii) data forwarding, (iii) weight updating, and (iv) depth threshold adaptation. During phase (i), each node calculates its weight on the basis of density of network and shares its weight and depth information with its neighbors using *hello* messages. At the same time, the courier nodes devise and start their preplanned tour from the bottom of the network up to the surface. During their tour, the courier nodes collect and aggregate data from the sensor nodes continuously, transmit them to the sink and restart their tour from the bottom. The courier nodes may change their mobility pattern to adapt to the changing network density. In phase (ii), the source node discovers the optimal forwarder among its neighbors by comparing their weights. The neighbor having the highest weight is elected as forwarder. After sensing the data, a node sends it toward the sink using the technique of carrier sense multiple access with collision avoidance (CSMA/CA). In phase (iii), the weights of the sensor nodes are revised according to the changing node density of the network. This phase is initiated (the node density changes) whenever the number of dead nodes increases by 2%. In the final phase, to adapt to the coverage problem that occurs due to the dying of nodes, the depth threshold is adjusted so that source/forwarder nodes may have more neighbors within their communication range.

There are a couple of disadvantages of the AMCTD protocol. First of all, since the protocol is depth-based, the communication structure is vertical (the communication direction is always upwards), that is, the protocol will fail for horizontal networks (shallow UWNs). Second, the protocol relies on the CSMA/CA technique to forward data packets. However, this technique has been shown to be ineffective in previous literature (Akyildiz et al., 2005).

5.2.15 The QDTR protocol

The adaptive and energy-efficient Q-learning-based DTN routing protocol, QDTR (Hu and Fei, 2010), assumes a layered network topology in a 3-D area, in which nodes are mobile in the current layers, and they can only communicate with the ones in the same layer and the neighboring layers (within the transmission range). Q-learning is a kind of reinforcement learning algorithm (Sutton and Barto, 1998) and is normally characterized by the Markov decision process (MDP).

Q-learning is based on a state-action function in which the value of an action at the current state depends on two things: (i) the direct reward and (ii) the value of the future states that the action would lead to. The direct reward function measures the behavior and performance of the system and is defined by $R = f(H, E, D)$, where the first factor (H) is the relative distance of a node from the sink, the second factor (E) is the residual energy of a node and the final factor (D) is the node density of the surrounding area. Each packet is associated with a state, and the node that holds the specific packet is defined by it. The value of a certain state indicates the likelihood of the packet getting delivered by the node associated with the state. In order to achieve good network performance and energy efficiency, the sensor nodes perform a single or multicopy forward at the beginning and spray packets among the neighbors of the destination at the end.

A drawback of QDTR is that it assumes a restrictive network structure where the sink is always situated on the topmost layer. In addition, the layered network structure also creates a restrictive communication pattern, which makes the application domain limited.

5.2.16 The VOM-DTN protocol

The vessel-based ocean monitoring delay tolerant networking (VOM-DTN) protocol (Wang et al., 2013) is a cost-saving DTN which utilizes the daily working voyages of existing vessels to transfer ocean monitoring data to the sink. In VOM-DTN, the underlying network is deployed in the vicinity of a vessel's travel path. A deployed node can exchange its sensed data with its neighbors directly. During its travel along the predefined route, the vessel can communicate directly with one or more sensor nodes when it comes near or crosses through the network region. During this communication opportunity, the vessel keeps collecting as much data as possible from the sensor nodes. When the vessel returns to the vicinity of the coast, it transfers all the carried data directly to the onshore sink node. The Spray and Wait (Spyropoulos et al., 2005) routing scheme is chosen for inter-node communication. VOM-DTN suffers from the same problem as that of the Spray and Wait protocol. For the Spray and Wait protocol, the key problem is choosing the number of message copies. If a large number of copies are used, the communication overhead increases. On the other hand, if a small number is used, the packet delivery ratio may decrease.

5.2.17   The DCT protocol

The distributed cooperative transmission (DCT) protocol (Tan et al., 2013) aims to enhance network performance by utilizing relay nodes for diversity gains. A diversity system can be considered as a system that receives two or more copies of the same transmitted signals from a transmitter. The protocol considers a cooperative scheme consisting of two relay nodes, one source and one destination. The literature considers a distributed UASN in the shallow ocean where the channel is heavily affected by multipath fading.

First, each node builds a list of neighbors and their corresponding distances (in terms of hop counts) from the sink by utilizing the periodic advertisement packet (ADV) from the sink node. Then the source node employs a Request to Send/Clear to Send (RTS/CTS) exchange (Tanenbaum, 2002) to discover the two suitable relays and the destination. To accomplish this, the source node first broadcasts a RTS packet. Upon hearing this packet, all neighbors that are closer to the sink (in terms of hop count) reply with a CTS packet. The CTS packet contains the neighbors' surrounding channel conditions in terms of signal to noise ratio. A node can sense its surrounding channel conditions by overhearing packets from its neighbors. After receiving all possible CTS packets, the source obtains a list of candidates along with three parameters (distance, time of arrival and channel condition) for each candidate. Based on these three parameters, the source chooses two relays and one destination in such a way that the source-relay-destination paths are considerably longer than the direct source-destination path. The source then transmits the actual message to the destination, which is received by the relay nodes first. The relay nodes decode the message in the first phase and forward it in the next phase. The destination then receives three copies of the message, one from direct source transmission and the other two from the relay nodes. The maximal ratio combing technique is applied to recover the original message.

The problem with this protocol is that two additional copies are needed for every message at every hop, creating a relatively larger network overhead. Furthermore, the protocol relies on periodic advertisement packets from the sink, which farther increases the network overhead.

5.2.18   The Ma-Sync protocol

Ma-Sync (Guo et al., 2013) is a mobile-assisted clock synchronization algorithm for UASNs. It exploits the use of a mobile beacon node to assist with clock synchronization. Ma-Sync utilizes the distance information between nodes to deal with the large propagation delay experienced by UASNs. In order to improve synchronization precision, Ma-Sync estimates the clock skew and offset by linear regression over a set of time information. In Ma-Sync, a mobile beacon node moves along a predefined trajectory and broadcasts synchronization packets periodically. The packets are time-stamped. The node that is to be synchronized receives several of these packets at different points in time and records these receiving times. Using the time stamps on the packets and their recorded receiving times, a node computes the clock skew and the clock offset. In a real environment, the speed of sound under water is affected by temperature, pressure, and salinity of the channel. To reduce the effect of propagation time variation, Ma-Sync uses linear regression over a set of different readings to compute the clock skew and offset. The main drawback of this protocol is its reliance on periodic synchronization packets, which creates a large network overhead.

5.3   A contemporary view of underwater delay-tolerant networks

As mentioned in Section 5.2, traditional delay-tolerant solutions for UANs either have their own limitations or rely explicitly on node mobility. The solutions (routing protocols) in this class follow a paradigm known as *store-carry-forward,* where participating nodes have to *store* and *carry* messages for some distance before encountering either (i) the desired destination and *forwarding* messages to it or (ii) a suitable node which can *store* and *carry* messages to the desired destination.

We have seen that, in a traditional DTN, a message is either replicated to increase delivery probability or forwarded based on some assumption about the underlying network topology or mobility pattern (contact schedule). But neither of these classes of solutions is perfect for an UASN, because UASNs do not behave like traditional DTNs. As both classes of solutions depend explicitly on node mobility, they are unsuitable for underwater sensor nodes that are stationary, and the entire UASN is static in nature (in a real-world scenario this indicates moored nodes and the absence of any underwater vehicle in the network), having no mobile nodes (AUVs). However, we believe that the concept of DTN can still be applied effectively to UASNs even in the absence of node mobility. A possible use of the DTN concept is outlined below.

5.3.1   *Observations and suggestions*

The peculiarities of the underwater acoustic channel, especially limited bandwidth and high and variable propagation delay, pose additional challenges in the design of MAC protocols. MAC protocols are crucial to any networking system where multiple nodes try to access the communication medium simultaneously. Traditional (and more generic) contention-based MAC protocols (e.g., MACA/MACAW/IEEE 02.11 (Tanenbaum, 2002)) that rely on handshaking techniques such as RTS/CTS exchange (Tanenbaum, 2002) are, in their current form, unsuitable for UASNs. This is because large link delays in the propagation of RTS/CTS packets lead to low channel utilization (Akyildiz et al., 2005). In recent times, a number of MAC protocols have been proposed specifically for UASNs (Rodoplu and Park, 2005; Hayajneh et al., 2009;

Guo et al., 2009; Noh et al., 2010; Zhou et al., 2012). However, each of these protocols has its own limitations. For example, the protocol in Rodoplu and Park (2005) only works well for low and uniformly distributed traffic, which is hard to predict in an underwater scenario; the orthogonal frequency division multiple access (OFDMA)-based MAC proposed by Hayajneh et al. (2009) is suitable for shallow water deployments only and, finally, MAC protocols such as propagation-delay-tolerant collision-avoidance protocol (PCAP) (Guo et al., 2009), delay-aware opportunistic transmission scheduling (DOTS) (Noh et al., 2010), and cooperative underwater multichannel MAC (CUMAC) (Zhou et al., 2012) depend on time synchronization, which is difficult to achieve in a delay-prone underwater acoustic network (Akyildiz et al., 2005; Shah, 2009), and this situation is compounded by the fact that even moored nodes may drift and move around with water currents. The lack of a suitable MAC protocol makes the underwater acoustic environment collision-prone and unreliable.

It is true that an UASN suffers from long link propagation delay, but this delay is not as extreme as space communication. Furthermore, when the UASN is static, most often the nodes in the network do not suffer from a lack of continuous end-t o-end connectivity, as is the case for most traditional DTNs. In such a scenario, in spite of having a valid and continuous multihop path from a source to a destination, a packet may still fail to reach the destination due to a collision-prone unreliable underwater acoustic environment which results from the lack of a suitable MAC protocol, as mentioned above. So the UASN can be characterized as a *disruption-prone network,* as well as a *delay-prone network.* Visually, the UASN looks like Fig. 5.2.

Fig. 5.2 reads as follows: an UASN possesses the inherent characteristic of long link propagation delay, which in turn affects the MAC protocol, which in turn causes the medium to become collision-prone and unreliable (disrupts communications in the network). The key differences between a traditional DTN and an UASN are listed in Table 5.1.

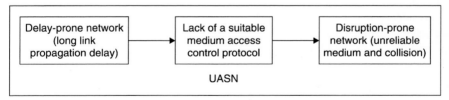

Fig. 5.2 Characteristics of an underwater acoustic sensor network.

Table 5.1 Key differences between a traditional DTN and an UASN.

Traditional DTN	UASN
1. Delay-prone due to the lack of continuous end-to-end connectivity or node mobility.	1. Delay-prone due to the inherent slow nature of the acoustic medium.
2. Disruption-prone due to the lack of continuous end-to-end connectivity or node mobility.	2. Disruption-prone due to delay-proneness, which implies lack of reliability.

So we see that delay is a property of an UASN, whereas disruption is a consequence of delay. Hence, steps should be taken to make the UASN more of a disruption-tolerant network rather than a delay-tolerant network. This in turn means reducing the number of collisions in a network which is inherently unreliable. A possible solution to this is described in the following section. For our solution, we remove the *carry* part from the store-carry-forward concept of DTN solutions (which essentially removes the node mobility requirement) and instead just use the store-and-forward concept. Rather than depending on node mobility or contact schedules (which is done by existing DTN solutions), our solution uses the concept in a more intelligent way, that is, the store-and-forward concept is utilized to assist the MAC protocol with a view to increasing the reliability of the underwater acoustic environment. This is explained in the following section.

5.3.2 A possible solution

As mentioned in Section 5.3.1, current underwater MAC protocols have their own limitations. On the other hand, although RTS/CTS-based MAC protocols suffer from low channel utilization in the underwater acoustic environment, they are generic in nature (meaning they can be deployed in any type of network, terrestrial or underwater) and they are able to control access to the shared communication medium, which in turn reduces the chance of packet collision and increases network reliability. Therefore, we use a simple RTS/CTS based protocol similar to multiple access with collision avoidance (MACA) (Tanenbaum, 2002) and use the DTN store-and-forward concept to improve the protocol's performance in terms of channel utilization. In order to understand how this is possible, we first look at the default behavior of the RTS/CTS scheme and examine how it (in its default form) yields a low channel utilization.

The default behavior of this MAC is very simple. Whenever a node wishes to send/forward a unicast packet (data or control) to its next hop, it first broadcasts a RTS packet locally. Hearing this RTS, every node (except the intended recipient) within the sender's communication range refrains from making any sort of transmission for a certain period. When the intended next hop receives this RTS, it replies with a CTS packet. Hearing this CTS, every node within the receiver's communication range abstains from making any sort of transmission for a certain period. Once the sender gets the CTS, it assumes that the medium is now free for packet transmission. This process takes place for every unicast packet and every time it is forwarded. According to Rappaport (1999), channel utilization (CU) can be thought of as the fraction of time a channel is used for data packet transmission. A network that uses the RTS/CTS protocol defines channel utilization as:

$$CU = \frac{Transmission\,Delay\,for\,DATA}{Transmission\,and\,1-hop\,Propagation\,Delay\,for\,(DATA+RTS+CTS)}$$

Transmission Delay for any packet is also known as *Packet Duration*. Furthermore, 1-hop *Propagation Delay for* (*DATA* + *RTS* + *CTS*) = 3 × 1-*hop Propagation Delay*. Hence, we rewrite the above equation as:

$$CU = \frac{Packet\,Duration(DATA)}{Packet\,Duration(DATA + RTS + CTS) + 3 \times 1 - hop\,Propagation\,Delay}$$

Let us consider the following network parameters:

- Channel Data Rate   = 100 kbps
- Data Packet Size   = 100 bytes
- RTS Packet Size   = 20 bytes
- CTS Packet Size   = 20 bytes
- 1-hop Distance   = 100 m (distance between the sender and receiver)
- Propagation Speed   = 1500 m/s (for acoustic signal)

This gives us:

$$Packet\,Duration\,(DATA) = \frac{100 \times 8\,bits}{100,000\,bps} = 0.008\,s = 8\,ms$$

$$Packet\,Duration\,(RTS) = \frac{20 \times 8\,bits}{100,000\,bps} = 0.0016\,s = 1.6\,ms$$

$$Packet\,Duration\,(CTS) = \frac{20 \times 8\,bits}{100,000\,bps} = 0.0016\,s = 1.6\,ms$$

$$1 - hop\,Propagation\,Delay = \frac{100\,m}{1500\,m/s} = 0.067\,s = 67\,ms$$

Substituting the aforementioned values into the CU equation gives us a channel utilization of $CU = \frac{8\,ms}{(8+1.6+1.6)\,ms + (3\times67)\,ms} = \frac{8\,ms}{212.2\,ms} \approx 0.038 = 3.8\%$. In the underwater acoustic environment. On the other hand, if we apply the same parameters to any terrestrial radio network where the propagation speed is 3×10^8 m/s and the 1-*hop Propagation Delay* is $\frac{100\,m}{3\times10^8\,m/s} \approx 0.00000033\,s = 0.00033\,ms$, then we get a channel utilization of $CU = \frac{8\,ms}{(8+1.6+1.6)\,ms + (3\times0.00033)\,ms} = \frac{8\,ms}{11.20099\,ms} \approx 0.714 = 71.4\%$.

This proves that the RTS/CTS scheme (for every data packet) is very inefficient under water.

An obvious way of improving CU in an underwater environment is by increasing the data packet size, which in turn will increase the *Packet Duration (DATA)* in the aforementioned CU equation. However, this is not always possible due to application requirements or restrictions. Therefore, we use the DTN store-and-forward concept to "virtually" increase the data packet size, that is, to create one long "virtual" data packet. This can be achieved by aggregating multiple data packets into one long virtual data packet and transmitting the packets back-to-back in one burst (Fig. 5.3).

The basic idea is as follows. Each sender node acts as a temporary data buffer. Once a node receives a data packet from the application or from some other node in the

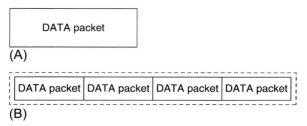

Fig. 5.3 Data packet—(A) original and (B) virtual.

network, it does not immediately forward the packet to the next hop node. Instead, the node stores the packet in its local buffer. The node continues to accumulate and store data packets until the number of data packets in the local buffer reaches a certain threshold. At this point, the sender node engages in a RTS/CTS exchange with the next hop receiver node (as explained before). The RTS/CTS exchange can be considered a channel reservation technique. Once the sender node successfully reserves the channel (all neighbors of the sender and all neighbors of the receiver are silent), it forwards all available data packets from its local buffer to the next hop receiver during the reservation time. Sending (forwarding) multiple data packets back-to-back (one long "virtual" data packet) using only one RTS/CTS exchange essentially increases the *Packet Duration (DATA)*, which in turn increases the CU using the aforementioned equation. For example, if (i) a sender (or forwarder) node stores (accumulates) up to 50 data packets in its local buffer, then (ii) reserves the channel using one RTS/CTS exchange and finally (iii) sends out (or forwards) all the 50 packets back-t o-back in one burst, then, using the same network parameters, the channel utilization will be

$$CU = \frac{50 \times 8\,ms}{(50 \times 8 + 1.6 + 1.6)\,ms + (3 \times 67)\,ms} = \frac{400\,ms}{604.2\,ms} \approx 0.66 = 66\%.$$

Thus, we have seen that the DTN store-and-forward concept can assist the RTS/CTS MAC protocol and make it more suitable for the underwater acoustic environment by increasing the performance of the MAC protocol in terms of CU. For comparison purposes, simplified versions of the default RTS/CTS MAC protocol and the DTN-assisted RTS/CTS MAC protocol are illustrated using the flowcharts in Fig. 5.4A and B, respectively.

5.4 Future trends

There are numerous areas of future research that can be explored:

- For solutions that rely on AUVs, optimization algorithms can be explored to design better mobility patterns so that these solutions can perform well in sparse, as well as dense networks.

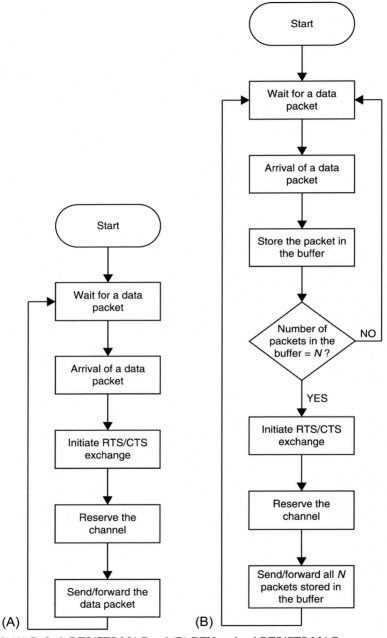

Fig. 5.4 (A) Default RTS/CTS MAC and (B) DTN-assisted RTS/CTS MAC.

- Artificial intelligence techniques need to be explored in different underwater DTN solutions (Section 5.2) in order to improve their adaptability in changing surroundings, which in turn may increase their applicability.
- Channel coding techniques can be integrated with DTN techniques to improve network performance in terms of reliability.
- Optimization algorithms need to be explored and applied to the contemporary solution described in Section 5.3.2 in order to determine the optimal value of N (number of packets stored in the buffer).
- The DTN solution outlined in Section 5.3.2 is suggested for static UWNs having moored nodes.

However, in real scenarios, these moored nodes will have passive mobility due to water currents.

Hence, mobility prediction techniques need to be explored to handle this passive mobility.

5.5 Conclusion

In this chapter, we have presented the current state-of-the-art solutions in underwater DTN. We have highlighted that the current solutions either rely on message replication or make some assumptions about the underlying network topology or mobility patterns (contact schedules). This indicates that current solutions are application (or network topology) dependent, and there is no one solution that will work on all network scenarios. Then we looked at the UAN from a different point of view and showed how the DTN concept can be applied to this different way of thinking. To this end, we have first characterized the UASN as an unreliable disruption-prone network due to the lack of an efficient MAC protocol. Then we have proposed our own idea as to how the DTN concept can be utilized to assist the MAC protocol and make it more efficient. Since our idea is still in its theory stage, for future work, we would like to implement and verify our idea in different underwater scenarios, both static and mobile. We would also like to compare the performance of our solution against various underwater delay-tolerant solutions.

References

Ahuja, R., Magnanti, T., Orlin, J., February 1993. Network Flows: Theory, Algorithms, and Applications. Prentice Hall, New Jersey, USA.

Akyildiz, I.F., Pompili, D., Melodia, T., 2005. Underwater acoustic sensor networks: research challenges. Ad Hoc Netw. 3 (3), 257–279.

Ayaz, M., Abdullah, A., Jung, L.T., June, 2010. Temporary cluster based routing for underwater wireless sensor networks. In: Proc. International Symposium in Information Technology (ITSim), Kuala Lumpur, Malaysia, pp. 1009–1014.

Chan, C., Motani, M., 2007. An integrated energy efficient data retrieval protocol for underwater delay tolerant networks. In: IEEE OCEANS 2007–Europe, Aberdeen, Scotland, pp. 1–6.

Climent, S., Capella, J., Bonastre, A., Orts, R., 2011. A new model for the ns-3 simulator of a novel routing protocol applied to underwater WSN. In: Proc. International Conference on Wireless Networks (ICWN'11), Las Vegas, NV, USA.

Climent, S., Capella, J.V., Meratnia, N., Serrano, J.J., 2012. Underwater sensor networks: a new energy efficient and robust architecture. Sensors 12, 704–731.

Domingo, M., Prior, R., 2007. A distributed clustering scheme for underwater wireless sensor networks. In: Proc. IEEE 18th International Symposium on Personal, Indoor and Mobile Radio Communications (PIMRC), Athens, Greece.

Fall, K., 2003. A delay-tolerant network architecture for challenged Internets. In: Proc. SIGCOMM '03, Germany, pp. 27–34.

Freitag, L., Grund, M., Singh, S., Partan, J., Koski, P., et al., 2005. The WHOI micro-modem: an acoustic communications and navigation system for multiple platforms. In: Proc. MTS/IEEE OCEANS, vol. 2, pp. 1086–1092.

Guo, Z., Colombi, G., Wang, B., Cui, J.-H., Maggiorini, D., et al., 2008. Adaptive routing in underwater delay/disruption tolerant sensor networks. In: Proc. 5th Annual Conference on Wireless On Demand Network Systems and Services, Garmisch-Partenkirchen, Germany.

Guo, X., Frater, M., Ryan, M., April 2009. Design of a propagation-delay-tolerant MAC protocol for underwater acoustic sensor networks. IEEE J. Ocean. Eng. 34 (2), 170–180.

Guo, Z., Wang, B., Cui, J.-H., 2010. Prediction assisted single-copy routing in underwater delay tolerant networks. In: IEEE GLOBECOM, Miami, FL, USA, pp. 1–6.

Guo, Y., Cui, H., Si, W., 2013. Large delay underwater sensor networks clock synchronization with mobile beacon. In: Proc. IEEE WCNC Workshop on Applications of Delay Tolerant Networking (A-DTN), Shanghai, China, pp. 211–215.

Hayajneh, M., Khalil, I., Gadallah, Y., 2009. An OFDMA-based MAC protocol for underwater acoustic wireless sensor networks. In: IWCMC'09, Leipzig, Germany.

Hu, T., Fei, Y., 2010. An adaptive and energy-efficient routing protocol based on machine learning for underwater delay tolerant networks. In: Proc. 18th Annual IEEE/ACM International Symposium on Modeling, Analysis and Simulation of Computer and Telecommunication Systems, Miami, FL, pp. 381–384.

Jafri, M.R., Ahmed, S., Javaid, N., Ahmad, Z., Qureshi, R.J., October 2013. AMCTD: adaptive mobility of courier nodes in threshold-optimized DBR protocol for underwater wireless sensor networks. In: Proc. 8th International Conference on Broadband and Wireless Computing, Communication and Applications ÇBWCCA'13), Compiegne, France.

Katz, I., 2007. A delay-tolerant networking framework for mobile underwater acoustic networks. In: Proc. 15th International Symposium on Unmanned Untethered Submersible Technology (UUST'07), Durham, NH, USA.

Luby, M., 2002. LT codes. In: 43rd Annual IEEE Symposium on Foundations of Computer Science, Vancouver, BC, Canada.

Magistretti, E., Kong, J., Lee, U., Gerla, M., Bellavista, P., et al., 2007. A mobile delay-tolerant approach to long-term energy-efficient underwater sensor networking. In: Proc. IEEE Wireless Communications and Networking Conference (WCNC), Kowloon, Hong Kong.

Merani, D., Berni, A., Potter, J., Martins, R., 2011. An underwater convergence layer for disruption tolerant networking. In: IEEE Baltic Congress on Future Internet and Communications, Riga, Latvia, pp. 103–108.

Noh, Y., Wang, P., Lee, U., Torres, D., Gerla, M., 2010. DOTS: a propagation delay-aware opportunistic mac protocol for underwater sensor networks. In: Proc. 18th IEEE International Conference on Network Protocols (ICNP), Kyoto, Japan, pp. 183–192.

Pompili, D., Akyildiz, I.F., 2009. Overview of networking-protocols for underwater wireless communications. IEEE Commun. Mag. 47 (1), 97–102.

Pompili, D., Melodia, T., Akyildiz, I.F., 2006. A resilient routing algorithm for long-term applications in underwater sensor networks. In: MedHocNet, Lipari, Italy.

Rappaport, T.S., 1999. Wireless Communications: Principles and Practice. Prentice Hall, New Jersey, USA.

Rodoplu, V., Park, M.K., 2005. An energy-efficient MAC protocol for underwater wireless acoustic network. In: Proc. of MTS/IEEE OCEANS'05, Washington DC, USA.

Shah, G.A., 2009. A survey on medium access control in underwater acoustic sensor networks. In: Proc. International Conference on Advanced Information Networking and Applications Workshops, Bradford, UK, pp. 1178–1183.

Smart, J.H., 2005. Underwater optical communication systems. Part 1: Variability of water optical parameters. IEEE Military Commun. 2, 1140–1146.

Spyropoulos, T., Psounis, K., Raghavendra, C.S., 2005. Spray and wait: an efficient routing scheme for intermittently connected mobile networks. In: Proc. ACM SIGCOMM Workshop on Delay-Tolerant Networking (WDTN'05), Philadelphia, PA, pp. 252–259.

Spyropoulos, T., Turletti, T., Obraczka, K., 2009. Routing in delay tolerant networks comprising heterogeneous node populations. IEEE Trans. Mob. Comput. 8 (8), 1132–1147.

Stojanovic, M., 1999. Underwater Acoustic Communication. Wiley Encyclopedia of Electrical and Electronics Engineering, New York, USA.

Sutton, R.S., Barto, A.G., 1998. Reinforcement Learning: An Introduction. The MIT Press, Cambridge, MA, USA.

Tan, D.D., Le, T.T., Kim, D.-S., 2013. Distributed cooperative transmission for underwater acoustic sensor networks. In: Proc. IEEE WCNC Workshop on Applications of Delay Tolerant Networking (A-DTN), Shanghai, China, pp. 205–210.

Tanenbaum, A., 2002. Computer Networks, fourth ed. Prentice Hall Professional Technical Reference, New Jersey, USA.

Wang, J., Li, D., Zhou, M., Ghosal, D., 2008. Data collection with multiple mobile actors in underwater sensor networks. In: 28th IEEE International Conference on Distributed Computing Systems Workshops, ICDCS'08, Beijing, China, pp. 216–221.

Wang, D., Hong, F., Yang, B., Zhang, Y., Guo, Z., 2013. Analysis on communication capability of vessel-based ocean monitoring delay tolerant networks. In: Proc. IEEE WCNC Workshop on Applications of Delay Tolerant Networking (A-DTN), Shanghai, China, pp. 200–204.

Yoon, S., Azad, A.K., Oh, H., Kim, S., 2012. AURP: an AUV-aided underwater routing protocol for underwater acoustic sensor networks. Sensors 12, 1827–1845.

Zhou, Z., Peng, Z., Cui, J.H., Jiang, Z., 2012. Handling triple hidden terminal problems for multichannel MAC in long-delay underwater sensor networks. IEEE Trans. Mob. Comput. 11 (1), 139–154.

Delay-tolerant networks (DTNs) for emergency communications[☆]

H. Chenji and R. Stoleru
Texas A&M University, College Station, TX, United States

6.1 Introduction

Natural or man-made disasters, whether sudden or predicted, cause untold loss of life and property. The power and communication infrastructure in the area is severely crippled and is rendered unusable. An organized response to this disaster is mounted by designated agencies. A disaster that spans one or few city blocks is usually handled by the local police force, the local fire department, a national level governmental agency (e.g., the FBI), or a combination of the above. However, when the disaster occurs on a large scale, such as the Japanese tsunami, several specialized agencies spanning international boundaries cooperate to make the response more effective.

A high degree of *situational awareness* is vital for effective disaster response. Reliable communication infrastructure is critical in ensuring situational awareness, since improved coordination between teams can lead to more rescued lives. Unfortunately, the local power (and hence the communication) infrastructure is severely crippled during a disaster. Oversubscription of remaining infrastructure, if any, hampers the communication capability of victims as well as emergency responders. Affected or rescued victims are unable to use their cell phones to contact the outside world. Emergency responders have to resort to using expensive and/or high-latency satellite phones to communicate. The lack of power infrastructure means that electronic devices need to be battery-powered. Diesel generators are not ubiquitous and are present only in important areas such as the emergency operations center (EOC).

This chapter utilizes the following disaster model as a motivating scenario. A large-scale disaster has affected a few cities or an entire geographical region, such as the earthquake in Haiti (USGS, 2010) and the tsunami in Japan (BBC, 2011). The communication infrastructure is disrupted (i.e., cellular networks are completely or partially damaged) for weeks if not months, there are serious shortages of power (i.e., power sources like nuclear reactors are damaged), surveying the disaster area for survivors under the rubble takes from days to weeks (with some inspiring examples of survivors emerging after tens of days) and the EOC is flooded with sensing and multimedia data from the field. This febrile, fast-paced environment lasts from one to several weeks, until the infrastructure is usable again.

☆ This chapter is a reprint of the chapter originally published in the first edition of Advances in Delay-tolerant Networks (DTNs): Architecture and Enhanced Performance.

Advances in Delay-tolerant Networks (DTNs). https://doi.org/10.1016/B978-0-08-102793-6.00006-0

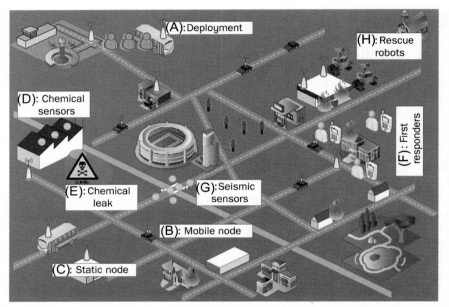

Fig. 6.1 Schematic of a large city in which a disaster has occurred. A national or international disaster recovery effort is underway. The particular points of interest, labeled A through H, illustrate how situational awareness can be improved. A detailed explanation is provided in Section 6.1.

A typical response to such a disaster is illustrated in Fig. 6.1. A national agency or an international coalition arrives with equipment (label A). Vehicles in the area, like ambulances or supply trucks, are outfitted with high-capacity wireless routers (label B). Certain important areas, such as collapsed buildings, are assigned static routers (label C). Rescue personnel deploy low-power wireless chemical sensors (label D) around a leak at a factory (label E), while others attempt to rescue victims from a burning building (label F). A bridge on the verge of collapse is monitored remotely using seismic sensors (label G), while rescue robots capable of recording HD video approach a gas station. The objective of this chapter is to explore computer networking architectures that can provide the data from such heterogeneous sources to first responders and rescue personnel.

In the United States, organized response to disasters is coordinated by the Federal Emergency Management Agency (FEMA). It has created several task forces (28 currently) that specialize in urban search and rescue (US&R) operations. US&R personnel perform the "location, rescue (extrication), and initial medical stabilization of victims trapped in confined spaces" especially in an urban/industrial setting (as opposed to ground, mountain, and battlefield search-and-rescue). During the Haiti earthquake, FEMA/US&R was responsible for 47 of 134 lives rescued. One can get an idea of the networking technologies currently used during disaster response by looking at the mandated and standardized equipment list for FEMA/US&R teams (FEMA, 2008). Each task force maintains its cache, containing over 16,000 items. The technical equipment details Project 25 (PTIG, 2014)-compatible two-way portable wireless radios. A 120 V AC-powered base station is also mentioned, along with battery-powered repeaters. Such radio systems have a large radio range capable of covering large areas and are securely encrypted. However, only voice and data channels are available on such systems, at very low data rates of 9.6 kbps (Motorola, 2011). Although reliable real-time long-range secure communication systems like Project 25 exist, they have several caveats, like cost, difficult integration with other systems, high power requirements, and physical bulk. The size of the equipment cache is about 66 tons. Mesh Dynamics (2008) has commercial offerings which accomplish network-centric warfare. To the best of our knowledge, these systems assume a powered, connected network. Additionally, they address spatially localized disasters or the triage area of disasters that occur over a large area.

6.2   Overview of proposed DTN solutions

This section presents an overview of proposed delay-tolerant networks (DTN) solutions for emergency responders. We first review several wireless networking systems and architectures for emergency response, motivate the need for DTN solutions, and present two mobility models that help DTN protocol and system modeling. Distress-Net, an emergency DTN, is then discussed. Two software-based solutions that help reduce energy consumption, as well as two that do the same in hardware, conclude the section.

6.2.1   Wireless networking systems and architectures for emergency response

6.2.1.1   WIISARD

The Wireless Internet Information System for Medical Response in Disasters (WIISARD) (Chipara et al., 2012; Brown et al., 2006; Arisoylu et al., 2005) is a 802.11-based wireless mesh network tailored to provide effective medical response in the event of a disaster. Digital tags on patients (Lenert et al., 2005) are read by medical personnel using personal digital assistants (PDAs) which roam the area while being connected to the Internet via backhaul connections (Killeen et al., 2006). Changes to such digital records are tracked and can be easily rolled back in case of conflict due to multiple simultaneous editing. It is to be noted that network connectivity is assumed to be persistent and highly available. Other components of the system include a blood pulse-oximetry device, a command center display that focuses on data quality, and a wireless device for first responders designed to enhance situational awareness. Chipara et al. (2012) propose the WIISARD communication protocol (WCP), which is a gossip-based protocol for data dissemination, and show that link properties vary between the different stages of the rescue drill. As stated by Chipara et al. (2012), the major contribution is not WCP itself, but the characterization of link quality and human mobility patterns during the medical triage phase.

6.2.1.2   RESCUE

Project RESCUE (Mehrotra et al., 2004) provides an overview of a wireless mesh network (WMN) for effective emergency response. Dilmaghani and Rao (2008) and Manoj and Baker (2007) argue for a WMN to be used in disaster response. They cite several shortcomings in several real-use cases that provide a baseline comparison to such systems. In (Dilmaghani et al., 2005), a hybrid WMN makes use of wireless wide area networks (WANs) as a backhaul link to access traditional networks. Several portable networked devices make use of routers affixed to lamp posts in order to achieve network connectivity.

6.2.1.3   SAFIRE

The SAFIRE project (Xing et al., 2009) deals with situational awareness for firefighters. Among the many problems dealt with are reliable data dissemination over ad hoc networks. Responders use a Wi-Fi-enabled tablet that uses a central push-pull method of data movement. The intended purpose is for use in a local emergency, and not a region-wide disaster. Based on the limited details available, the system offers robust middleware based on 802.11 and/or WiMax-based networking. To the best of our knowledge, these systems assume an AC-powered connected network and do not offer integration of low-power smart devices.

6.2.1.4   AID-N

Another academic effort that improves disaster recovery in the medical triage area is the Advanced Health and Disaster Aid Network (AID-N) (Gao et al., 2007). It presents the design of an electronic triage system for use on victims. Low-power IEEE 802.15.4-based hardware replaces colored paper tags that are used to identify the severity of a victim's situation. A tiered architecture is proposed: level 1 contains an ad hoc network of embedded 802.15.4-based devices, level 2 consists of 802.11-based PDAs and laptops used by personnel to access victim information, and level 3 is a central high-capacity server that stores information. Level 1 devices form an ad hoc mesh network that uses the Flows routing protocol (Gao et al., 2007) which was proposed as part of the CodeBlue project (Gao et al., 2007). It is a spanning tree-based protocol that uses hop-by-hop acknowledgments. Level 2 devices connect to an 802.11g-based wireless router, which in turn connects to the level 3 central server. To summarize, AID-N provides a tiered multi-PHY architecture that uses mesh networking.

6.2.2   Why delay-tolerant networks for emergency communication?

The above solutions require fixed infrastructure such as cell phone towers or a high-power Wi-Fi access point. When it is impractical to deploy such equipment, due to either lack of power sources or cost prohibition, ad hoc mesh networking can be used. Wireless mesh networking utilizes several fixed access points that serve clients that connect to the mesh. As the size of the affected area increases, the amount of fixed infrastructure that is needed to cover the entire area for wireless connectivity increases exponentially, resulting in increased cost and complexity. For this reason, it might not be efficient to use mesh networking. DTNs form a sparsely connected network that is capable of tolerating large link delays and low link availability. Mobility in the area is leveraged to mule data on vehicles capable of carrying networking equipment.

One advantage of the DTN architecture is that it can be implemented purely in the application layer or above—as a result, it is capable of integrating many heterogeneous subarchitectures. For example, the use of wireless protocols like 802.11 (Wi-Fi), 802.15.4 (ZigBee), and cellular networks have been proposed in recent research that looks at architectures for emergency response. Each of these technologies has its own advantages and disadvantages. For example, 802.15.4, is optimized for low-power sensing—devices implementing 802.15.4 typically have low energy footprints but are unable to offer bit rates of more than a few hundred Kbps. Wi-Fi is ubiquitous and can be found on most commercial off-the-shelf (COTS) devices; it requires significantly more energy and it is able to offer communication rates of a few tens of Mbps. Both Wi-Fi and ZigBee have radio ranges that can be measured in tens of meters. Cellular networks, on the other hand, require the most energy but are able to offer constant connectivity to clients at bit rates comparable to Wi-Fi while being spread over a few kilometers.

A comprehensive description of motivating factors for a new delay-tolerant architecture can be found in (Fall, 2003) and RFC 4838 (Cerf et al., 2007). Interoperability between devices with heterogeneous capabilities and functionality is emphasized. Specialized low-power devices should be able to seamlessly integrate with other classes of devices, different network stacks notwithstanding. Since there may exist different types of data with different priorities and useful lifetimes, programmatic support for prioritizing data and specifying lifetimes is provided. Nodes may have a combination of many characteristics, like low energy resources and limited storage capacity. In the event that such nodes find that their resource requirements hamper functionality, the node needs to be able to delegate the responsibility of ensuring data flow to another suitable node in the network. The Bundle Protocol is presented in RFC 5050 (Scott and Burleigh, 2007). The primary data unit is called a "bundle." Application layer implementation ensures network abstraction. The actual implementation of a DTN gateway will need to have support for "convergence layers," which are essentially interfaces to one or more network interfaces present. The custody transfer feature ensures that nodes can delegate responsibility to another node. Acknowledgments can be requested either on a per-hop or on an end-to-end basis.

6.3 Mobility models for emergency DTNs

Mobility models are essential to mathematically characterize node movement patterns during disaster recovery. Many algorithms, especially routing protocols, need information about node mobility in order to make better, optimal decisions.

6.3.1 Postdisaster mobility model

The postdisaster mobility (PDM) model (Uddin et al., 2009) uses two main components to functionally model the interaction between survivors and rescue workers: "mobile agents" (MAs) and "Centers." Centers are fixed, static areas on a city map and represent important areas such as the EOC or an evacuation center such as a stadium. PDM is a map-based mobility model as opposed to a random waypoint: each MA moves from one point to another based on a predefined set of roads. There are different categories of MAs, like patrol cars, ambulances, volunteers, and supply vehicles, each with its own movement model, as described below. It is assumed that a networking device is present at each Center as well as on each MA. The PDM model is described formally as follows. There are multiple Centers C, of which two are special and compulsory: the EOC and the Triage. An illustration is shown in Fig. 6.2. The categories of MAs are: volunteers, supply vehicles, ambulances, and patrol cars. Each category has its own min and max speeds, and an agent belonging to that category chooses a speed uniformly between min and max at random, for each leg of travel. In the volunteer movement model (D), each volunteer is placed at a randomly assigned home center $c_H \in C$ initially. Next, every volunteer individually chooses a random

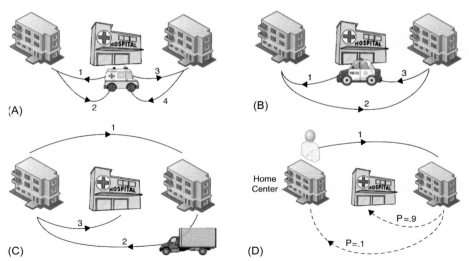

Fig. 6.2 Illustrating the postdisaster mobility model. There are three centers: the EOC on the left, the triage (hospital) in the middle, and another center. The mobile agents represented are (A) ambulance, (B) patrol car, (C) supply vehicle, and (D) volunteer.

point within the entire map with 90% probability or chooses c_H with 10% probability, and travels to it along the shortest path. The process is repeated upon reaching the point. In the supply vehicle movement model (C), each SV is placed at a randomly assigned home center $c_H \in C$ initially. Then each SV individually chooses a center $c_d \in C$ at random and travels to it along the shortest path. The process is repeated upon reaching c_d. In the patrol car movement model (B), each car has a predefined list of centers $c_1, c_2, c_3, \ldots c_n \in C \times C \times C \ldots C$, and is placed at c_1 initially. Next, it travels to c_2 along the shortest path, and the process is repeated by choosing the next center in the list. In the ambulance movement model (A), each ambulance is always assigned to the Triage initially. Next, each ambulance chooses a center (including the triage) at random to travel to, following which the ambulance always returns to the triage. The process is repeated, resulting in a series of alternating centers and triages.

6.3.2   Enhanced postdisaster mobility model

As explained in the introduction of this chapter, US&R personnel are essential when the area affected by the disaster is an urban area. The movement patterns of these personnel are not included in the PDM model. Chenji et al. (2013a,b) propose an enhancement to the PDM model: a new category of mobile agents called the "urban search and rescue worker (USAR)" is proposed. An illustration is shown in Fig. 6.3. These USARs operate in an area of a fixed radius around a center and move using random waypoint within that area. In the USAR movement model, each USAR member is placed at a predefined home center $c_H \in C$, initially. Then, every USAR agent individually randomly chooses a point uniformly within radius r of its c_H, and travels to it

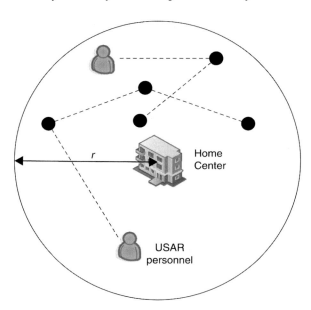

Fig. 6.3 Enhancing the PDM model with the USAR movement model. Several USAR personnel move around a home center within a radius r, using the random waypoint mobility model.

along the shortest path possible in the map. After reaching the point, the process is repeated. USARs need not visit c_H compulsorily.

6.4 DistressNet

DistressNet (Chenji et al., 2012) is a system that utilizes DTNs to address the needs of US&R personnel during an emergency response. It is not designed for use in the medical triage area, as many of the above-described systems are, but is instead meant to be deployed over the large disaster-struck area. The hardware and software architecture incorporates heterogeneous hardware, open-source software, and COTS devices. Networking equipment carried by vehicles and mobile personnel are used to mule data from one geographical point to another. Wireless sensors, which could be present on victims trapped under rubble or installed by US&R personnel on buildings to monitor chemical activity, periodically generate data that is then ferried to the EOC using these vehicles. The large amount of data generated by a wide range of devices, ranging from video to air quality data, can be accessed via a cloud computing-like interface called a "fog" (Chenji et al., 2013a,b). Rescue personnel can use their smartphones to access and visualize high-quality data. DistressNet middleware and technical contributions include an infrastructure placement module (Chenji et al., 2012), a QoS-aware DTN routing protocol (Chenji et al., 2013a,b), and an energy-saving scheme that operates in a Pareto-optimal fashion (Chenji and Stoleru, 2014).

6.4.1 Motivating scenario

DistressNet addresses the needs of the disaster recovery process in the USAR application domain, as opposed to the medical triage area of a large disaster, as illustrated in Fig. 6.4. When a disaster hits an urban metropolitan area that spans tens of square miles (2011 Joplin tornado) or hundreds of square miles (2011 Japan earthquake), power and communication infrastructure are rendered unusable. A situation report about the 2011 Joplin tornado (Missouri DPS, 2011), 3 days after the disaster, offers a glimpse into the situation: electric services are still being restored, a few cell sites have been restored, cell phones are being distributed and satellite telephone has been set up. In this kind of environment, the presence of broadband Internet access cannot

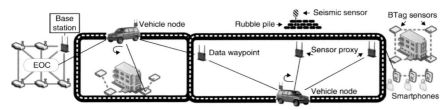

Fig. 6.4 Schematic of a fog computing deployment showing all components. Data generated by BTag and seismic sensors are ferried to the base station using vehicle nodes. A data waypoint improves the data transfer process by creating a contact opportunity between two vehicle nodes.

be assumed, and blanketing a large urban area with battery-powered communication hardware is near impossible. Providing data to USAR responders at high spatial and temporal resolution, with only tens of routers, becomes a challenge. We assume that in such an environment there are multiple collapsed buildings (buildings in Fig. 6.4) or rubble piles in the affected area ("Rubble pile" in Fig. 6.4), and the emergency operations center ("EOC" in Fig. 6.4) is situated tens of miles away from the affected area. Limited Internet connectivity is available only at the EOC. There is some mobility in the area ("Vehicle node" in Fig. 6.4) as medical supplies and rescued victims are transported from the field to the EOC.

6.4.2 DistressNet applications

We now describe the conceptual working of seven new motivating applications designed to satisfy some of the emergency responders' requirements. After describing the hardware architecture in the next section, the implementation details of these new apps are discussed.

The app-oriented architecture of DistressNet enables third parties to add piecewise functionality without worrying about lower-layer functions like routing or storage. As an example, a file storage and social networking service for first responders can be very useful and improve situational awareness. A traditional file storage service in the cloud has the following functionality and properties: (i) The ability for a client to upload a file to the cloud, without specifying a destination node; (ii) The ability for a client to retrieve a file, without specifying the location of the file; and (iii) Data robustness due to intelligent replication performed by the cloud service back end. The file-sharing application allows authenticated users to share data with other users, or groups of users. As an example, one can imagine a team of responders sharing the layout of an explored building, along with current hazards, with other nearby teams. An important feature that emphasizes the need for such a service is that the destination for data sent is unknown—it is simply stored in the mobile cloud environment, which we call *Fog*, and accessed by anyone who connects to the fog computing infrastructure, in the disaster area.

The *File Sharing App* can be seen as a client/front end, designed to be used by a first responder, that connects to a *File Sharing Service* server/back end and authenticates itself. The clients can then use the fog application programming interface (API) to ADD/DELETE/MODIFY files which they own (as determined after authentication), as well as to specify availability metrics depending on the importance and criticality of the data file. They can also share selected files with other clients or users. Examples of files that users can upload include video taken using a smartphone's camera. Users can also specify whether they want to back up these files to an external storage provider like Amazon S3 or Flickr. If a client wishes to use their own external account, an encrypted query, using the provider's API (S3's API or Flickr's API), is sent to a special device that has the file sharing service as well as Internet access. The HTTP header and body are intercepted by the file sharing service and sent to the Internet Gateway using the Fog's underlying network.

The fog API insulates the user, first responders in this case, from the inner workings of the underlying network. It aims to provide a service similar to those offered by cloud storage services like Dropbox and Amazon S3. Three primitives called ADD, DELETE, and MODIFY provide an interface into the fog. The ADD primitive uploads a file into the Fog. This file could consist of a tweet as produced by the social networking app (to be presented next), or a photo from the user's phone as produced by the file sharing app. The MODIFY/DELETE primitives allow users to modify or delete files that they own.

When an ADD request is sent from the file sharing app to a nearby device running the file sharing service, the file is first transferred locally to the device. Then, the file is replicated on multiple fog devices according to the criticality of the data. The actual fog devices that are chosen as endpoints depend on the output of several algorithms to be presented later in this dissertation. A MODIFY operation causes the file sharing service to send the difference between the current version of the stored file and the new incoming file to the fog devices which contain the original. These devices will locally modify their copy of the file and push it to another user's file sharing app upon connection. The DELETE operation simply sends a low-overhead message to the Fog device which says that the local copy of the file on the device should be deleted. The synchronization between the devices running the file sharing app and file sharing service occurs as follows: when an app discovers a service nearby, it can supply a list of files stored locally and ask for changes to those files. The service then replies with a list of changes, which the App can apply to its local copy.

API-level security is available if the user chooses external service providers—thus providing encryption on an end-to-end basis for the user. Users need not disclose their existing external credentials in order to use the Fog. We consider Amazon S3 as an example of an external representational state transfer (REST) cloud storage service provider. Files are uploaded to S3's servers using a published API which offers both a REST and a simple object access protocol (SOAP) interface using XML. When a user signs up, a secret key is assigned. This secret key is then used alongside a keyed-hash message authentication code (HMAC) Secure Hash Algorithm 1 (SHA1) in order to authenticate all HTTP (optionally, HTTPS) requests. Whenever a user wishes to place a file in S3, a challenge string is first constructed based on a predefined ruleset, which then serves as the "message" in HMAC-SHA1. The output, which is a base64 encoded string, forms part of the HTTP request header. The entire HTTP header and body (if applicable) are then sent to Amazon by the client.

The *Social Networking App* is a user client that provides services like interteam messaging, and allows victims to publicly tweet their status. Users can communicate with other users, or mass message other teams or team members. They can also choose to post messages to an external Twitter account. It is important to note that such messages will be available in the Fog, as well as on their Twitter profiles—thus, an external service like Twitter is mirrored in the Fog. All data from Twitter is pulled regularly by an Internet gateway and sent to the fog (Fig. 6.5) so that the data stays synchronized. Primitives specific to social networking, such as "following" a user or replying to a message, are handled by a *Social Networking Service* that runs on the same device as the file sharing service, such that the fog API is used by the service to synchronize data.

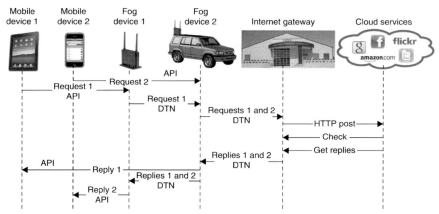

Fig. 6.5 Sequence diagram for accessing cloud and social networking services (e.g., Twitter).

The FEMA US&R equipment cache list (FEMA (2008)) mentions delsar life detection sensors: a steel spike (Fig. 6.6) that is driven into rubble, which responders can then monitor for voices or knocks from victims. Upon manually probing the rubble at different places, the victim can be localized and rescue operations can commence.

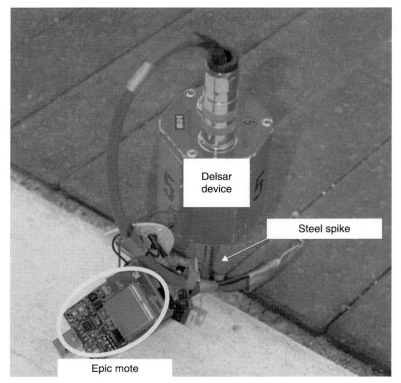

Fig. 6.6 The delsar life detector attached to a Sensor class device.

However, there are sources of noise like footsteps and vibration from nearby vehicles which are also picked up. The goal of the *Vibration Sensing App* is to automatically detect and classify the source of vibration. To profile these sources, the steel spike of the sensor was driven into a small wedge in a pavement outside our building on campus. Three sources of noise/data were profiled: a stone dropped from a height, footsteps of pedestrians, and a knock made by a hammer on a pavement. The fixed-point in-place 1024-bin fast Fourier transform (FFT) is shown in Fig. 6.7.

It is important to note that the amplitude of the signal alone cannot be used to classify a source. Hence, two features were extracted from the FFT: (i) average value of the frequencies weighted by their respective amplitudes (f_1) and (ii) the mean amplitude of the frequencies (f_2). We then used these features in a simple KNN (k-nearest neighbor) classifier. Suppose that we have g different types of data $G_1 \dots G_g$ and n samples for each group, for a total of gn samples $s_1 \dots s_{gn}$. Let each sample be a vector consisting of two features $[f_1, f_2]$. The KNN classifier first needs to be trained using these samples. Training consists of storing each sample and its corresponding group in memory. Now, given a new sample $S = [F_1 F_2]$ that needs to be classified, Algorithm 6.1 can be used to calculate the group G that S belongs to, thus identifying the source of the vibration.

Algorithm 6.1: K-NN Classifier
1. **for** each $s_i \in s_1 \dots s_{gn}$ **do**

2. Compute $d_i \leftarrow \sqrt{\left(F_1 - S_i^{f_1}\right) + \left(F_2 - S_i^{f_2}\right)^2}$

3. **end for**
4. $r_1 \dots r_k \leftarrow$ the k smallest d_i
5. *groups* $\leftarrow$ Union of groups that each of $r_1 \dots r_k$ belong to
6. G $\leftarrow$ most common group in *groups*

FEMA uses X-codes to denote the search status of buildings during urban search and rescue. An example X-code is shown in Fig. 6.8A. This includes information like the number of survivors inside, the location of chemical hazards, if any, and the most recent date/time that the structure was searched (Guven and Ergen, 2011). The primary motivation for the *Building Monitoring App* was the fact that X-codes are currently painted on walls (Fig. 6.8B). These X-codes are likely to remain constant and not change very often at the rate of several times per second. A *Building Monitoring Service* runs on low-power sensors and stores the data programmed into it by using a client front end like the Building Monitoring App. One or more of these sensors can be deployed in or around a building. Vehicles in the vicinity can automatically gather data from these sensors and mule it using other vehicles to the EOC. Thus, an electronic X-code is possible, and the data stored can be picked up by vehicles and transferred to the EOC for usage and further processing.

6.4.3   Hardware and software architecture

There exist three distinct classes of devices in DistressNet: sensors, smartphones, and routers. We now describe the hardware used to implement these three classes.

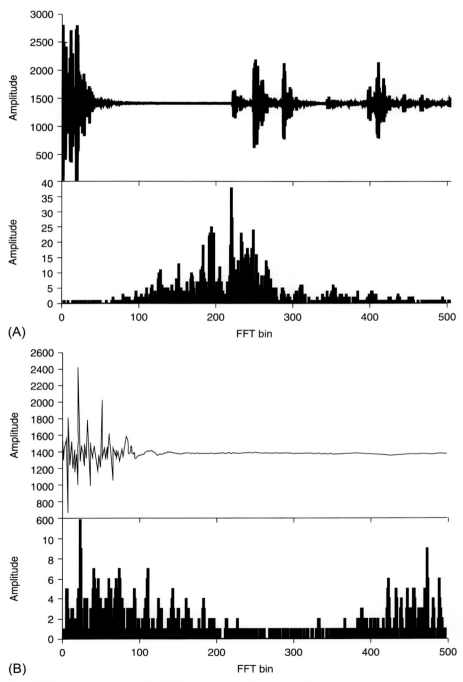

Fig. 6.7 Spectrum and signal of (A) stone drop, (B) footstep, 3

(Continued)

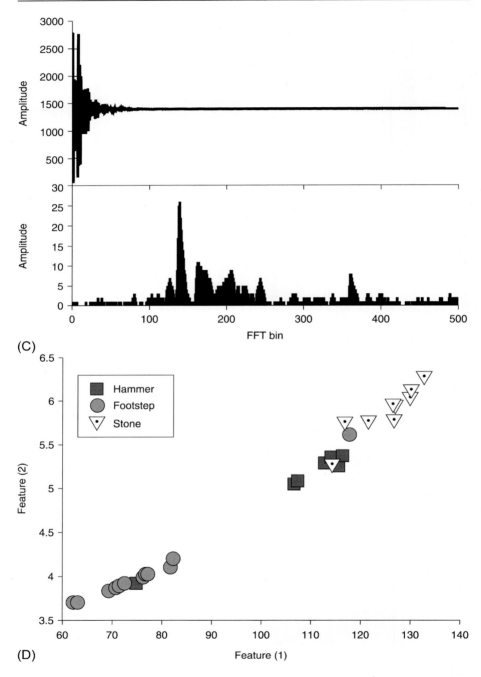

Fig.6.7, Cont'd (C) hammer strike, (D) shows the classifier results based on two features.

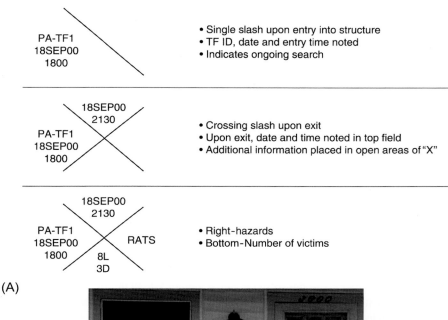

(A)

(B)

Fig. 6.8 (A) FEMA X-code format. The four major pieces of information are the ID of the rescue team, time and date of last search, number of alive and dead people, and hazards present. (B) An X-code seen in New Orleans during the Hurricane Katrina rescue effort. (A) This image was taken unmodified from FEMA, 2006. Rescue Field Operations Guide US&R-23-FG. Available from: http://www.fema.gov/pdf/emergency/usr/ usr_23_20080205_rog.pdf. (Accessed 15 August 2013).

Fog sensors are typically low-power, battery-powered wireless sensor network platforms such as an Epic mote. The data sensed may be either mission-critical or informational. Being heavily duty-cycled, they are designed to last for several weeks with a single charge. 802.11 support is rarely found on these devices, with 802.15.4 or no networking being more common. Applications deployed on such devices include sensing and services (building monitoring service, sink election app, vibration sensing app, and self-localization app in Fig. 6.9A). The BLIP stack provides user datagram protocol (UDP) connectivity over the IPv6 provided by 6lowpan (Montenegro et al., 2007). In fog computing, we use RPL (Winter et al., 2012), an IPv6 routing protocol for low-power and lossy networks, as the default routing protocol. Fog sensors can upload data into the fog using a fog router as a proxy (to be described shortly).

Smartphones refer to popular network-centric consumer electronics like tablets, smartphones, and laptops, which have networking capabilities, but have limited resources. Fog smartphones are primarily used by first responders to access data stored in the fog. They provide a rich interface to the data collected in the field, while also providing some functionality themselves. Not as resource-constrained as fog sensors, most devices have 802.11 capability and are designed to last a few days on a single charge. Smartphones also have various sensors such as global positioning systems (GPS), cameras, microphones, and accelerometers. They are usually incapable of routing or advanced networking capabilities and have limited, but not scarce, resources. Apps installed on these devices are more data consumers than data generators. The hardware platform used was the iPod touch as well as the iPad. We were limited to the application layer since the software development kit (SDK) does not allow non-trivial modifications to the operating system for security reasons. The network stack on these devices consists of TCP/UDP over IPv4/v6 and 802.11. The fact that 802.11 IBSS mode was readily supported out of the box made us choose iOS over Android.

Fog computing apps and services that run on fog smartphones include the building monitoring app (building monitoring app in Fig. 6.9B) that is used to program fog sensors running the building monitoring service once they are deployed with relevant information. The file sharing and social networking apps (as shown in Fig. 6.9) are also implemented on smartphones, as is the team separation detection app.

Routers are portable, battery-powered devices that provide basic wireless networking functionality and are deployed in the field. An example is a common 802.11 router found in most homes today. They can be assumed to have expansion ports to provide additional functionality like persistent storage or cellular connectivity. These can be either static or deployed inside a vehicle. The hardware platform used is the Mikrotik RB433UAH routerboard, which has three MiniPCI slots and two USB 2.0 ports, allowing for two 802.11abgn wireless cards configured for 2.4 and 5 GHz, respectively. It also has 512 MB of Not-AND gated (NAND) flash and 128 MB of RAM. Open-WRT (OpenWRT Dev Team, 2014) is an open-source GNU/Linux-based operating system compatible with this router, which was chosen because of the openness and the wide range of software and support available. The USB port can be used to provide functionality such as a new physical layer such as 802.15.4 (to communicate with fog sensors), enhanced storage like a USB flash drive, or both. The networking stack used is 802.11abgn below IPv6/v4 and UDP (Fig. 6.9C). Since most COTS Wi-Fi-compliant devices support only the 2.4 GHz band, we decided to use the 5 GHz interface exclusively for routing. Dynamic host configuration protocol (DHCP) is provided on the 2.4 GHz interface for clients to connect. All routers have statically assigned IPs—router n has an IP of 192.168.50.n for its 5 GHz interface and

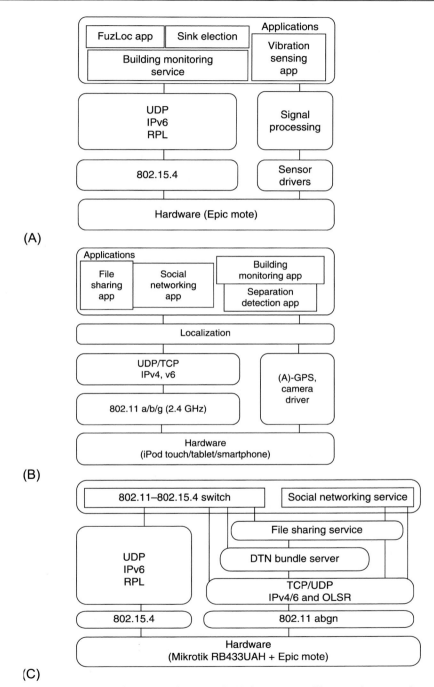

Fig. 6.9 The hardware/software architecture of (A) fog sensors, (B) smartphones, and (C) routers.

192.168.*n*.1 for the 2.4 GHz interface. Each router can handle 255 end-user devices—they are assigned IPs in the 192.168.*n*.0/24 range.

Fog routers employ delay and disruption networking. For implementing DTN functionality, we used the IBR-DTN (Schildt et al., 2011) implementation, which is readily available as a package for OpenWRT. A "Bundle" (RFC 5050) is the primary data unit in DTN. Each DTN node is identified by a URI like dtn://dn.zigbeegateway1. A client application with an ID of motel can connect via an API to the local "Bundle Server." Then, any traffic intended for this app will simply need to be addressed to dtn://dn. zigbeegateway1/mote1. Examples of functions provided by the bundle server API include registering the application name (e.g., "motel"), setting a destination, requesting encryption or authentication or custody transfer, and setting the lifetime. All communication in the DTN layer can be encrypted and authenticated, as defined in RFC 6527. Each DTN node first generates, predeployment, a 2048-bit Rivest, Shamir, and Adleman (RSA) public/private key pair. The public keys of all nodes are aggregated and shared among all DTN nodes. This data is used for bundle encryption in a public-key cryptography fashion. It is to be noted that bundle encryption is always between source and destination, and authentication is always on a single-hop basis. Bundle authentication uses the HMAC-SHA1 message authentication cipher, which encrypts a message based on a key. In this case, the key is a preshared plaintext key of arbitrary length that is different from the public/private keys. DTN is implemented as an overlay network of nodes where multiple local clients can connect to a local DTN server (bundle server in Fig. 6.9) in the application layer. A special DTN app on the router, which can talk to 802.15.4-based fog sensors as well as the DTN server, provides DTN proxying functionality ("802.11–802.15.4 Switch" in Fig. 6.9). The file sharing and social networking services have the capability to interface with fog smartphones by acting as a Fog API provider, while simultaneously talking to the bundle server in order to replicate data on other fog routers.

6.5   Routing protocols for emergency DTNs

Traditional DTNs are somewhat more resource-rich when compared with a DTN used for emergency response. Existing infrastructure is crippled or unusable after disaster strikes. Rescue personnel have to transport and deploy hardware to provide basic infrastructure like power and communication. As a result, certain assumptions made during the design of algorithms and protocols for traditional DTNs may not hold for emergency DTNs. Briefly speaking, emergency DTNs are *energy-deficient*. Most DTN routing protocols typically aim to achieve the best performance, in terms of the delivery delay, delivery ratio, or other metrics. The number of transmitted messages, which is a good indicator of the energy consumed at the routing layer, affects performance. The replication factor for multicopy routing protocols in turn affects the number of transmitted messages. In this section we discuss intercontact routing, which makes use of the fact that recurrent contacts occur in an emergency DTN, to reduce message replication and thus save energy.

Another issue in emergency DTNs is the heterogeneous nature of data routing protocols are expected to handle. Data could be mission-critical, informational, or somewhere in between. Simultaneously, the contact bandwidth in a DTN is limited—leading to fragmentation of data during a contact. It could be the case that mission-critical data needs to be delivered as a single block to the EOC within a short time, so that decisions can be taken with complete knowledge of a situation. Informational or debug-level data would need to be delivered quickly to the EOC, but, due to inherent redundancy introduced by the sampling interval, need not be delivered as a single block. In other words, different data streams have different *quality of service* (QoS) requirements. How can rescue personnel tune a routing protocol so that certain data is delivered quickly while other data is delivered contiguously? We discuss Raven, a DTN multicopy routing protocol that allows the DTN user to tune for different QoS requirements.

6.5.1 Intercontact routing for emergency DTNs

Intercontact routing (Uddin et al., 2013) takes advantage of the fact that the movement of vehicles and personnel during disaster recovery is not *entirely* random. For example, as explained above in the section on the PDM model, an ambulance always returns to the hospital or triage area after visiting a Center. The choice of the next center to travel to from the hospital is random, but it is guaranteed that it will return to the hospital after visiting a center. Intercontact routing aims to exploit this recurrence, in order to find optimal paths for packet delivery. The key idea here is that packet replication can be reduced by choosing paths that have a high probability of delivery—and reducing packet replication leads to reduced energy consumption.

DTNs have inherently high intercontact times, due to sparse mobility and low node densities in the area of deployment. Therefore, a packet is likely to spend a lot of time in an on-node message buffer, which increases its delivery delay. We can easily observe that the delay experienced by a packet in a DTN depends on the previous contact as well as the next contact. Uddin et al. (2013) propose the concept of *intercontact delay* to capture this observation. The network is modeled using an *intercontact graph*, where each vertex represents an encounter between two nodes, and the edge weight between two vertices is the intercontact delay between those two contacts. For example, in Fig. 6.10A, a mobile agent A1 visits a center (called building B1) at time $t = 0$, moves to center B2 after 10 min ($t = 10$), and, after visiting other Centers, returns to B1 at $t = 60$. The intercontact graph for this scenario is shown in Fig. 6.10B. Each encounter represents a vertex: there are two encounters between A1–B1 and A1–B2. The edge from A1–B1 to A1–B2 has a weight of 10, the time taken to move from B1 to B2. But, due to the geographic position of the Centers, it takes $t = 60 - 10 = 50$ min to visit B1 again *after visiting B2*. This asymmetry is captured using directional edges, as shown in Fig. 6.10B. Further, each edge stores not only the average intercontact delay, but also the variance of the delay. Similarly, these parameters can be computed for a path in the intercontact graph using linearity of expectations. The probability of delivery within a timeout or a deadline along a particular path can then be computed using this mean and variance.

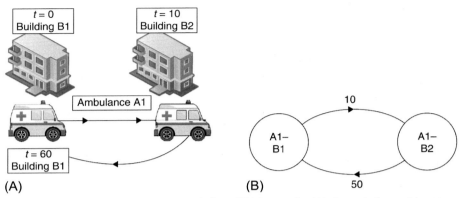

Fig. 6.10 Building the intercontact graph for a PDM scenario. (A) An ambulance A1 moves between two centers B1 and B2, and (B) the corresponding intercontact graph.

The routing protocol stores the mean and variance of the "minimum cost path" (cost here is defined as the 95th percentile of the delay) to a given destination. A separate *intercontact delay table* is constructed to contain the mean and variance per every neighbor pair; values are updated after every contact using the new intercontact delay sample generated from the contact. Whenever two nodes meet, they update entries in each other's table in a distance-vector manner. For reasons of notational consistency with (Uddin et al., 2013), let there be three nodes i,j,k. An encounter between nodes i and j is denoted as vertex ij in the intercontact graph. $\delta(ij \rightarrow ik)$ and $\sigma^2(ij \rightarrow ik)$ represent the mean and variance of the edge between encounters ij and ik. A path in the intercontact graph is denoted as $(ij \rightarrow k)$: a path from vertex ij to some different vertex that contains k. Its delay is denoted by $d(ij \rightarrow k)$.

A packet, when created, is assigned an initial number of copies L, a delivery deadline, and a time to live (TTL) initialized to the deadline. Let node i contain a packet intended for k. During a contact with node j, i computes delivery probability P for each of its neighbors x using the intercontact graph as follows:

$$p_x = P\{0 < \text{delay} < \text{TTL} | \text{delay} > 0\}$$

$$= \frac{\Phi\left(\dfrac{TTL - \text{delay}_x}{\sqrt{\text{var}_x}}\right) - \Phi\left(\dfrac{-\text{delay}_x}{\sqrt{\text{var}_x}}\right)}{1 - \Phi\left(\dfrac{-\text{delay}_x}{\sqrt{\text{var}_x}}\right)}$$

where Φ is the cumulative distribution function (CDF) of the normal distribution, and ($d(ij \rightarrow k)$ is the value of $d(ij \rightarrow k)$ calculated by node i):

$$\text{delay}_x = \delta(ij \rightarrow ix) + d_i(ij \rightarrow k)$$

$$\text{var}_x = \sigma^2(ij \rightarrow ix) + \sigma_i^2(ij \rightarrow k)$$

Each neighbor x is then assigned $p_x L$ copies to distribute to other nodes. The budgeted number of messages L is decremented after every transfer of $p_x L$ copies. In case the probabilities are zero or cannot be computed, the protocol uses Spray and Wait (Spyropoulos et al., 2005) as a fallback.

6.5.2 Raven: Quality of service-aware routing for emergency DTNs

The Raven (Risk AVersE routing in dtNs) routing protocol introduces the concept of risk-aversion to DTN routing. In research related to finance and traffic engineering, decision making in the presence of uncertainty is called risk-aversion. A risk-averse user will prefer a strategy whose reward has lower variance (i.e., more predictable) but a higher mean (i.e., lower reward). Similarly, in an emergency DTN, data is routed such that either the packet delivery delay or the variance of the packet delivery delay (commonly called *jitter* in traditional networks) is lower. Raven models the emergency DTN using the enhanced PDM model. This EPDM model is represented graphically using *stochastic multi-graphs*. When compared with a traditional graph, stochastic multigraphs have a random variable with a known distribution as edge weights instead of a scalar quantity. Further, multiple edges are possible between two vertices.

An example PDM scenario is depicted in Fig. 6.11A. Following a disaster, the EOC has been set up, a collapsed building (RUBBLE) has been identified for search and rescue operations and a medical TRIAGE area has been set up, resulting in three

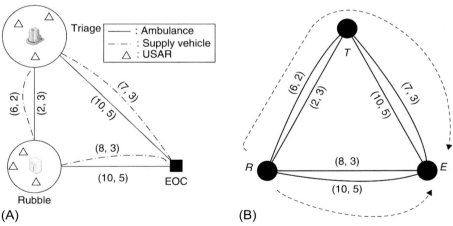

(A)   (B)

Fig. 6.11 (A) A simple scenario with three centers and three mobile agents: ambulances, supply vehicles, and USARs. Numbers next to a path indicate the (mean, variance) of the travel delay in minutes along that path, for the category of mobile agent represented by the line type (*solid* vs *dashed*). (B) Stochastic multigraph for this scenario. Vertices R, T, and E correspond to the rubble, triage, and EOC centers, respectively. The edges represent mobile agents and have the same weights.

centers. For simplicity, only three categories of MAs are shown: ambulances, supply vehicles, and USARs. USARs move around the triage and rubble in an area of fixed radius shown by the dotted line. The scenario in Fig. 6.11A is represented using a stochastic multigraph in Fig. 6.11B. A stochastic multigraph in this context is a graph whose vertices represent centers and an edge represents a MA category that visits the two incident centers. Assuming that the travel time distributions are Gaussian, the delay distribution of a path is the sum of the distributions of its constituent edges.

The key idea behind constructing this stochastic multigraph is that the travel time incurred by a MA when traveling between two centers is essentially a distribution of travel times. Each MA has some element of randomness in its mobility model: for example, an ambulance can choose a center at random. This distribution, computed using simulation, is assigned as the edge weight of an edge (MA) incident between two vertices (two centers). Since multiple MAs can travel between a pair of Centers (barring patrol cars, which loop between a given list of centers), multiple edges are created between any pair of vertices. Furthermore, since data is muled on vehicles (i.e., MAs), the packet delivery delay is proportional to the travel time incurred by a vehicle in moving from the source center to the destination center. Thus, the mean and variance of the distribution assigned to the edge (path) represent the mean and variance of the packet delivery delay along the edge (path). As mentioned above, risk-aversion in this context refers to the interplay between the mean and variance of the packet delivery delay.

How can one choose a path in this stochastic multigraph such that the mean (or variance) of the path weight is reduced? First, risk is quantified as a linear combination of mean and variance, using a risk-aversion coefficient $\rho \geq 0$:

$$\text{risk} = \text{mean} + \rho^*\text{variance}$$

The edge weights in the stochastic multigraph are reduced to scalars (i.e., the "risk" is computed) using ρ. For a high value of ρ, the paths chosen have a low variance of the delivery delay, while a ρ of 0 chooses paths with a low mean delay.

Algorithm 6.2: K-Safest Paths Algorithm (KSfP)
Input: Stochastic multigraph δ, K, source s, dest d
Output: $\mathbb{K}$, a set of K paths in δ between s and d
1. for edge e in $edges(\delta)$ **do**
2. $weight(e) \leftarrow \mu_e + \rho \times \sigma_e^2$
3. end for
4. $\delta' \leftarrow \delta$ with edge weights as above
5. $\mathbb{K} \leftarrow$ Apply K shortest paths on δ' with src/dest s/d
6. return $\mathbb{K}$

In order to choose K least risk paths so that *multicopy routing* can be performed, we adopt the K-shortest paths (KShP) problem in deterministic graphs to stochastic graphs. A "safe" path of two paths p_1 and p_2 is the one with the lower risk min (R_{p1}, R_{p2}). The objective of the K-Safest Paths Algorithm (KSfP) is to choose the K safest paths of a stochastic graph S, given a source node and a destination node.

K is a natural number. Existing algorithms for KShP (classical version) include a modified Bellman-Ford algorithm that stores the top K shortest paths at each pass instead of storing only the shortest (JGraphT library), and Yen's algorithm (Yen, 1971). The stochastic shortest path problem (KSfP with $K = 1$) is a nonconvex combinatorial problem (Lim et al., 2012). A dynamic programming approach is incorrect, since subpaths of optimal paths are not optimal. The risk of a path is not a linear combination of the risks of the edges, but is, in fact, nonlinear, as seen above ($R_p \neq \sum R_e$). We, therefore, propose the use of variance instead of the standard deviation for simplicity. While dimensional homogeneity is not present, due to the use of variance, which is the square of the standard deviation, the implementation of KSfP becomes straightforward and simple. The algorithm is shown in Algorithm 6.2. The stochastic graph is first converted into a deterministic graph (Steps 1–3). The edge weight is computed using the modified risk formula $R_p = \mu_p + \rho^* \sigma^2_p$. Each edge e in the stochastic graph S is assigned a deterministic edge weight (Step 2). The modified graph S' is now completely deterministic (Step 4). Any KShP algorithm can now be applied (Step 5). The result is a set of paths K that have the least risk.

6.6 Minimizing energy consumption in emergency DTNs

We now focus on hardware-based techniques to reduce energy consumption in an emergency DTN. In the previous section, software-based techniques were presented: they aimed to reduce energy consumption by reducing the number of transmitted messages. One could think of a solution where the radio is turned off during the large inter-contact times characteristic of DTNs. However, such solutions are hampered by another characteristic of opportunistic DTNs: *random mobility*. The dynamic and fast-paced environment of disaster recovery means that mobility cannot be predicted accurately in the future. Thus, we need solutions that are able to predict mobility and duty-cycle the radios appropriately.

The system performance of an emergency DTN is equally important: suboptimal duty-cycling can lead to exacerbated packet delivery delay and/or packet delivery ratio. Thus, it becomes important to ensure that the system performance is maximized while energy consumption is minimized. There is an inherent tradeoff between performance and energy consumption; the low node density and internode contact rates mean that missing a single encounter with another node, due to duty-cycling, can adversely impact packet delivery delay. We now explore solutions to this problem.

6.6.1 Duty-cycling using mobility prediction

Recent research has proposed the use of statically placed hardware, called throwboxes (Zhao et al., 2006) or data waypoints (Chenji et al., 2012), to improve the performance of a DTN. These additional nodes create artificial contact opportunities, increasing the capacity of the DTN. However, in an energy-constrained DTN such as one used in disaster response, continuous forwarding of packets may decrease the node's lifetime—nullifying performance improvements. Banerjee et al. (2007)

propose a multitiered architecture for reducing energy consumption for these throw-boxes. The key idea is that a long-range, low-bit rate (i.e., low-power) radio can wake up a short-range, high-bit rate (i.e., high-power) radio whenever a contact is in range.

Because of the difference in radio ranges, a contact may not enter the high-power radio's range even though it enters the longer range. In such cases, the energy spent by the node in waking up for the contact is proved useless. Thus, a mobility prediction scheme is necessary. Let r be the radius of the long-range radio (for consistency with (Banerjee et al., 2007)). Construct a square of side $2r$ such that the node is located at the center of the square, as shown in Fig. 6.12. This larger square can be divided into smaller squares numbered $0 \ldots k$. Under Markovian assumptions, entry T_{ij} of a learned probability transition matrix T is the probability that a mobile node (the contact) when present in cell i will move to cell j. The contact always broadcasts its previous trajectory when in range of the node, so that the matrix T can be constructed and maintained. D_1 is the distance (and D_1/v, the time) the contact has to travel before entering the range of the high-power radio. D_2/v is the expected time the contact will spend in the range of the node, assuming the contact continues on a straight path. If the time taken to wake up the high-power radio is not greater than D_1/v, the node calculates Pr, which is the probability that the contact will enter the radio range of the node ($Pr = T_{AB} \times T_{BC}$). More details can be found in (Banerjee et al., 2007). Such a scheme is lightweight and can be implemented in as little as 10 KB of memory on an 8 MHz processor (the MaxStream XTend radio is used as the long-range radio).

The authors then present a scheme to maximize the number of transferred bytes while meeting an average power constraint. First, it is shown that the problem of choosing the optimal subset of contacts to participate in is NP-Hard. The problem

Fig. 6.12 A square of side 2r such that the node is located at the center of the square.

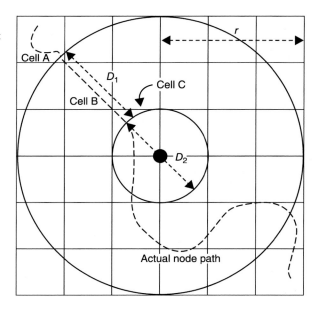

can be modeled as follows. Suppose that there is a set of contact events $E_1 \ldots E_n$ which occur in time $[0,t]$. During each event E_i, b_i can be transferred at an energy cost e. The problem is now to find a subset of events such that the total energy spent does not exceed a threshold Pt while the total number of bytes transferred is maximized. The decision version of this problem, where the node is required to forward at least k bytes of data, is shown to be NP-Complete. The proposed suboptimal solution is called the Token Bucket Scheduler, and is shown in Algorithm 6.3.

Algorithm 6.3: Token Bucket Scheduler
Input: Avg. power constraint P, number of tokens m, present contact data: energy cost E_p, bytes b_p
  1. **if** $m \cdot P < E_p$
  2. Do not wake up the high power radio
  3. **else**
  4. $b_k \leftarrow$ mean byte size of last k contacts
  5. $E_k \leftarrow$ mean energy of last k contacts
  6. $T_m \leftarrow$ inter-arrival time of contacts
  7. **if** $m \cdot P - E_p + T_m \cdot P < E_k$ and $b_p < b_k$
  8. Do not wake up the high-power radio
  9. **else**
10. Wake up the high-power radio
11. **end if**
12. **end if**

First, energy tokens, which represent the amount of energy that can be spent by the node, are generated at the rate of P every second. The node uses the mobility prediction scheme described above to decide whether to stay awake for the current contact: it can do so only if it has accumulated enough tokens (Steps 1–3). In case it does have the required number of tokens, the node now decides whether to stay awake for the current contact or the next one. Estimates about the next contact are made using data from previous contacts (Steps 4–6). After estimating the number of tokens it can accumulate between now and the next contact, it computes whether staying awake for both contacts is possible (Steps 7–9). If not, it chooses the larger of the two contacts (Steps 10–11). The running time of this algorithm is linear in the number of contacts, and space complexity is $O(k)$.

6.6.2 Characterizing the performance-energy consumption tradeoff

One of the key observations in (Chenji and Stoleru, 2014) is that simply reducing radio usage may not be the most optimal way to save energy. For example, in DistressNet, Mikrotik RB433UAH devices serve as fog routers, which also function as data waypoints/throwboxes. Here, the radio consumes 0.72 W, whereas the baseboard consumes 3 W. Therefore, the device has to be completely shut down to save energy. Recent research has focused on ways to wake up the router by correctly predicting contacts. However, the research question proposed in this work is different: if these throwboxes are completely excluded from the routing process instead of being woken up at a contact, what is the impact on performance? In other words, can the system performance vs energy consumption tradeoff be characterized? If so, does a Pareto front exist? Can the system be made to operate at any of these Pareto optimal points at the behest of the user? Chenji and Stoleru (2014) propose a technique not just to reduce the energy consumption, but to do so in a Pareto-optimal manner by quantifying the effect on system performance simultaneously. The energy consumption can be reduced further compared with state-of-the-art methods by accepting the lowest possible system performance penalty for the amount of energy to be saved. Now, whether a node generates or consumes data depends on the flows in the network. Only nodes that are neither consumers nor producers of data can be completely shut down. In DistressNet, to characterize data flows in a cloud-like environment, the concept of potential and firm flows is used. A firm flow is one whose source and destination are well known (such as IP addresses, host names, or a DTN URI, for example). A potential flow's source is well known, but its destination is simply the Fog. It is up to the Fog service to convert each potential flow into one or more firm flows. The exact number of firm flows created depends on the potential flow's *availability metric*, which ranges from 0% to 100%. It denotes the importance of the data: an availability of 100% means that data will be available on all Fog devices—a firm flow is created from the source to every other fog device. An availability of 25% means that data will be available on a quarter of all Fog routers: a subset of devices is chosen and firm flows are created to each of them. For each firm flow, Raven (Chenji et al., 2013a,b), which has been discussed previously, is chosen as the underlying routing protocol. Therefore, K paths need to be chosen for each source-destination pair for each firm flow.

The disaster area consists of c centers $C_1 \ldots C_c$ and is represented by the stochastic multigraph S. There are f firm flows $F_1 \ldots F_f$ and their sources/dests $\{F_i^S\}/\{F_i^D\}$. There are p potential flows $P_1 \ldots P_p$, their availabilities $A_1 \ldots A_p$ and sources $\{P_i^S\}$ but no destinations. Thus, the total number of flows is $h = f + \sum_{i=1}^{p}[cA_i]$. Each of these h flows has an associated parameter $k_1 \ldots k_h$ that is used by Raven (the number of paths to compute). The user has specified a global Raven parameter K: as a result, $1 \le k_i \le K$. When Raven computes paths for each of these h flows, let the number of unique Centers in the union of these paths be $\overline{c} < c$.

Now, since S is stochastic, each path in this graph has an associated mean and variance (equal to the sum of means/variances of the constituent edges), and is hence a distribution. The path weight represents the packet delivery delay because the physical travel delay in a DTN is a major component of the packet delivery delay (Zhu et al., 2010). When a packet is sent on k paths simultaneously, the expected delay is the minimum of the delays of the k individual paths; it follows that the per-flow packet delivery delay is the minimum of k normally distributed random variables. For the flow numbered i with paths parameter k_i, the delay is $D_i = \min\{D_{i1}, D_{i2} \ldots D_{iki}\}$ where each D is a distribution and not a scalar. A closed-form expression for this minimum of several random variables is not trivial. For h flows, the overall packet delivery delay distribution is $D = \frac{\sum_{i=1}^{h} D_i}{h}$, where, to re-emphasize, all quantities except h are normally distributed random variables with a mean and a variance. Using the mean-risk probability model, the risk of this distribution D is $risk(D) = E[D] + \rho^* \sqrt{V[D]}$.

Given the above notation, the objective is to minimize the delay (which is called risk when variance is taken into account, i.e., risk$(D):=E[D]$ when $\rho = 0$) as well as $\bar{c}$. There are three parts to this problem: (1) the conversion of potential flows to firm flows, (2) applying Raven to each of the firm flows so that the delay distribution D can be estimated, and (3) tuning Raven's K parameter so that energy and risk(D) are minimized. The above three problems are solved in a single optimization problem as follows. To convert a potential flow P_i into $[cA_i]$ firm flows, consider a binary vector V of length c. The ith element V_i corresponds to Center C_i: $V_i = 0$ if the Center is not chosen as a destination, and $V_i = 1$ otherwise. The sum of this bit vector should be $[cA_i]$. Repeating this procedure for p potential flows, the length of V becomes pc. Because Raven needs a K parameter for each flow, V is augmented with h more integers. Thus, the vector V of length $(pc + h)$ can now be used as input. It is a dual objective nonlinear program:

$$\min_{V} \ \mathrm{RISK}(V), \mathrm{UNIQ}(V) \tag{6.1}$$

$$s.t. \sum_{j=1}^{c} V_{(i-1)c+j} = [cA_i], \quad i = 1...p \tag{6.2}$$

$$1 \leq V_i \leq K, \quad i = (pc+1)...(pc+h) \tag{6.3}$$

Deals with potential flows. It stipulates that the total number of firm flows created (by setting a Center's bit) for each potential flow equals $[cA_i]$. Makes sure that the K parameter needed by Raven cannot exceed the global value of K specified by the user. Involves two procedures, RISK and UNIQ, which calculate risk(D) and $\bar{c}$ respectively.

Algorithm 6.4 details the calculation of the two objectives. Steps 1–7 involve the creation of firm flows from potential flows: for each potential flow (Step 1), if the bit corresponding to a Center is set (Step 3), a new firm flow is created (Step 4). Now that all h firm flows have been created, Raven is applied to each of these (Step 9) along with the Raven parameter K (Step 10). The resulting set of k_i paths are collected and duplicates are removed (Step 11). The number of unique vertices in these paths is nothing but the number of unique centers (Step 13), since each vertex in the stochastic multigraph corresponds to a center. The delay distribution, which is the path weight of each of these paths, is calculated (Step 14), averaged (Step 15) and the risk is calculated (Step 16).

Algorithm 6.4: RISK(V) and UNIQ(V)
   1. **Input**: vector V, firm flows $\{F\}$, K, p
   2. **for** $i := 1 ... p$
   3. **for** $j := 1 ... c$
   4. **if** $V_{(i-1)c+j} = = 1$
   5. $F \leftarrow F \cup$ (a new firm flow between P^S_i and C_j
   6. **end if**
   7. **end for**

8. **end for**
9. *paths* ← *Φ*
10. **for** $i := 1 \dots h$
11. $Q_k \leftarrow$ (Raven with s/d F_i^S and F_i^D, k-parameter $V_{pc + i}$
12. *paths* ← *paths* ∪ Q_k
13. $D_i \leftarrow \min (Q_1, Q_2 \dots Q_K\}$
14. **end for**
15. UNIQ(V) = $\bar{c}$ ← unique nodes in *paths*
16. RISK$(V) \leftarrow \sum_{t=1}^{h} E[D_i] + \rho^* \sqrt{V[D_i]}$
17. **return** UNIQ(V), RISK(V)

Is a dual objective, nonlinear optimization problem. Because of its complexity, a stochastic optimization approach is preferred, as opposed to a deterministic one. Evolutionary algorithms are especially suited to solve multiobjective problems—and genetic algorithms (GAs) are the most popular variety of evolutionary algorithms. We use the NSGA-II algorithm to solve, due to its speed and low complexity. It is also able to handle disconnected Pareto fronts.

The input to the algorithm, vector *V*, is referred to as a "chromosome" in GA parlance. It is a string of numbers, either binary or real-valued (in this case, integer-valued). Through multiple GA operations like crossover, selection, and mutation, new candidate solutions are generated in a stochastic fashion. The current set of candidate solutions (the "population") is evaluated (the "fitness" is calculated) and filtered to retain only nondominated solutions. NSGA-II gives preference to Pareto optimal points that are situated far away, so as to capture both extremes of the front. The crossover operator used is simulated binary crossover (SBX), since the chromosome is real-valued. Mutation occurs according to the polynomial operator. Selection happens in a binary tournament fashion.

6.7   Conclusions and future trends

This chapter has summarized the use of DTNs during the disaster recovery process. Starting with a motivating scenario consisting of a natural disaster that occurs over a large geographical area, the industry state of the art was reviewed. The US&R task force of FEMA was introduced, along with a description of their equipment cache.

The academic state of the art has proposed several solutions for the problem of improving situational awareness during disaster recovery. Some solutions use mesh networking and mobile ad hoc networks (MANETs) to build systems capable of improving situational awareness. Examples include WIISARD, RESCUE, SAFIRE, and AID-N. Then, the use of DTN for disaster response was motivated by showing the shortcomings of existing systems. DTNs can function in spite of low node densities and limited bandwidth, which is common when disasters affect large areas spanning tens of square miles. Next, the post disaster mobility model was introduced as being able to model mobility during disaster response using a variety of submodels, and the enhanced PDM model added the USAR movement model to the above set. DistressNet, a DTN for disaster response, was then discussed. It proposes a new

"Fog Computing" paradigm where the data generated by sensors is presented to first responders using a cloud computing-like interface. Heterogeneous battery-powered devices, which use open source software and COTS hardware, comprise the networking backbone.

The fact that emergency DTNs are resource-constrained while being mission-critical motives the need for a new class of DTN routing protocols. The intercontact routing protocol is able to reduce energy consumption by reducing the number of transmitted messages while still maintaining a high delivery rate. It does so by exploiting the recurrence present in mobility patterns during disaster rescue. By choosing paths with a high probability of delivery only, packet replication—and thus the number of transmitted messages—is reduced. Raven is a DTN routing protocol that enables the DTN's users to specify QoS requirements. By tuning a single parameter Q, the user is able to specify whether a low packet delivery delay or a low variance of delay is preferred. Raven is able to accomplish this goal by constructing a stochastic multigraph for the PDM model.

Next, the chapter explored hardware-based means to reduce energy expenditure in an emergency DTN. First, a multitiered architecture for DTN throwboxes was discussed. A long-range, low-bit rate, low-power radio wakes up the main tier, which is an 802.11-based shorter-range, higher-bit rate, higher-power radio. One problem with the different radio ranges is that a node may not enter the main radio's range after being detected by the low-power radio. A mobility prediction scheme was proposed, which is then used by the low-power radio to wake up the main tier. Additionally, a token bucket lifetime scheduler deals with the problem of meeting an average power constraint while ensuring that a minimum number of bytes are forwarded by the throwboxes. Another energy-saving scheme determined that more energy can be saved by excluding some devices from the routing process. Such an approach has a marked effect on the system's performance, and this energy-performance tradeoff was investigated.

6.7.1 Future trends

Security is an important concern in emergency DTNs. Malicious adversaries could delay the victim rescue process, or sensitive information about victims could be leaked. Solutions such as encryption increase energy consumption both at the central processing unit (CPU) as well as the network layers. The contact time in DTNs is limited; encrypting data could increase the overhead and decrease goodput, affecting the DTN's performance. The Bundle Security Protocol, recently standardized as RFC 6257, provides a good start for researchers who want to investigate this issue.

The performance of algorithms that require some input from the DTN, such as intercontact routing and raven, depends on the quality of the input. Errors introduced during the computation of the input metric could cascade and adversely affect the performance of the DTN. Mathematical techniques such as sensitivity and uncertainty analysis can help calculate the error in the output, given the uncertainty in the input. Thus, using such theoretical techniques can lead to better algorithms that are better equipped to handle varying levels of uncertainty in the input.

Disaster rescue is a complex process and involves the coordination of many teams, which sometimes span international boundaries. The networking systems used by these teams could have different algorithms running at different layers, depending on their specialization and requirements. For example, the emergency DTN used by Team A could use intercontact routing, while Team B uses Raven. When these teams are required to coordinate, different coexisting protocols compete for the same resources while attempting to maximize performance. Thus, the study of the coexistence of multiple algorithms and protocols becomes very important. Game theory could possibly provide mathematical frameworks for researchers to analyze the coexistence of protocols in an emergency DTN.

References

Arisoylu, M., Mishra, R., Reao, R., Lenert, L.A., 2005. 802.11 wireless infrastructure to enhance medical response to disasters. AMIA Ann. Symp. Proc. 2005, 1–5.

Banerjee, N., Corner, M., Levine, B., 2007. An energy-efficient architecture for DTN Throwboxes. In: INFOCOM 2007. 26th IEEE International Conference on Computer Communications, Anchorage, Alaska, USA, 2007.

BBC, 2011. Japan Hit by Tsunami After Massive Earthquake. Available from: http://www.bbc.co.uk/news/world-asia-pacific-12709850. (Accessed 15 August 2013).

Brown, S.W., William, M., Griswold, G., Demchak, B., Leslie, B., et al., 2006. Middleware for reliable mobile medical workflow support in disaster settings. AMIA Ann. Symp. Proc. 2006, 309–313.

Cerf, V., Burleigh, S., Hooke, A., Torgerson, L., Durst, R., et al., 2007. Delay-Tolerant Networking Architecture. Available from: http://tools.ietf.org/html/rfc4838. (Accessed 15 August 2013).

Chenji, H., Stoleru, R., 2014. Pareto optimal cross layer lifetime optimization for disaster response networks. In: Sixth International Conference on Communication Systems and Networks (COMSNETS), Bangalore, India, 2014.

Chenji, H., Zhang, W., Won, M., Stoleru, R., Arnett, C., 2012. A wireless system for reducing response time in urban search & rescue. In: IEEE 31st International Performance Computing and Communications Conference (IPCCC), Austin, Texas, 2012.

Chenji, H., Smith, L., Stoleru, R., Nikolova, E., 2013a. Raven: energy aware QoS control for DRNs. In: IEEE 9th International Conference on Wireless and Mobile Computing, Networking and Communications (WiMob), Lyon, France, 2013.

Chenji, H., Zhang, W., Stoleru, R., Arnett, C., 2013b. DistressNet: a disaster response system providing constant availability cloud-like services. Ad Hoc Netw. 11, 2440–2460.

Chipara, O., Griswold, W.G., Plymoth, A.N., Huang, R., Liu, F., et al., 2012. WIISARD: a measurement study of network properties and protocol reliability during an emergency response. In: Proceedings of the 10th International Conference on Mobile Systems, Applications, and Services (MobiSys), Low Wood Bay, Lake District, UK, 2012.

Dilmaghani, R., Rao, R., 2008. An ad hoc network infrastructure: communication and information sharing for emergency response. In: Proceedings of the 2008 IEEE International Conference on Wireless and Mobile Computing, Networking and Communications (WiMob), Avignon, France.

Dilmaghani, R.B., Manoj, B.S., Jafarian, B., Rao, R.R., 2005. Performance evaluation of rescue mesh: a metro-scale hybrid wireless network. In: Proceedings of IEEE WiMesh, Santa Clara, CA, USA, 2005.

Fall, K., 2003. A delay-tolerant network architecture for challenged internets. In: Proceedings of the 2003 Conference on Applications, Technologies, Architectures, and Protocols for Computer Communications (SIGCOMM'03). ACM, New York, NY, USA.

FEMA, 2008. US&R Task Force Equipment List. Available from: http://www.fema.gov/task-force-equipment. (Accessed 15 August 2013).

Gao, T., Massey, T., Selavo, L., Crawford, D., Chen, B.-R., et al., 2007. The advanced health and disaster aid network: a light-weight wireless medical system for triage. IEEE Trans. Biomed. Circuits Syst. 1 (3), 203–216.

Guven, G., Ergen, E., 2011. Identification of local information items needed during search and rescue following an earthquake. Disaster Prev Manag 20 (5).

Killeen, J.P., Chan, T.C., Buono, C., Griswold, W.G., Lenert, L.A., 2006. A wireless first responder handheld device for rapid triage, patient assessment and documentation during mass casualty incidents. AMIA Ann. Symp. Proc. 2006.

Lenert, L.A., Palmer, D.A., Chan, T.C., Rao, R., 2005. An intelligent 802.11 triage tag for medical response to disasters. AMIA Ann. Symp. Proc. 2005, 440–444.

Lim, S., Sommer, C., Nikolova, E., Rus, D., 2012. Practical route planning under delay uncertainty: stochastic shortest path queries. In: Proceedings of Robotics: Science and Systems. Sydney, Australia, July 2012.

Manoj, B., Baker, A.H., 2007. Communication challenges in emergency response. Commun. ACM 50 (3), 51–53.

Mehrotra, S., Butts, C., Kalashnikov, D., Venkatasubramanian, N., Rao, R., et al., 2004. Project rescue: challenges in responding to the unexpected. Proc. SPIE 5304, 179–192.

Mesh Dynamics, 2008. Network-Centric Warfare and Wireless Communications. Available from: http://www.meshdynamics.com/documents/MD_MILITARY_MESH.pdf. (Accessed 15 August 2013).

Missouri DPS, 2011. Joplin Tornado Situation Report 6 a.m. May 24. Available from: http://sema.dps.mo.gov/newspubs/SRTemplate.asp?ID-09110049. (Accessed 15 August 2013).

Montenegro, G., Kushalnagar, N., Hui, J., Culler, D., 2007. Transmission of IPv6 Packets over IEEE 802.15.4 Networks. Available from: http://tools.ietforg/html/rfc4944. (Accessed 15 August 2013).

Motorola, 2011. Project 25 Interoperable Communications for Public Safety Agencies. Available from: http://www.motorolasolutions.com/web/Business/Solutions/Business%20Solutions/Mission%20Critical%20Communications/ASTRO%2025%20Trunked%20Solutions/_Document/Project%2025%20Whitepaper.pdf. (Accessed 15 August 2013).

OpenWRT Dev Team, 2014. OpenWRT Wireless Freedom. Available from: https://openwrt.org/. (Accessed 15 August 2013).

PTIG, 2014. Project 25 Technology Interest Group. Available from: http://www.project25.org/. (Accessed 1 July 2014).

Schildt, S., Morgenroth, J., Pöttner, W.B., Wolf, L., 2011. IBR-DTN: a lightweight, modular and highly portable bundle protocol implementation. In: Electronic Communications of the EASST. vol. 37, pp. 1–11.

Scott, K., Burleigh, S., 2007. Bundle Protocol Specification. Available from: http://tools.ietf.org/html/rfc5050. (Accessed 15 August 2013).

Spyropoulos, T., Psounis, K., Raghavendra, C.S., 2005. Spray and wait: an efficient routing scheme for intermittently connected mobile networks. In: Proceedings of the 2005 ACM SIGCOMM Workshop on Delay-tolerant Networking, Philadelphia, Pennsylvania, USA.

Uddin, M., Nicol, D., Abdelzaher, T., Kravets, R., 2009. A post-disaster mobility model for delay tolerant networking. In: Proceedings of the 2009 Winter Simulation Conference (WSC), Austin, TX, USA.

Uddin, M., Ahmadi, H., Abdelzaher, T., Kravets, R., 2013. Intercontact routing for energy constrained disaster response networks. IEEE Trans. Mob. Comput. 12 (10), 1986–1998.

USGS, 2010. M7.0—Haiti Region. Available from: http://comcat.cr.usgs.gov/earthquakes/eventpage/pde20100112215310060_13. (Accessed 1 July 2014).

Winter, T., Thubert, P., Brandt, A., Hui, J., Kelsey, R., et al., 2012. RPL: IPv6 Routing Protocol for Low-Power and Lossy Networks. Available from http://tools.ietf.org/html/rfc6550. (Accessed 1 July 2014).

Xing, B., Mehrotra, S., Venkatasubramanian, N., 2009. RADcast: enabling reliability guarantees for content dissemination in ad hoc networks. In: INFOCOM 2009. IEEE, Rio de Janeiro, Brazil.

Yen, J.Y., 1971. Finding the K shortest loopless paths in a network. Manag. Sci. 17 (11).

Zhao, W., Chen, Y., Ammar, M., Corner, M., Levine, B., et al., 2006. Capacity enhancement using throwboxes in DTNs. In: 2006 IEEE International Conference on Mobile Adhoc and Sensor Systems (MASS), Vancouver, BC, Canada, October.

Zhu, H., Fu, L., Xue, G., Zhu, Y., Li, M., et al., 2010. Recognizing exponential intercontact time in VANETs. In: Proceedings of the 29th Conference on Information Communications, INFOCOM'10, San Diego, California, USA.

Environment friendly green data broadcasting in delay-tolerant opportunistic networks[☆]

Sanjay K. Dhurandher[a], Jagdeep Singh[b], Isaac Woungang[c], Jahanavi Mishra[b], and Tarun Dhankhar[b]
[a]Department of Information Technology, Netaji Subhas University of Technology, New Delhi, India, [b]Division of Information Technology, Netaji Subhas Institute of Technology, University of Delhi, New Delhi, India, [c]Department of Computer Science, Ryerson University, Toronto, ON, Canada

7.1 Introduction

By design, OppNets (Huang et al., 2008) typically incur high delay, intermittent connectivity, and no assurance of an end-to-end route between the sender and destination nodes. These nodes are not permanently connected. The network topology may change based on the mobility of nodes, node activation, or node deactivation. OppNets are a subclass of delay-tolerant networks (DTNs) (Mauve et al., 2001), which themselves aid in countering the communication problems and delays during the routing of data. The dynamic store-carry-and-forward technique is used for routing in OppNets. This technique helps overcome the network partition problem along with reducing the load on the communication channel, leading to low overhead and high delivery ratio (Fig. 7.1).

The OppNet finds usage in various scenarios where the flexibility of the network in terms of freedom of nodes, and ease of setup is desired. The network keeps on varying in terms of the nodes that participate in relaying the messages because nodes keep on moving and may enter/exit the network at intermittent moments. The messages hop from one moving node to another until they reach their destinations. One of the primary drawbacks of OppNets is the power failure of nodes, which occur if a node does not possess sufficient battery power to forward the messages to the desired destination. This may lead to a failure of a particular route and an overall reduction in the network reliability. The multihop nature and varying participation of nodes in an OppNet make it susceptible to attacks that can curtail the network from ensuring the delivery of messages and ensuring that the messages that are delivered are indeed correct. Malignant nodes may choose to drop or alter the messages they are supposed to forward toward destination. This behavior may go undetected because well-meaning nodes also drop messages for various reasons and the nodes are not usually sophisticated enough to

[☆] Fully documented templates are available in the elsarticle package on CTAN.

Advances in Delay-tolerant Networks (DTNs). https://doi.org/10.1016/B978-0-08-102793-6.00007-2

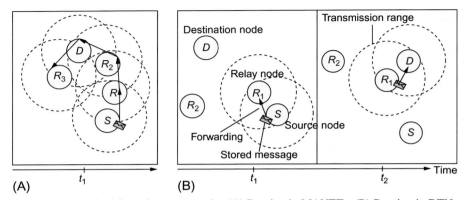

Fig. 7.1 Routing in delay-tolerant networks. (A) Routing in MANETs. (B) Routing in DTNs.

verify the authenticity of messages. Therefore, the message can be lost during transmission. The reasons for this to occur are power failure of nodes, short-range wireless communication, and mobility of nodes. Alongside this, uncertain mobility patterns, restriction on battery power and storage, added to damaged links, are other known drawbacks of OppNets.

Energy consumption (Buzzi et al., 2016) is a problem when a node sends or receives a packet or when multiple copies of one packet are sent. To maximize the chance of delivering the packets while increasing the battery life of nodes, it is required to avoid unnecessary transmission of packets repeatedly and to route only through those nodes that have sufficient remaining battery power. The reason for this is that replication will lead to the worst possible performance when there are finite resources. Each node consists of three basic subsystems: (1) a sensing subsystem, which is used to obtain the data; (2) a processing subsystem, which is used to process the data; and (3) a wireless communication subsystem, which is used for communication between the nodes.

Every node includes a power source, which is a battery with limited energy; and it is a difficult task to recharge these batteries owing to their huge number or the hostility of the terrain where they are deployed in. Therefore, it is necessary to prolong the life of nodes by applying different energy reducing techniques and providing energy harvesting (EH) mechanisms for the nodes to increase the network lifetime.

7.2 Issues related to energy efficiency in OppNets

In OppNets, the energy utilized by the nodes is a pivotal problem, which occurs repeatedly if a node exchanges a packet and when it sends a packet several times to various other nodes. Therefore, in order to save the node's energy and battery life, hence, increasing the probability of successful delivery of packets, it is essential: (1) to prevent the transmission of packets which have already been delivered to the destination and (2) to transmit the packets only through those nodes that have sufficient remaining battery power, in such a way that the ratio of delivery of packets is maximized.

In OppNets, there are numerous issues related to energy efficiency. In the following, we discuss some of them.

7.2.1   Energy efficiency issues in geocasting

Geocasting is a technique where messages are sent to specific locations instead of the devices identified explicitly. Using it to transmit the data in an OppNet has some benefits since it helps in achieving a successful delivery of packets to all nodes within a specified area rather than via arbitrary nodes. Lu and Liu (2014) proposed a routing protocol called GeoOPP, which provides a geocasting service to OppNets. This scheme combines the flooding and unicasting techniques, by transmitting a message to a particular geographic region, then flood it to all the nodes within that specified region. To transmit a message toward a region, GeoOPP utilizes the greedy algorithms for routing in the network, the use of the progress within radius metric (PWRM) to determine the progress and efficiency of a node in forwarding the message toward its destination based on one of the futures visited areas. In doing so, the probability of a node visiting a region and having a contact inside that region is estimated using the Chebyshev's inequality. One of the main drawbacks of this protocol is that all the nodes are assumed to have an unlimited amount of battery power, which is nonrealistic in practice.

7.2.2   Energy efficiency issues related to security in OppNets

One of the areas in a network where energy is extensively utilized is managing its security and privacy. A large amount of energy is spent in recognizing the attackers and in employing the mechanisms in countering the attacks or recovering from them. One of the primary reasons for threats related to privacy and security in OppNets is that even if hard security drills are performed before including a device in the network, this would not help in excluding the malicious devices. In an OppNet, two ways of defense should be employed: (1) a preventive mechanism – for instance, blocking the harmful devices and (2) a reactive mechanism – by implementing a strategy for detecting the harmful devices.

An OppNet faces various challenges when it comes to security and privacy. These challenges include: providing a secure routing mechanism, ensuring the privacy of the network, detecting intrusions, dealing with specifically dangerous attacks, maintaining privacy of data and ensuring its integrity, and authenticating the nodes, helpers, honeypots, and honey farms.

There are various hard and soft mechanisms that have been employed for maintaining the security and privacy of OppNets (Zou et al., 2013). Sending multiple copies of a message guarantees that at least one copy will reach the target node successfully. This technique, however, leads to high-energy utilization due to redundancy and duplication. In doing so, it must be ensured that no packet is dropped to avoid reduced efficiency. Also, there must be a limit on the number of times that a device can request for help so as to not pay heed to multiple false alarms. Another solution consists of utilizing tags on packets with signal prints and matching rules in order to detect denial of service attacks using MAC addresses. One can also reduce the load on each node in order to save the energy and maintain the network connectivity. Also, all nodes can monitor the nodes adjacent to them for ID spoofing. Furthermore, some protocols can be used for the validation of the source address. The techniques employed for maintaining the security and privacy of networks are varied. However, it is often suggested to employ soft mechanisms since they tend to utilize less energy. On the other hand, hard security mechanisms require larger amounts of energy, but they offer a much better security. Hence, there is a trade-off between energy efficiency and the reliability of the security mechanism in a network.

7.3   Related work

Vullers et al. (2010) proposed energy sustainable methods and paradigms for future networks. Suitable management techniques to reduce the environmental footprint and operational expenditure of 5G networks by using EH hardware are reviewed. Several open issues such as the need of accurate data profiles, small inefficient transfers in current wireless transferring techniques, and lack of accurate energy consumption and transmission models are identified. According to the reported findings, in order to get the proper control on energy consumption for a self-sustainable 5G network system, it is recommended to implement a set of procedures such as device-to-device communication, cell sleeping and zooming, and wireless transfer of energy.

In Sivakumar and Manoharan (2017), the ERP routing protocol is proposed, which is more practical and efficient than the BubbleRap and EABubble counterpart. ERP is an energy-aware routing protocol for mobile OppNets, in which the dynamic changes in the energy of nodes are taken into consideration during the selection of relay nodes to carry the message toward its destination. From simulation results, it is proved that ERP outperforms the existing protocols in terms of cost, node survival ratio, and delivery ratio.

In Zhao et al. (2016), some critical issues in the design of routing protocols for dense lossy networks are investigated, which include power usage, reliable, and efficient delivery of data. Traditional protocols deliver data along some predefined paths with high power consumption. In this scheme, a reliable and energy-aware opportunistic routing protocol (named REOR) is proposed, which is shown to increase the lifetime and reliability of the network. This protocol delays the death of nodes up to a significant extent; it also ensures an undisturbed connectivity by selecting a suitable power transmission and set of forwarder nodes. Through simulations, it is proved that this protocol is 50% more reliable and 30% more energy efficient than the IPv6 protocol. In REOR, a recovery method is also included to detect and resend the lost data efficiently and an overhearing-based coordination method is employed to avoid duplications.

Zhou et al. (2015) proposed an energy efficient routing protocol for underwater wireless networks in Internet of Underwater Things (IoUT). In this scheme, the sensor nodes sense and store the current data about the underwater surroundings, then they communicate with each other to forward sensor data to the sink nodes. In doing so, to reduce the expenditure in forwarding the packets as well as the power consumption, an enhanced and efficient version of the Channel-Aware Routing Protocol (E-CARP) is implemented, which is also proved to reduce the cost in communication while increasing the network performance to a significant level.

In underwater acoustic sensor networks, achieving an efficient and reliable transmission is a challenge due to the complex underwater surroundings and the complexity in analyzing and monitoring the environmental changes. In investigating this issue, Wang et al. (2015) proposed an energy-aware packet transmission scheme called efficient grid routing using three-dimensional (3D) cubes (EGRC), which use the node's residual power and end-to-end delay in finding the next-hop node to make the transmissions reliable. In this scheme, the network is divided into several small clusters, each of which participates in the clustering process. Precisely, an algorithm for cluster-head selection is developed, which selects the node having the highest remaining residual energy and shortest distance to the base station as cluster head. This cluster-head node in each cluster is in charge of the data aggregation and transmission to its cluster members. Through simulations, it is proved that the proposed EGRC protocol outperforms the standard protocols in terms of power consumption and reliability.

In UWSNs, multipath effects, high and variable latency, low bandwidth, fading, and bit errors, are the main causes of channel impairments, which reduce the reliability in data transmission and lead to poor quality-aware data collection. Thus, designing a quality of service (QoS)-aware data collection protocol to analyze, monitor, and explore the oceans in underwater conditions is a challenge. In Sivakumar and Manoharan (2017), a multiobjective evolutionary routing protocol called MERP is proposed for various UWSNs applications. This protocol uses the natural evolution of the multiobjective genetic protocol to achieve a reliable information collection in UWSNs. The proposed routing mechanism maintains a high stability and quality of links in a set of cluster heads, leading to robustness in terms of data delivery in UWSNs. By design, this scheme also avoids loops in the data paths and unnecessary multihop transmission of packets by keeping the up-down movement of the data in the network under control. Through simulations, it is proved that MERP reduces the power consumption as well as the number of corrupted data packets while increasing the network life span to a significant extent. It is also proved that during the analysis and monitoring of the underwater surroundings, the MERP protocol quickly achieved its target compared to traditional UWSNs routing protocols.

Rani et al. (2017) proposed an energy-aware chain-based protocol for USWNs (called E-CBCCP). In this scheme, a distance-based communication utilizing the concept of location informed nodes is considered in the analysis and monitoring of the domains during the steady state. To achieve communication among the relay nodes, a hop-to-hop mechanism is implemented; and to select the suitable relay nodes, the confidence level of the sensor nodes is determined. Through simulations, it is proved that E-CBCCP outperforms the channel-aware routing protocol for underwater acoustic wireless networks (CARP) protocol (Basagni et al., 2015) in terms of energy consumption and data packets transmission.

Piovesan et al. (2018) discussed the role that energy will play in the design of future generation mobile networks, stressing on the use of renewable energy as a key enabler to reduce the environmental footprint of 5G technology deployments while maintaining acceptable QoS levels. First, a comprehensive review of management techniques is provided and several issues in the design of cost-effective mobile architectures are discussed, along with proposed solutions in the literature such as edge/fog computing, software-defined networking, and C-RAN solutions. Second, from an energy efficiency standpoint, available energy sustainable paradigms and techniques to boost the energy efficiency of networks featuring renewable energy sources are reviewed, along with related challenges and open issues.

Borah et al. (2017) proposed the energy-aware versions of some traditional benchmark routing protocols for OppNets: PRoPHET, PRoWait, and EDR. In each of these schemes, the best suitable hop for message transmission is selected based on the node's remaining energy. Through simulations, it is proved that E-PRoPHET, E-EDR, and E-PRoWait have less power usage in comparison to PRoPHET, EDR, and PRoWait, respectively.

7.3.1 Comparison of energy efficient routing protocols for OppNets

The following table provides representative energy efficient routing protocols for OppNets along with their characteristics, tools, techniques, and limitations.

Energy efficient routing protocols

Reference	DTN technique	Tools	Features	Limitations
Borah et al. (2017)	E-Prophet, E-EDR, and E-PRoWait	ONE simulator	These techniques are designed by incorporating some energy-related parameters that constrain the nodes to forward the message to only those encountered relay nodes that have enough remaining battery power	Performance of the proposed protocols using real mobility traces is not evaluated

Continued

Reference	DTN technique	Tools	Features	Limitations
Khalid et al. (2016)	AEHBPR	ONE simulator	This technique incorporates some energy factors in a utility function in order to select the best next-hop node to transmit the message	AEHBPR is compared only with two benchmark protocols: HBPR and AEProphet, which limits the scope of the results
Pan et al. (2017)	Group-Purchase Scheme (GPSCH)	ONE simulator	This technique allows a message to choose the next path depending upon the communication probability between the message and its subscribers so that the message forwarding delay in the network can be minimized	Delivery ratio in GPSCH is not stable in the beginning of the simulations and energy efficiency is not considered
Patel and Chaudhary (2015)	Volatile spray and wait routing in DTNs	ONE simulator	The proposed routing algorithm uses the Spray-and-Wait mechanism to boost the message spreading while increasing the efficiency of the routing process. In terms of performance evaluation, VSWR outperforms the Spray-and-Wait, epidemic, and prophet protocols	Proposed algorithm is not evaluated for heterogeneous networks

Continued

Continued

Reference	DTN technique	Tools	Features	Limitations
Chilipirea et al. (2013)	Energy-efficient social-based routing protocol for OppNets	UPB 2012	The delivery cost is improved by using some social-based parameters for routing	Proposed protocol does not yield better results in comparison to BUBBLE Rap in terms of performance
Borah et al. (2018)	Energy-aware Location Prediction-based Forwarding for Routing using Markov Chain (ELPFR-MC)	ONE simulator	A Markov chain-based technique is implemented to control the energy consumption of nodes	Performance of ELPFR-MC is not evaluated using real-trace mobility models

7.4   Environment friendly green data broadcasting techniques

In the recent years, green communication (Han et al., 2011) has gained a considerable attention from the research community. It is essentially the selection of energy-efficient communication and networking techniques in order to reduce the harm to the environment. The present scenario exploits the green communication technologies, raising the carbon footprint and associated environmental degradation issues. In the sequel, we discuss novel protocols to resolve the energy efficiency issues.

7.4.1   Energy efficient PRoPHET routing protocol

The energy efficient PRoPHET routing protocol proposed in Khalid et al. (2016) relies on the idea that a node is more likely to interact or visit a location in the near future if it has interacted with or visited that location in the past. This technique incorporates some energy-related parameters, where the nodes only forward the messages to encountered relay nodes with sufficient battery power, yielding the E-Prophet protocol. Initially, the residual energy of the neighboring nodes of each source node is calculated. Then, those neighboring nodes whose residual energy is greater than that of the source node are considered. The message is forwarded only to the subset of these nodes whose delivery probabilities are greater than that of the source/ intermediate nodes.

7.4.2 Energy efficient PRoWait routing protocol

In the energy efficient PRoWait routing protocol proposed in Khalid et al. (2016), initially, the residual energy of each neighboring node to the source node is calculated, and based on this, only those nodes whose residual energy are greater than that of the source/intermediate node are considered. Then, the message is forwarded only to the subset of these nodes whose delivery predictability values (obtained using the PRoWait rules) are greater than that of the source/intermediate nodes.

7.4.3 Energy efficient EDR routing protocol

In the energy efficient EDR routing protocol (Khalid et al., 2016), the selection of a suitable next-hop forwarder of a message is determined by the so-called forwarding parameter, which is determined by jointly minimizing the distances between the neighboring nodes and destination and by maximizing the number of encounters with the destination for every packet delivery. Initially, the residual energy of the residual energy of the source node is calculated, then that of the neighboring nodes are calculated; and only those intermediate nodes whose residual energy is greater than that of the source nodes are considered as candidate relay nodes for forwarding the message to its destination. In doing so, the average of the forwarding parameters for each sender-receiver set is considered and the ideal group of nodes is selected to act as relay nodes.

7.4.4 Energy efficient ATDTN routing protocol

Dhurandher et al. (2019) proposed an energy-aware version of the altruism-dependent trust-based data forwarding protocol (ATDTN) for OppNets (Kumar et al., 2017). In this protocol, each node is associated with a dynamically changing altruism value, which represents its trust in the network. This value is then used to decide on whether the node should participate or not in the message forwarding. When doing so, the residual energy of the source node's neighboring nodes is calculated, along with that of the intermediate nodes, and those nodes whose residual energy is greater than that of the source node are considered as potential message forwarders based on the altruism parameter.

7.4.5 Geocasting techniques

Rajaei et al. (2016) proposed the Geocasting Spray-And-Flood (GSAF) routing protocol for OppNets. Its design relies on a flexible approach that consists of taking random walks toward the destination cast, where the cast definition is given by the sender. Those messages that follow the directions away from the cast are discarded when the node's buffer capacity is exhausted. That way, some spaces in the buffer are free up for new incoming messages that are to be delivered. In this protocol, the message is disseminated as follows: a node that receives a message needs only to check if that message is originated from the cast defined in it; and that node carries and forwards the message to other selected relay nodes according to a ticketing mechanism (Rajaei et al., 2016). When a cast receives a message, a controlled flooding is utilized to

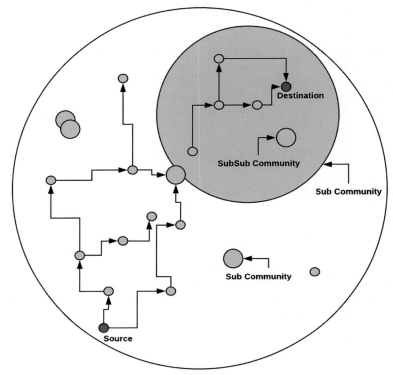

Fig. 7.2 Example of geocasting.

disseminate it in such a way that the message cannot exit the borders of the destination cast. When doing so, those messages that have expired are deleted (Fig. 7.2).

The following techniques have been proposed to achieve geocasting (Rajaei et al., 2016) in OppNets.

7.4.5.1 Floating content

In OppNets, the geocasting technique is essential to preserve the privacy of content and to prevent any data leaks during transmission. The region for content sharing is defined in the form of a circle using its center and radius. Each region has two circles defined for it and the content is supposed to float inside the circle $P(x, y)$, R. Users inside this circle can share content using a flooding technique. A larger circle $P(x, y)$, A is designed to enable the retrieval of the outgoing content to the main circle. Based on some priorities, the users between both circles ($R < r < A$) can retain the content in their buffers, then pass on the content to users moving toward the main circle. On the other hand, the users moving out of the larger circle ($r > A$) must delete the content in order to empty their buffers, and this helps decreasing the probability of leaking the data outside the defined region. The assumption that the content can only be created locally and then distributed is a limitation for the distributor although it helps improving the privacy (Fig. 7.3).

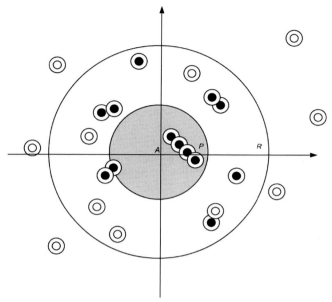

Fig. 7.3 Example of floating content.

7.4.5.2 GeoOPP

Lu and Liu (2014) proposed the so-called geocasting for opportunistic networks (GeoOPP), a two-phase scheme which involves forwarding a message and flooding it. Typically, the network is divided into tiny nonoverlapping cells. Then, the rate of reaching a destination cell through each cell in the network is computed, and the message is forwarded to a neighbor only if its future destination has a greater probability of delivering the message compared to its current location. Unicasting and flooding are combined in the sense that the message is first forwarded to the predefined geographical region, then flood to all nodes in that region. In addition, a geographic greedy routing is adapted in such a way that the nodes can only choose those neighbors who can carry the message closer to its destination; and this proximity is measured by means of the so-called radius metric. In this process, the Chebyshev's inequality is used to estimate the probabilities that a node visits a region and establish contact with nodes in that region.

7.4.5.3 Expected visiting rate

In the opportunistic geocast technique (Ma and Jamalipour, 2011), routing decisions are often based on the expected visiting rate (EVR) of the users to the destination region, that is, the probability of each user visiting the destination cast to deliver the messages to the destination region, which is computed based on the historical data of nodes visited in the network. In this technique, the message is an ascending gradient of EVR until reaching the destination region. Thereafter, a controlled broadcast is used to identify all the members of the intended audience in that region.

7.4.5.4 Geocasting in mobile partitioned networks

In Piorkowski (2008), a geocast service for mobile partitioned networks is proposed, with the objective of minimizing the delivery latency. In its design, the time stability of the node's mobility pattern is exploited and the collective mobility pattern is represented by a mobility map. Assuming that in the partitioned subregions, the users are continuously connected to each other and all the nodes are confined within their own subregions, the nodes cannot move between the subregions and a mobility-assisted forwarding strategy is implemented to decide on suitable relay nodes to carry the message to its destination. In doing this, the goal is to minimize expected message delay and maximize the message delivery rate. One of the drawbacks of this scheme is that the membership of a node is static in unicasting and multicasting.

7.5   Evaluation

In this section, the performance of E-PRoPHET, E-ProWait, E-EDR, PORON, and E-ATDTN is evaluated using the ONE simulator (Keranen et al., 2009) using real mobility data traces (Scott et al., 2009). Each line of this database describes a connection event and has five fields as represented in Table 7.1.

Energy efficient routing protocols

Reference	DTN technique	Tools	Features	Limitations
Dong et al. (2015)	Broadcasting combined with multi-NACK/ACK protocol (BCMN/A)	OMNET++	Multiple acknowledgement and negative acknowledgement are taken into consideration	Energy efficiency is not considered
Wang et al. (2017)	DTN congestion avoidance routing algorithm (DCAR)	ONE simulator	This algorithm uses fixed sink stations as centers for data collection and publishing. These sink stations manage the token allocation and collection in the network. A node can transmit packets only after it obtains a token for transmission. By using this technique, the algorithm manages the concurrent transmissions of packets	Energy efficiency is not considered

Continued

Reference	DTN technique	Tools	Features	Limitations
Dhurandher et al. (2009a)	E2-SCAN	GloMoSim simulator	The routing and data forwarding operations are protected by means of the SCAN protocol. The security of the network is enhanced through the detection and reaction to malicious nodes	When nodes are highly mobile, only localized broadcasting of TREV packets may not be helpful. Overhearing the channel constantly in E2-SCAN may lead to a large amount of energy consumption
Dhurandher et al. (2014)	GAER	ONE simulator	This protocol uses the genetic algorithm to select a better next hop among a group of neighbor nodes for message routing toward the destination. With this technique, the selection of the best possible node as next hop is dependent on its remaining battery power	GAER is compared only with Epidemic, PROPHET, and Spray-and-Wait routing protocols
Dhurandher et al. (2009c)	An energy-aware routing protocol for ad hoc networks based on the foraging behavior in ant swarms	GloMoSim simulator	Power consumption in routing and multipath transmission attributes of ant swarms are exploited	Compared to some benchmark schemes, the efficiency of the protocol in terms of packet delivery rate and energy per packet in high mobile conditions is established and is proved by simulations

Continued

Continued

Reference	DTN technique	Tools	Features	Limitations
Zhou et al. (2013)	Contact probing process based on the random waypoint (RWP) model	MATLAB	Random WayPoint model is applied	The trade-off between contact opportunities and energy efficiency is investigated, leading to a method for quantifying the detecting probability. In general, it is hard to obtain the exact expression of such probability

Energy efficient routing protocols

Reference	DTN technique	Tools	Features	Limitations
Dhurandher et al. (2017)	EHBPR	ONE simulator	The EHBPR protocol addresses the energy constraints in HBPR	The performance of EHBPR was not compared against that of other energy efficient routing protocols
Obaidat et al. (2010)	DEESR	Global mobile information system simulator	The decision to select the next-hop node to carry the message toward its destination is based on various parameters such as longevity, battery power, distance, and link quality	The detection times increase with the increase in the number of nodes. Security aspects can be improved

Continued

Reference	DTN technique	Tools	Features	Limitations
Yang et al. (2013)	OIA based on antenna selection (AS) and singular value decomposition (SVD)	ONE simulator	This technique considers the antenna selection and the singular value decomposition	The scaling law for the number of users is not discussed, and energy efficiency is not considered
Wong and Jia (2013)	Efficient Social Relation Opportunistic Routing (SROR) algorithm	NS-2 simulator	SROR is beneficial to social networks such as dynamic networks	The packet delivery probability can be improved by finding the best forwarding node when performing the routing. Energy efficiency can be improved
Dhurandher et al. (2018)	DEEP	ONE simulator	The selection of the best next-hop forwarder for a message relies on the node's energy	Security aspects are not incorporated
Dhurandher et al. (2009b)	EEAODR	GloMoSim simulator	This method balances the energy load among nodes so that a minimum energy level is maintained	Clustering and directional antenna can also be considered

Table 7.1 Syntax of standard event reader format.

Time	Action	First node	Second node	Type

For our simulations, the considered performance metrics are residual energy, dead nodes and delivery probability, under varying Time-to-Live (TTL) and message generation interval (MGI). The simulation parameters are given in Table 7.2. For comparison purpose, the initial battery power is 45–50 K Joules for E-PRoPHET, E-EDR, E-PRoWait, PORON, and E-ATDTN. The execution of the ONE simulator interface for the relevant protocols is captured in Fig. 7.4 and the creation, interaction, and updates of various objects are shown in Fig. 7.5, where the incoming and outgoing

Table 7.2 Simulation parameters.

Parameter	Value
Real mobility data trace	Cambridge-haggle-one-infocom 2006
Trace format	Standard events reader
Trace fields in each line	5
Communication interface	Bluetooth
Transmission range	10 m
Number of nodes	98
Number of contacts	170,601
Simulation time	337,418 s
Transmission speed	250 kbps
Message size	500 kb up to 1 Mb
Buffer capacity	5 Mb
Movement model	Shortest path
Message time-to-live (TTL)	100–300 min

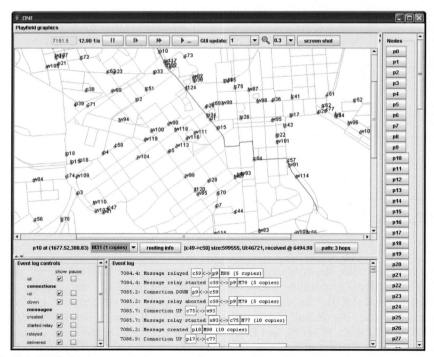

Fig. 7.4 ONE simulator execution screenshot.

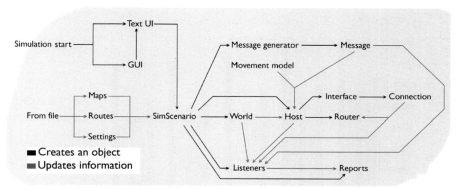

Fig. 7.5 Interaction of objects in ONE simulator.

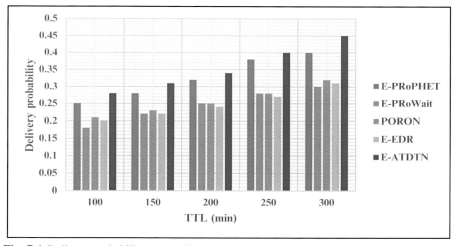

Fig. 7.6 Delivery probability versus TTL.

arrows represent the interconnections from the start of the simulation to the creation of the report. The various objects in between explain the paths followed.

7.5.1   Simulation results

First, the delivery probability is studied under varying TTL, and the results are captured in Fig. 7.6. It can be seen that when the TTL is increased, the delivery probability of each protocol is decreased. The message delivery probability is also decreased on the account of the time period increase of each message allotment with the TTL increment. Indeed, as the TTL increases, the time duration allotted to each message gets increased, and more messages get stored in the node's buffer, leading to a decrease in the message delivery probability. It is clear that E-ATDTN outperforms all the other studied protocols.

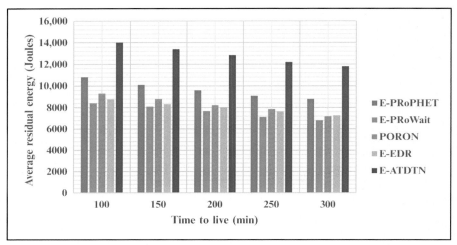

Fig. 7.7 Average residual energy versus TTL.

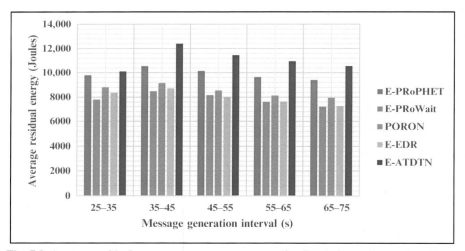

Fig. 7.8 Average residual energy versus message generation interval.

Second, the TTL is varied and the average residual energy is investigated. The results are captured in Fig. 7.7. It is observed that when the TTL is increased, the average residual energy is decremented. Indeed, in terms of residual energy when the TTL is increased, E-ATDTN is 32.80%, 68.9%, 55.32%, and 60.83% better than E-PRoPHET, E-PRoWait, PORON, and E-EDR protocols, respectively.

Third, the message generation interval is varied and the effect of this variation on the residual energy is investigated. The results are given in Fig. 7.8. It can be seen that initially, when the message generation interval is small, the average residual energy is high. As the message generation interval is further increased, a drop in the average residual energy is observed. Therefore, the message generation interval and average residual energy have an inverse relationship.

Fourth, the TTL is varied and the number of dead nodes is studied. The results are captured in Fig. 7.9. It is observed that as the TTL is increased, the number of dead nodes is also increased.

Fifth, the initial energy used is 50 kJ; the TTL is varied and the effect of this variation on the residual energy is studied for the GSAF, F-GASF, and CGOPP protocols. The results are depicted in Fig. 7.10. It is observed that when the TTL is increased, the

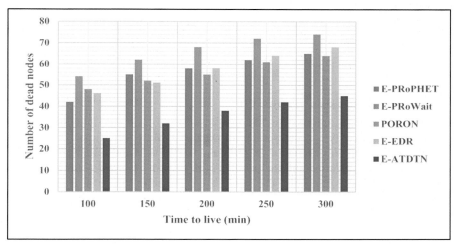

Fig. 7.9 Number of dead nodes versus TTL.

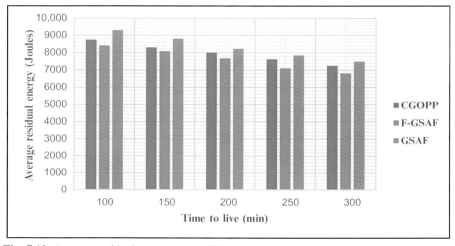

Fig. 7.10 Average residual energy versus TTL.

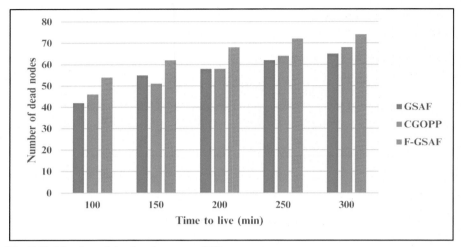

Fig. 7.11 Number of dead nodes versus TTL.

average residual energy is decreased for all protocols. Moreover, the performance of GSAF is 4.32% superior to that of CGOPP and 9.05% superior to that of F-GSAF.

Sixth, the TTL is varied and the number of dead nodes is studied for the GSAF, F-GASF, and CGOPP protocols. The results are captured in Fig. 7.11. It is clear that as the TTL is increased, the number of dead nodes is also increased.

7.6   Conclusion

In this chapter, we have discussed on energy efficiency issues in OppNets related to geocasting and security areas. A review of energy-aware routing protocols for OppNets in these areas, including the features, tools, and limitations, has been proposed and the ONE simulator has been used to achieve a performance evaluation of few energy efficient protocols, namely E-PRoPHET, E-ProWait, PORON and E-EDR, and E-ATDTN. Under varying TTL, it has been shown that E-ATDTN outperforms the other studied protocols in terms of average residual energy, number of message drops, and number of dead nodes. Finally, it has also been shown that the performance of GSAF is 4.32% superior to that of CGOPP and 9.05% superior to that of F-GSAF in terms of the number of dead nodes and residual energy. Further, the authors try to enhance the performance of energy efficient protocols on other real mobility traces such as the ones provided in Rhee et al. (2011) as future work.

References

Basagni, S., Petrioli, C., Petrocia, R., Spaccini, D., 2015. CARP: a channel-aware routing protocol for underwater acoustic wireless networks. Ad Hoc Netw. 34, 92–104.

Borah, S.J., Dhurandher, S.K., Tibarewala, S., Woungang, I., Obaidat, M.S., 2017. Energy-efficient Prophet-PRoWait-EDR protocols for opportunistic networks. In: GLOBECOM 2017—2017 IEEE Global Communications Conference, IEEE, pp. 1–6.

Borah, S.J., Dhurandher, S.K., Woungang, I., Kandhoul, N., Rodrigues, J.J.P.C., 2018. An energy-efficient location prediction-based forwarding scheme for opportunistic networks. In: 2018 IEEE International Conference on Communications (ICC), IEEE, pp. 1–6.

Buzzi, S., Chih-Lin, I., Klein, T.E., Poor, H.V., Yang, C., Zappone, A., 2016. A survey of energy-efficient techniques for 5G networks and challenges ahead. IEEE J. Select. Areas Commun. 34 (4), 697–709.

Chilipirea, C., Petre, A.-C., Dobre, C., 2013. Energy-aware social-based routing in opportunistic networks. In: 2013 27th International Conference on Advanced Information Networking and Applications Workshops, IEEE, pp. 791–796.

Dhurandher, S.K., Misra, S., Ahlawat, S., Gupta, N., Gupta, N., 2009a. E2-SCAN: an extended credit strategy-based energy-efficient security scheme for wireless ad hoc networks. IET Commun. 3 (5), 808–819.

Dhurandher, S.K., Misra, S., Obaidat, M.S., Bansal, V., Singh, P.R., Punia, V., 2009b. EEAODR: an energy-efficient ad hoc on-demand routing protocol for mobile ad hoc networks. Int. J. Commun. Syst. 22 (7), 789–817.

Dhurandher, S.K., Misra, S., Obaidat, M.S., Gupta, P., Verma, K., Narula, P., 2009c. An energy-aware routing protocol for ad-hoc networks based on the foraging behavior in ant swarms. In: 2009 IEEE International Conference on Communications, IEEE, pp. 1–5.

Dhurandher, S.K., Sharma, D.K., Woungang, I., Gupta, R., Garg, S., 2014. GAER: genetic algorithm-based energy-efficient routing protocol for infrastructure-less opportunistic networks. J. Supercomput. 69 (3), 1183–1214.

Dhurandher, S.K., Sharma, D.K., Woungang, I., Saini, A., 2017. An energy-efficient history-based routing scheme for opportunistic networks. Int. J. Commun. Syst. 30 (7), e2989.

Dhurandher, S.K., Borah, S.J., Woungang, I., Tibarewal, S., Barolli, L., 2018. DEEP: distance and encounter based energy-efficient protocol for opportunistic networks. J. High Speed Netw. 24 (2), 119–131.

Dhurandher, S.K., Woungang, I., Singh, J., Borah, S.J., 2019. Energy aware routing for efficient green communication in opportunistic networks. IET Netw. 8 (4), 272–279.

Dong, M., Ota, K., Liu, A., Guo, M., 2015. Joint optimization of lifetime and transport delay under reliability constraint wireless sensor networks. IEEE Trans. Parallel Distrib. Syst. 27 (1), 225–236.

Han, C., Harrold, T., Armour, S., Krikidis, I., Videv, S., Grant, P.M., Haas, H., 2011. Green radio: radio techniques to enable energy-efficient wireless networks. IEEE Commun. Mag. 49 (6), 46–54.

Huang, C.M., Lan, K.-C, Tsai, C.Z., 2008. A survey of opportunistic networks. In: Proceedings of IEEE International Conference on Advanced Information Networking and Applications Workshops, Okinawa, Japan, 25–28 March 2008, pp. 1–6.

Keranen, A., Ott, J., Karkkainen, T., 2009. The ONE simulator for DTN protocol evaluation. In: Proceedings of SIMUTools, Rome, Italy, March 2–6, pp. 1–9.

Khalid, K., Woungang, I., Dhurandher, S.K., Barolli, L., Carvalho, G.H.S., Takizawa, M., 2016. An energy-efficient routing protocol for infrastructure-less opportunistic networks. In: 2016 10th International Conference on Innovative Mobile and Internet Services in Ubiquitous Computing (IMIS), IEEE, pp. 237–244.

Kumar, A., Dhurandher, S.K., Woungang, I., Obaidat, M.S., Gupta, S., Rodrigues, J.J.P.C., 2017. An altruism-based trust-dependent message forwarding protocol for opportunistic networks. Int. J. Commun. Syst. 30 (10), 1–11.

Lu, S., Liu, Y., 2014. Geoopp: geocasting for opportunistic networks. In: 2014 IEEE Wireless Communications and Networking Conference (WCNC), IEEE, pp. 2582–2587.

Ma, Y., Jamalipour, A., 2011. Opportunistic geocast in disruption-tolerant networks. In: IEEE GLOBECOM.

Mauve, M., Widmer, J., Hartenstein, H., 2001. A survey on position-based routing in mobile ad hoc networks. IEEE Netw. 15, 30–39.

Obaidat, M.S., Dhurandher, S.K., Gupta, D., Gupta, N., Asthana, A., 2010. DEESR: dynamic energy efficient and secure routing protocol for wireless sensor networks in urban environments. J. Inf. Process. Syst. 6 (3), 269–294.

Pan, D., Sun, L., Jiao, J., 2017. Group-purchase scheme for content dissemination in opportunistic networks. IEEE Access 5, 6060–6074.

Patel, V.G., Chaudhary, H.L., 2015. Volatile spray and wait routing in delay tolerant network. In: 2015 Second International Conference on Advances in Computing and Communication Engineering, IEEE, pp. 72–77.

Piorkowski, M., 2008. Mobility-centric geocasting for mobile partitioned networks. In: IEEE International Conference on Network Protocols, October 19–22, Orlando, FL, USA, pp. 1–10.

Piovesan, N., Gambin, A.F., Miozzo, M., Rossi, M., Dini, P., 2018. Energy sustainable paradigms and methods for future mobile networks: a survey. Comput. Commun. 119, 101–117.

Rajaei, A., Chalmers, D., Wakeman, I., Parisis, G., 2016. GSAF: efficient and flexible geocasting for opportunistic networks. In: 2016 IEEE 17th International Symposium on a World of Wireless, Mobile and Multimedia Networks (WoWMoM), IEEE, pp. 1–9.

Rani, S., Ahmed, S.H., Malhotra, J., Talwar, R., 2017. Energy efficient chain based routing protocol for underwater wireless sensor networks. J. Netw. Comput. Appl. 92, 42–50.

Rhee, I., Shin, M., Hong, S., Lee, K., Kim, S.J., Chong, S., 2011. On the Levy-walk nature of human mobility. J. IEEE/ACM Trans. Netw. (TON) 19 (3), 630–643.

Scott, J., Gass, R., Crowcroft, J., Hui, P., Diot, C., Chaintreau, A., 2009. CRAWDAD dataset Cambridge/Haggle (v. 2009-05-29), CRAWDAD wireless network data archive.

Sivakumar, T., Manoharan, R., 2017. ERP: an efficient reactive routing protocol for dense vehicular ad hoc networks. Turk. J. Elec. Eng. Comput. Sci. 25 (3), 1762–1772.

Vullers, R.J.M., Van Schaijk, R., Visser, H.J., Penders, J., Van Hoof, C., 2010. Energy harvesting for autonomous wireless sensor networks. IEEE Solid-State Circuits Mag. 2 (2), 29–38.

Wang, K., Gao, H., Xu, X., Jiang, J., Dong, Y., 2015. An energy-efficient reliable data transmission scheme for complex environmental monitoring in underwater acoustic sensor networks. IEEE Sensors J. 16 (11), 4051–4062.

Wang, H., Lv, H., Wang, H., Feng, G., 2017. DTN congestion avoidance routing algorithm based on tokens in an urban environment. J. Sensors. DCAR 2017, 1–9.

Wong, G.K.W., Jia, X., 2013. A novel socially-aware opportunistic routing algorithm in mobile social networks. In: 2013 International Conference on Computing, Networking and Communications (ICNC), IEEE, pp. 514–518.

Yang, H.J., Shin, W.-Y., Jung, B.C., Paulraj, A., 2013. Opportunistic interference alignment for MIMO interfering multiple-access channels. IEEE Trans. Wireless Commun. 12 (5), 2180–2192.

Zhao, M., Kumar, A., Chong, P.H.J., Lu, R., 2016. A reliable and energy-efficient opportunistic routing protocol for dense lossy networks. IEEE Wireless Commun. Lett. 6 (1), 26–29.

Zhou, H., Zheng, H., Wu, J., Chen, J., 2013. Energy-efficient contact probing in opportunistic mobile networks. In: 2013 22nd International Conference on Computer Communication and Networks (ICCCN), IEEE, pp. 1–7.

Zhou, Z., Yao, B., Xing, R., Shu, L., Shengrong, B., 2015. E-CARP: an energy efficient routing protocol for UWSNs in the internet of underwater things. IEEE Sensors J. 16 (11), 4072–4082.

Zou, Y., Wang, X., Shen, W., Hanzo, L., 2013. Security versus reliability analysis of opportunistic relaying. IEEE Trans. Veh. Technol. 63 (6), 2653–2661.

Part Two

Improving the performance of delay-tolerant networks (DTNs)

Assessing the Bundle Protocol (BP) and alternative approaches to data bundling in delay-tolerant networks (DTNs)[☆]

W.M. Eddy
MTI Systems, Greenbelt, MD, United States

8.1 Introduction

The delay-tolerant networks (DTN) Bundle Protocol (BP) (Fall, 2003) is one of the main approaches in recent research involving internetworking in "challenged" environments. Based on the promise of this approach, groups have produced descriptions and specifications for BP-based DTN protocols, including informational and experimental requests for comments (RFCs) produced via the Internet research task force's DTN Research Group (DTNRG) (Cerf et al., 2007; Scott and Burleigh, 2008), and in addition worked towards international standards in the consultative committee for space data systems (CCSDS) (CCSDS, 2012). As there is a great deal of inherent flexibility in BP, several extensions have been proposed, and many of them are implemented to some extent. These have supported widely different approaches to building networks based on the DTN architecture via the BP as an internetwork overlay protocol for relaying across the challenged hops.

In addition, besides the BP, there have been other proposals for solving similar problems to those the DTN architecture is aimed at. These proposals have also been implemented and tested to some extent. This chapter focuses on comparison between some alternative approaches to challenged networking and different profiles of BP features.

Although the idea of delay tolerance and the DTN architecture is inspired by space communications scenarios and the Interplanetary Internet concept, one recent survey of DTN applications categorized 11 different areas of terrestrial use for which DTN has been studied and over a dozen different performance studies, simulation tools and emulation test beds in use (Voyiatzis, 2012). This breadth in types of challenged networks and diversity in platforms and metrics that researchers have used makes it difficult to compare and correlate results empirically. For instance, not all platforms implement the same DTN features (e.g., reactive fragmentation, security, different routing algorithms, etc.), and individual experimenters enable or add features relevant

---

☆ This chapter is a reprint of the chapter originally published in the first edition of Advances in Delay-Tolerant Networks (DTNs): Architecture and Enhanced Performance.

Advances in Delay-tolerant Networks (DTNs). https://doi.org/10.1016/B978-0-08-102793-6.00008-4

to their environment. In addition, while several demonstrations and real-world trials have been conducted, due to the scale and expense involved in these, many researchers have also based results and evaluations on simulation work. This has led to a mixture of data collected with different methodologies and at differing levels of fidelity and detail. Even though some of this data is available, it is difficult to correlate or confirm results between projects, even for variations of DTN BP configurations alone, without regard to non-BP alternatives (Walker et al., 2010).

For each approach discussed in this chapter, consideration is taken of the unique challenges and metrics relevant to the target deployment environments. There are no typical or well-known model scenarios for DTN, as there are other fields of networking (e.g., dumb-bell, parking lot topologies, etc.). Specific challenges and metrics are explained in individual papers, but there are also some generalized discussions in the literature (Kapadia et al., 2011), which include:

- Potential challenges:
  - o Node encounter schedules may be known in advance (e.g., via an oracle), calculable (e.g., based on trajectory knowledge), community-based, or even completely random or unknown to individual nodes.
  - o Network connectivity between two or more nodes may fall anywhere between relatively static and highly dynamic, and the nodes may be homogeneous or vastly heterogeneous in their capabilities.
  - o The network capacity itself may range from high-bandwidth links between some nodes (either sporadic or highly available) to very low-bandwidth links (e.g., acoustic underwater networking).
  - o Storage resources at forwarding nodes may be either constrained or relatively unconstrained versus the demand of the traffic model. Storage capabilities may be fairly homogeneous or more widely heterogeneous between nodes.
  - o Energy concerns will generally apply, though how much of a concern energy is and how badly this impacts the contact schedule between nodes is closely dependent on other facets of the deployment environment.
- Potential metrics:
  - o Message delivery ratio (MDR) can be used to assess whether data eventually arrives at the destination in a given scenario. MDR-based metrics do not assess timeliness, resource usage, etc., but only whether data eventually reaches its destination.
  - o Message Delay can be used to assess how timely a given approach is in its data delivery; however, the irony of using delay minimization as a metric of evaluation in any "delay-tolerant" networking approach is well understood.
  - o Number of Replicas can be minimized to save on power, bandwidth/capacity, and storage resources; however, minimizing replication of data is also known to impact MDR and other metrics in the opposite direction, due to the way such things are measured in current configurations.
  - o Energy/power usage is always a concern for wireless communications, and, in the case of challenged networks, the implications of protocol design and configuration on power needs are not well understood. In space exploration, it may be simply accommodated through mission-planning in power budgets and duty cycles or scheduling. In other cases, where beaconing and multiple node (not point-to-point) datalinks are in use, the convergence layer adapters, routing mechanisms, and other software configuration may make energy usage more significant as a metric.

Selecting from subsets of these challenges and the applicable metrics has driven exploration of dozens of routing and data delivery mechanisms, including concepts such as epidemic routing, direct-contact forwarding, one-hop relay forwarding, use of oracles, location-based routing, gradient-based forwarding, controlled replication, network coding, and many others. It is unfair, and of dubious value, to try to compare these with one another without regard to the intended deployment environment and operational considerations of each. Even the detailed configuration of a single feature (e.g., fragmentation of bundles) can be studied in detail and produce results that vary widely (Magaia et al., 2011), so comparison of the detailed configuration of DTN BP to an alternative should only be made when both are tailored for a similar environment. In this chapter, we discuss specific DTN alternatives proposed for challenged networks, along with specific DTN profiles of features used in similar environments. This provides the closest and most fair comparison of DTN to other approaches, and also highlights the fact that DTN has been applied to numerous differing challenged network scenarios, whereas other approaches have, to date, generally been more limited in their intended scope. We do not attempt direct performance or other hard quantitative comparisons here, as this is not helpful to do, based on the current literature with data from incomparable scenarios. Our view on this is consistent with other recent surveys on DTN (Khabbaz et al., 2012).

8.2 DTN architecture and Bundle Protocol implementation profiles

The concept of a bundle layer in order to support DTN is described in RFC 4838 (Cerf et al., 2007), and is the basic element of the DTN architecture. The architecture does not exactly define the protocol, but focuses on describing the challenges that need to be overcome and the features that the protocol needs to support. It was clear at the time of publication of RFC 4838 that some facets of the technical approach would need to be very flexible in order to accommodate the wide variety of underlying networks and connectivity scenarios envisioned. As testing and experimentation with DTN advanced, a number of implementations sprang up, each focusing on particular aspects of the DTN architecture, and developing specific mechanisms to meet its needs within the general DTN architecture. Although these may have grown several noninteroperable features, they do still interoperate for very basic bundle exchanges (e.g., using small bundles unicast to endpoint identifiers (EIDs) of the "DTN" schema, with manual routes over a user datagram protocol (UDP) convergence layer, without custody transfer, or need for reactive fragmentation), and they all basically conform to the BP specification, demonstrating its flexibility to multiple operational environments.

Some of the most important DTN architecture concepts that were kept flexible and loosely defined include:

- Endpoint Identifier (EID) formats—Uniform resource identifiers (URIs) are described as the format for EIDs in bundles; however, the actual EID schemes are not standardized across the architecture. Endpoints can utilize any format of URI that may seem useful for their individual constraints on design, implementation and operation. However, nodes in the network need to be able to effectively forward bundles received, and generally (though not necessarily) utilize the destination EID for this. Thus, in actual implementation and deployment experience, specific EID schemas have been selected and used in different environments. Also, some later attempts to apply DTN technology for information-centric networking also utilize other metadata for forwarding and storage, de-emphasizing forwarding based on the destination EID.

- Reliability and Custody Transfer—Custody transfer was recognized as an optional feature that only certain applications might make use of, and also custody transfer support within forwarding nodes is not strictly required. Thus, implementations have differed in terms of support for custody transfer, and particularly in regard to how other BP features are (or are not) supported in conjunction with custody transfer. For instance, multicast delivery requires additional extensions to support custody transfer (Symington et al., 2009), and these are not implemented in all cases. The architecture also views reliability primarily as a matter of retransmission, and does not specifically address detection and correction of corrupted data. Multiple methods of detecting corrupted bundles exist, though they are not universally implemented or even commonly used in research so far (Eddy et al., 2009; Symington et al., 2011).

- Routing and Forwarding—The concept of routing over a time-varying multigraph was developed, but no specific algorithms have yet been strongly or universally advocated by the DTNRG, either for establishing adjacencies and distributing routing information between nodes, or for defining EID schemas with specific aspects (e.g., aggregation) used to make forwarding determinations. An enormous number of routing mechanisms were developed (Ali et al., 2010) and forwarding has generally occurred based on either (1) lookup of "flat" destination EIDs, (2) lookup within hierarchical groupings of EIDs (e.g., "regions" were an early proposal for grouping EIDs and simplifying routing and forwarding), or (3) "intentional naming", in which specific (canonical) destination EIDs are eventually resolved by intermediate nodes based on more general EIDs that had been used by the sender.

- Fragmentation and Reassembly—The architecture describes both proactive and reactive fragmentation mechanisms. Because reactive fragmentation is not fully compatible with several other extensions (e.g., for security or multicasting), its support and usage seem to have been limited to date. While proactive fragmentation is compatible with most other extensions, efficient and effective use of proactive fragmentation requires pre-knowledge of contact durations and capacities, which few scenarios accommodate outside those with scheduled satellite or space links.

- Congestion and Flow Control—Applicability of congestion control to the bundle layer was largely left as a research topic by the architecture, and has been relatively undeveloped to any common basis in implementations. Flow control was presented as a matter of managing storage at the nodes, and has been primarily handled through node-local resource-management mechanisms, specific to given implementations. End-to-end or even node-to-node flow and congestion control is not generally supported within either homogeneous or heterogeneous deployments of DTN software today, aside from one mechanism implemented in recent revisions of the Interplanetary Overlay Network (ION) code.

- Security—Because of resource-scarcity in challenged networks, security for DTNs is needed at least in terms of providing authentication and access control to node resources. However, the basic BP does not provide any security mechanisms. After the BP specification, the Bundle Security Protocol (BSP) later defined BP extensions for authentication and encryption. To date, there has been no compelling work on the key management issues, or linking of keying material and security policies to EIDs. The BSP has, however, been implemented in multiple codebases and shown to interoperate in at least some basic configurations. Limited experience using the BSP suggests that making it more automated is a key operational concern in making DTN security practical in a large network. In small-scale use, manual configuration is palatable, though this is understood to involve not just initial configuration, but also periodic rekeying, managing node reputations, and management of policies for storing and routing bundles on behalf of other nodes. In larger networks, the administrative burden of doing all this manually, given existing software and protocols, would be prohibitively expensive, though for smaller networks it might be feasible.

BP experimental deployments have selected "profiles" of features to implement, specific to their constraints and requirements. These profiles have differed widely, and we provide an overview of several of them in the remainder of this section.

8.2.1   DTN/BP—DINET profile

The ION software implementing BP was used in the Deep Impact NETwork (DINET) experiments conducted by the Jet Propulsion Laboratory (JPL) in 2008 (Jones, 2009). The software employed supports a number of different protocol features that can be configured, primarily intended for use in support of space communications and as flight software. In order to select the profile of DTN options and extensions to be used, the DINET experiments only exercised a profile intended to be close to what might actually be used for space operations, based on the contemporary concepts for space internetworking.

The DINET experiments ran over the course of 2 months and validated the improvements possible with DTN technologies applicable to space operations. DTN software was run on a number of ground systems with emulated disruptions as well as on the flight system of the EPOXI spacecraft in a deep-space orbit with actual delays and disruptions to the connectivity between it and the Deep Space Network (DSN) used to relay data between flight and ground systems.

The ION software configuration used for DINET included: (1) the "IPN" style of endpoint identifiers, (2) basic custody transfer, (3) LTP over the space links, and transmission control protocol (TCP) terrestrially for convergence layer adapters, (4) contact graph routing (CGR) and prioritized forwarding of bundles, (5) proactive fragmentation, and (6) use of a subset of the BSP for security (via the bundle authentication block with particular ciphersuites).

8.2.2   DTN/BP—UK-DMC profile

Another use of BP between space and ground assets was in the 2008 NASA Glenn Research Center (NASA GRC) testing with the UK-DMC satellite operated by Surrey Satellite Technology Ltd. (SSTL) (Ivancic et al., 2010a). Multiple ground stations (in England, Alaska, and Hawaii) and the experiment operations center (in Cleveland, Ohio) were hosting nodes running the open source DTN2 software with minor modifications. The UK-DMC spacecraft was an operational asset of SSTL, used for its commercial imaging business. For this experiment, it ran slightly modified flight software in order to create DTN bundles encapsulating the normal Earth-imaging data captured by the spacecraft sensors.

The UK-DMC experiments demonstrated and validated that both proactive and reactive fragmentation was useful in supporting large file transfers over multiple ground stations with contacts too short to individually support the needed data transfers (Ivancic et al., 2010b). The flight software components were minimized in order to fit within limitations of the real system and not negatively complicate or impact pre-existing spacecraft operations. This was a highly minimalist implementation.

The profile of BP features in the flight code and DTN2 version used to support the UK-DMC experiments include: (1) "DTN" EIDs, (2) no custody transfer, (3) use of Saratoga (Wood et al., 2007) as a convergence-layer for space-finks and TCP overground links, (4) forwarding based on destination EIDs only, with manually configured static routes, (5) both proactive and reactive fragmentation, and (6) no specific security extensions present, but MD5-based checksums over the data were computed in order to detect corruption.

8.2.3   DTN/BP—N4C profile

The European N4C project (N4C, 2011) operated DTN networks in several different parts of Europe during trials running for longer terms (multiple months or years) in order to provide connectivity to communities in remote regions with little infrastructure. The systems deployed supported e-mail, web-caching, weather data, and animal tracking applications. The considerations for these applications differed substantially from the spacecraft networking scenarios demonstrated by JPL and NASA GRC.

The data mule approach chosen for relaying in N4C is inherently different from the scheduled connectivity between nodes that the space scenarios used, driving different routing configuration. The router platform design for the N4C DTN nodes took more consideration of power/energy constraints, for instance. It also had significant design effort in order to support "gatewaying" or providing adaptations of normal Internet (nonnative-DTN) applications (Farrell et al., 2009).

The profile of DTN features included in DTN2 demonstrated during the N4C program include: (1) "dtn" EIDs derived from hierarchical Internet Protocol (IP) addresses and domains, (2) custody transfer, (3) TCP and UDP convergence layer adapters, (4) static routing table configuration, forwarding during opportunistic contacts, as well as PRoPHET routing protocol in some cases (Lindgren et al., 2004), (5) both proactive and reactive fragmentation, (6) security in applications, but no use of DTN/BP-layer security.

8.2.4   DTN/BP—SPINDLE profile

For the most part, the ION and DTN2 deployments described previously use destination EIDs set by the bundle source, in order to specify the intended destinations of bundles. In these systems, the in-network storage is primarily utilized by intermediate bundle nodes only when a path to the destination EID is not immediately available. The SPINDLE system developed by BBN Technologies (Krishnan et al., 2007) significantly differs from this approach, and makes more advanced use of the storage capacity at intermediate nodes. In SPINDLE, the in-network storage is also applied for caching data that might be commonly requested, permitting later retrieval within pockets of connectivity, even when the original source is not actually reachable. This caching also reduces latency and network load even in a well-connected (non-challenged) network, and is very closely linked to some of the concepts that have recently become more popular under the information-centric networking (ICN) title.

Field experiments with the SPINDLE code were conducted at US military facilities in Hawaii and Virginia. These tests involved mobility patterns, data volumes, and connectivity disruptions that differed substantially from the N4C data mule scenarios, and drove some of the more advanced concepts for routing and forwarding that the implementation included.

The profile of DTN features included in SPINDLE and demonstrated as part of the defense advanced research projects agency (DARPA) DTN program include: (1) support for intentional names that are progressively resolved as bundles traverse through nodes within the network, (2) beyond just custody transfer, actually caching data within nodes for time beyond that needed to complete a single delivery, (3) TCP, UDP, and negative acknowledgment (NACK) oriented reliable multicast (NORM)-based convergence layer adapters, (4) flexible support for many routing systems and inclusion of novel routing algorithms such as prioritized epidemic (PREP) and anxiety-prone link state (APLS), (5) reactive fragmentation, (6) optional use of BSP blocks for security. BBN has also experimented with alternative mechanisms for expressing bundle lifetimes as relative values rather than absolute timestamps compatible with RFC 5050.

8.2.5   Summary of DTN profiles

The BP profiles used by different demonstrations described in this section are selected from among many others that are described in the literature. These are examples selected for comparison with specific BP alternatives proposed, and are not intended to be all-encompassing of the variety of BP implementations and configuration options.

The DTN architecture has proven to be adaptable to a wide range of networking requirements, including deep-space communications, Earth-science sensor networking, providing remote regions with access to Internet-like services, and military operations scenarios described in this section, along with several others not discussed here, such as vehicular DTN (VDTN) using the BP to support disrupted connectivity in vehicular networking (Rodrigues et al., 2011; Rogerio et al., 2012). There are substantial differences between the four BP configurations described in this section. Although all were based on the BP itself, the profiles of options and extensions utilized were distinct enough that they would not be capable of interoperating with one another. For assessment of the BP and DTN architecture against alternatives in the next section, we provide side-by-side comparisons of the alternatives against the most similar DTN BP profile described in this section (see Fig. 8.1).

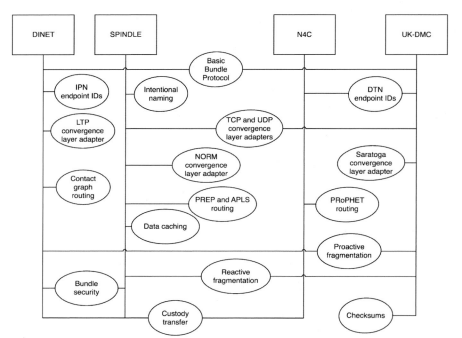

Fig. 8.1 DTNBP feature profiles.

8.3   Alternative approaches

The generalized "challenged networks" for which DTN has been intended, cover many areas of research that have also been pursued as their own specialties. For instance, Mobile Ad-Hoc Networking (MANET) pre-dates DTN as a popular research topic, but MANETs can be viewed as a special class of DTN in which delays may not necessarily be large within pockets of connectivity, but the connectivity is simply dynamic and the potential for temporary partitioning of the network exists. Disruption of links and reachability is the concern, rather than propagation delay. So, while DTN might be applicable to MANET scenarios, protocols developed for MANET are generally not at all suitable for meeting the goals of the DTN architecture.

This section considers several other areas where specialized solutions have been proposed, and assesses them within their niches, in comparison to DTN-based approaches using the BP in its closest demonstrated profile of features.

8.3.1   Haggle and opportunistic networking

Chaintreau et al. identify "opportunistic networking" as a subclass of both DTN and MANET (Chaintreau et al., 2006). Through analysis of several human-mobility datasets, they found that an approximate power law holds for the intercontact times of nodes in all datasets examined. Some existing forwarding algorithms perform poorly in this case. For instance, many algorithms previously described in the litera-ture are designed and evaluated for exponential distributions of intercontact times. This has often been based on assumptions of either node locations that are uniformly distributed within a bounded region, or random waypoint and Brownian motion models. Both of these assumptions have been shown to differ from sets of real traces.

The Haggle project introduced the concept of pocket switched networking (PSN) (Hui et al., 2005; Scott et al., 2006), in which mobile users with personal devices use (1) neighborhood connectivity to other local devices, (2) infrastructure connectivity to the global Internet, and (3) physical user mobility, in order to support data transfer between devices.

Comparing Haggle with DTN/BP as deployed by the N4C project, the challenges and metrics each attempted to address can be summarized as:

- Node Encounter Schedules: PSN uses statistical analysis of contact event traces in order to avoid needing a deterministic contact graph or assuming that a fixed forwarding structure will suffice. PSN must work with a highly dynamic and unpredictable contact schedule. N4C used opportunistic contacts to transfer data, but static routes through mules, some (such as helicopters) with predictable contact times.
- Network Connectivity: PSN assumes multiple possible connectivity technologies will exist per node, maximizing the potential for forwarding data within contacts. PSN originally focused on power-law distributions of intercontact times. N4C primarily used Wi-Fi for con-nectivity during opportunistic contacts with infrastructure and data mule nodes; however, the stack could have adapted to other technologies as well.
- Node Capacity: PSN is intended to accommodate a wide range of node capabilities and capacities between nodes. Since N4C primarily utilized Wi-Fi links, typical node capacities were in a relatively uniform range. Asymmetry in links is not specifically addressed in either Haggle/PSN or the N4C work.
- Storage Resources: Management of storage is a major concern for PSN, since users them-selves comprise much of the network-forwarding resources. Storage resources for N4C were provided by network infrastructure and mules rather than end-user platforms as in PSN, so were somewhat less of a concern.
- Energy Concerns: Energy usage is a major concern for PSN, as mobile devices still have highly limited energy storage and users would not tolerate the system if it depleted batteries while forwarding data for others. Energy was also a concern for N4C; however, solar recharging or mains power was available for much of the forwarding infrastructure, so the specific concerns differ slightly from PSN. N4C tests demonstrated that the system would be disabled by users if it seemed to impose excessive energy costs on users (e.g., hikers were very conscious of remaining battery life).
- Message Delivery Ratio: Delivery success ratios were studied in simulation for several dif-ferent forwarding algorithms (Boldrini et al., 2006) in Haggle. None achieved full 100% delivery, so user expectations need to be assessed with such a system. Delay tolerance of the user can greatly increase delivery ratio. For the N4C experiments, delivery ratios of 90% for particular deployments have been achieved (N4C, 2011) and the PRoPHET routing protocol was utilized.
- Message Delay: Delays can be reduced in PSN, due to ability to cache in the network. How-ever, due to node storage limitations, this may not be very easy to exploit to high benefit. N4C enabled applications that did not work due to prior lack of real-time connectivity. N4C did not make special attempts to reduce latency beyond providing services in the first place.

- Number of Replicas: Algorithms for Haggle forwarding take as much context into account as possible and reduce the replication significantly in comparison to epidemic algorithms (Hui et al., 2006). Similar algorithms could be used with DTN/BP, but were not part of the N4C project.

Haggle and PSN are focused entirely on human mobility. To the DTN architecture, this is just one type of supported scenario. Concrete Haggle protocol specifications have not been as well-defined and demonstrated as for DTN. Through Haggle, forwarding and routing strategies appropriate for PSN scenarios have been explored that could be adapted into DTN BP-based systems. With proper algorithms and configuration, the flexibility of the DTN architecture should enable BP to function comparably to Haggle in PSNs. The N4C experiments targeted mule-based infrastructure differently than the pure PSN, but in some aspects, such as PRoPHET routing, the N4C capabilities demonstrated are similar to what Haggle aims for.

For networks with contacts between humans providing a primary means of relaying, the current state of research is that there are no known "magic bullet" algorithms that work well across all metrics. The best algorithms depend on the characteristics of the social graph between nodes and other aspects of the relationships between nodes (Zhu et al., 2013). Thus, whatever the protocol in use (DTN/BP, Haggle protocols or others), flexibility in the routing system to adapt algorithms (not just policies) for different applications and graphs appears to be beneficial.

8.3.2   File-based space operations

File-based space operations were proposed as an advancement beyond the simple packet and frame-based systems that have been used by many space missions. The basic idea is to take advantage of standards that the space operations community developed for file transfer in the CCSDS File Delivery Protocol (CFDP) and advances in onboard storage, communications and file management systems (Haddow et al., 2012). The CFDP protocol can optionally be used as a relay layer to transfer files iteratively across the temporarily connected portions of an end-to-end path. File transfers need to be scheduled in conjunction with the space link availability. Relaying like this makes use of CFDP's "extended procedures" (CCSDS, 2007), which support store-and-forward operations via indication of waypoints that the file transfers are to pass through. Relaying proceeds based on the waypoints specified by the file source (i.e., as a form of strict source-routing).

This is obviously a much simpler and more limited architecture than that enabled by DTN, since the routing and forwarding mechanism is less flexible and does not clearly facilitate any future advancements in security, prioritization of traffic or scalability of the network. While file-based operations using CFDP has sometimes been described as an alternative to DTN (Flentge, 2011), other formulations include DTN as part of the underlying file transfer system, rather than just CFDP alone (Blake, 2012). This could allow file-based operations to overcome the limitations of CFDP extended procedures alone in terms of routing flexibility, pre-knowledge of file sizes and contact schedules, etc. A DTN capability to support for file-based operations was incorporated into the Space Internetworking Strategy Group (SISG) Solar System Internet concept (Interagency Operations Advisory Group, 2010).

Comparing pure file-based operations with DTN/BP as deployed in the DINET experiment makes this clearer.

- Node Encounter Schedules: For file-based operations node encounter schedules can be predicted in advance; file-sources can anticipate waypoints that will need to be used to reach the destination, but there is no included way to automate their knowledge of changing conditions or infrastructure. For DINET, node encounter schedules were predicted in advance and conveyed using the CGR protocol, adding a level of automation not present for pure file-based operations.
- Network Connectivity: In space operations, connectivity is generally point-to-point based on orbits of nodes, and scheduled periodically in order to meet the control and science needs of the mission. The assumptions of file-based operations versus the DINET DTN/BP profile do not differ in this regard.
- Node Capacity: In space communications, both high-rate and low-rate links may be present (from low bits per second to several gigabits per second). High degrees of asymmetry are generally present between the uplink/command channel and the downlink channel for telemetry and science data. Protocols that scale to high data rates and asymmetries are essential. DINET demonstrated high utilization over such links.
- Storage Resources: Storage resources on board spacecraft are growing, but are still limited in comparison to the volume of science data and measurements, often requiring periodic contact with the network in order to downlink bulk science data and purge onboard storage so that it can be used for upcoming data collection. Storage concerns for DINET were identical to what they would be for file-based operations.
- Energy Concerns: Energy limitations for a space mission can be factored into the contact schedule, and do not necessarily need to impact the other properties of the system in any way, since highly directional, point-to-point, scheduled links are generally used (i.e., not multinode wireless local area networks, beaconing links, etc.). Energy concerns for DINET were similar to what they would be for a mission using file-based operations.
- Message Delivery Ratio: Nearly 100% message delivery is required for space operations. Uplink/command data is not allowed to be lost in the network, and, if downlink mission data is lost, it can reduce the value of the mission. DINET demonstrated that this delivery ratio is possible, and used prioritization of data to express relative importance.
- Message Delay: Optimizing delay does not seem to have been a concern for the file-based operations proposals; however, use of relays is known to reduce delay versus waiting for end-to-end paths to be contemporaneously available. Delivery acceleration ratio was computed for DINET in order to validate that prioritization of bundles would result in speedup of delivery for high-priority data.
- Number of Replicas: No replication of data during the forwarding operations seems to be anticipated for file-based operations. Forwarding algorithms used in DINET did not result in replication of data.

DINET using the BP and the DTN architecture demonstrated features beyond what simple file-based operations are capable of (such as security and automation of the forwarding decisions based on dynamic data conveyed through CGR). The cost is in terms of software and system complexity in order to provide that automation. In general, file-based operations without use of DTN underneath them may be feasible for small-scale space operations scenarios, but the value of including DTN becomes evident for larger, more complex scenarios, where automation becomes necessary to enable manageability.

8.3.3   HTTP-based DTN architecture

Use of the Internet's HTTP as a hop-by-hop protocol to relay data, similar to the way that the DTN BP operates, was proposed by Wood and Holliday (2011). HTTP headers can be used to convey data similarly to the fields in the BP headers. Multi-purpose internet mail extensions (MIME) can be used to encapsulate and describe the application data. Making this approach work for challenged networks involves decoupling HTTP from TCP (Wood et al., 2009b), similarly to the way that BP is decoupled from its underlying convergence layer adapter protocols. HTTP has been supported as an application protocol proxied on top of DTN/BP by other projects (Ott and Kutscher, 2005), but this should not be confused with using HTTP as the relaying/ overlay layer itself, as proposed by HTTP-DTN.

HTTP-DTN adapts existing protocols (MIME, HTTP, etc.) in order to implement ideas from the DTN architecture, without the BP. The HTTP-DTN approach should not be confused with merely a semantic rewording of the DTN BP, however, as there are some novel aspects that BP work has not yet investigated. One example is that, in addition to pushing data between nodes, HTTP can request specific objects and support pull-based forwarding models, where the node caching some data en route might not even need to make routing determinations itself based on the destination EID. The concept of delegating that task to peers who know their transitive forwarding abilities and interests allows HTTP-DTN to more directly support models of operation closer to ICN proposals. ICN approaches utilizing the BP have, in fact, used information other than the bundle destination EID in order to make caching and forwarding decisions.

Some of the HTTP-DTN proponents were also members of the UK-DMC experiment team. Thus, the HTTP-DTN approach has features that would allow it to work well in scenarios similar to the UK-DMC demonstrations. The HTTP-DTN approach has been sketched and discussed, but not fully worked out into an operational system with available software in quite the same way as Haggle, for instance. It needs to be considered here, though, as an alternative to BP-based DTN, because it demonstrates that adaptation and extension of existing widely used Internet protocols may be possible to meet the challenges that BP designers saw as warranting a new architecture and new overlay protocol (the BP).

The challenges and metrics addressed by HTTP-DTN versus the UK-DMC profile of DTN/BP are:

- Node Encounter Schedules: As described, HTTP-DTN makes no assumption about encounter schedules, but does not attempt to describe forwarding decisions and routing data dissemination. If contacts are scheduled, the HTTP transfers to be completed could be scheduled as well. The UK-DMC network was stable except for the scheduled space links. The bundle transfers were scheduled and commanded in conjunction with the link availability, similarly to the way they would be with HTTP-DTN.
- Network Connectivity: HTTP-DTN does not depend on or make assumptions about node connectivity. Support for proactive fragmentation is trivial, and could be receiver-driven rather than just sender-driven, for instance, by using HTTP's ability for requesting ranges of an object. Reactive fragmentation is also possible, though it needs further definition to

be robustly supported. The UK-DMC experiment had a well-connected ground network, but would have also functioned over disruptions in the ground network. As evidence, one bundle relayed over the ground network experienced reactive fragmentation when there were issues with the virtual private network connectivity between ground station and destination nodes, causing the TCP connection between the convergence layer adapters to be closed, but when connectivity was regained the transfer completed successfully and the complete bundle was recreated.

- Node Capacity: HTTP-DTN should work with non-TCP transports below HTTP in order to accommodate wide ranges of capacity and asymmetry hop-by-hop. The UK-DMC experiment had high asymmetry and several Mbps data rate in the space link, with thinner links available on the ground. The benefit of a bundle agent providing rate-buffering between different stretches of capacity in the network was demonstrated.
- Storage Resources: No specific consideration of storage resources has been made yet for HTTP-DTN. Storage resources, other than on board the UK-DMC, were not an issue in the UK-DMC DTN demonstrations.
- Energy Concerns: No specific consideration of energy has been made yet for HTTP-DTN. The UK-DMC, like other spacecraft, disables subsystems during periods when they are not needed, and subsystems are scheduled according to availability of power and sunlight to the solar arrays to recharge batteries. Contacts in the UK-DMC experiments were scheduled in order to fit within the power budget of the spacecraft and its orbit.
- Message Delivery Ratio: All data were successfully delivered in the UK-DMC experiments. HTTP-DTN would have performed similarly in the test scenario.
- Message Delay: Using short contacts over multiple ground stations reduced delay compared with waiting for longer contacts, and can enable partial analysis of data to begin sooner. HTTP-DTN would have performed similarly in the test scenario.
- Number of Replicas: Data was not replicated by the forwarding algorithms used in UK-DMC experiments. HTTP-DTN would have performed similarly in the test scenario.

HTTP-DTN represents a very interesting approach to enhancing existing Internet standards in order to implement ideas from the DTN architecture. Conceptually, it may prove to be simpler to implement, while performing similarly against metrics in many scenarios, particularly those similar to the UK-DMC demonstrations with an Earth-observing satellite, but likely others as well. Other proposals have been drafted to modify or extend standard protocols; for instance, extending IP is possible and may provide some benefits, like compatibility with IPsec and network management systems (Neufield, 2010). To date, such proposals are not fleshed out completely enough to definitively say whether or not they could technically contend with DTN/BP in all aspects, and in flexibility across the full range of challenged networks for which BP has been demonstrated. These topics remain interesting for future work.

8.3.4 Information-centric networking

ICN has recently been defined to synthesize the concepts being explored in multiple different projects pursuing content-based caching and forwarding of data through a diversity of approaches. The common thread among them is that, rather than being sent from sources to destination addresses, data is published and disseminated based on the particular content and user interest in it.

The similarity is obvious to some DTN-based work that has utilized caching within the network, though the routing and forwarding of data in ICN systems uses more of a publish-subscribe model or expression of interest in order to drive data to destinations. Among the BP-based work discussed here, DTN systems using caching generally build heuristic improvements on flooding or epidemic forwarding, based on projected opportunistic contacts. Compared with ICN, SPINDLE is the closest DTN system, as it incorporates not just these types of mechanisms, but also the intentional naming capability, which resolves named data into specific destination EIDs as it moves across the network.

- Node Encounter Schedules: For ICN, nodes might have constant communications possibilities rather than a schedule of encounters with other nodes. SPINDLE is designed to operate even when encounter schedules are not known in advance.
- Network Connectivity: Much ICN work assumes a well-connected network. SPINDLE was shown to work over networks that were largely partitioned into disconnected groups with limited, temporary interconnections.
- Node Capacity: ICN generally involves at least some nodes with a large network capacity between them, though it may be heterogeneous. SPINDLE does not rely on particular capacities between nodes, and should handle at least limited asymmetry in link rates.
- Storage Resources: ICN generally involves at least some nodes with a large storage capacity. To get the benefit out of SPINDLE caching, a similar level of storage resources is necessary.
- Energy Concerns: ICN work does not frequently analyze energy concerns directly. SPINDLE may save radio lifetime overall, in comparison with other approaches, since it can shed load from longer-haul links and localize most traffic; however, it may imply that nodes manage the energy they expend on behalf of other nodes.
- Message Delivery Ratio: ICN techniques generally result in all content being eventually delivered. This may not be true of all ICN approaches, but it is the case at least when ICN is providing an optimization for normal server-based distribution. SPINDLE testing shows that delivery ratio could be very high, even with many disconnections, and was superior to typical MANET protocols (e.g., ad hoc on-demand distance vector (AODV)).
- Message Delay: ICN techniques minimize delay by utilizing local copies of content, rather than end-to-end transfers from the original source. SPINDLE exploits in-network storage in almost the exact same way as ICN techniques.
- Number of Replicas: ICN does not attempt to minimize replicas, in general, since replication among many caches is the key to performance in other metrics. SPINDLE, similarly to ICN, does not limit replication, though its means for fanning out data can be driven by modules that could potentially implement this as a policy goal.

In summary of the comparison between ICN and the SPINDLE instantiation of DTN, many of the same concepts were implemented in SPINDLE prior to the advent of current ICN work. SPINDLE is using the basis of BP, common to many other projects, and highly extensible, while ICN projects are often creating new protocols, not interoperable with or similar to existing systems. Whether or not some of the ICN protocols being proposed in the community are either inherently able to work in challenged network environments, or able to do so through some modification, remains an interesting field of research. To date, ICN work has not considered the challenged networking scenarios as much as normal terrestrial Internet ones, and the capability of some profile of the BP to serve as an ICN protocol given Internet-like connectivity is also not completely explored.

8.3.5 Overall assessment

It has been shown that the BP and DTN architecture are highly flexible and, indeed, applicable to a wide range of challenged network environments, as intended. The architecture accommodates this through the use of tailored convergence layer adapters specific to the network environment underneath the BP overlay. Diversity in the BP implementations themselves was also anticipated, and accommodated through aspects of the BP design, such as generic URIs for EIDs, extension blocks, not defining a specific forwarding algorithm, etc.

Other alternative proposals generally focus on singular subtopics within the scope of challenged networks, but are not applicable to the full range. For any given type of challenged network scenario, it may be possible to find alternative approaches that are more efficient or higher-performing in some aspects than the BP for that particular scenario. General solutions, like BP, are not always optimal, or intended to be optimal.

The benefits to the DTN/BP approach in a given scenario are (1) potential reuse of existing, documented, tested and debugged protocols and code robust for the core BP functionality and (2) potential reuse of existing routing, security, prioritization, and other features developed either generically or for other similar scenarios. The logic for investing in BP is that these should reduce costs, development schedule and operational issues in deployments versus more customized solutions. Achieving these benefits can only happen if the core protocols and software are, indeed, reusable across types of challenged networks, and if a particular environment shares similarity with others that have already had investment made in tailoring the BP to them. Whether or not these conditions hold in practice remains to be seen, and will be a matter of future study as individual organizations consider deploying infrastructure or applications to serve challenged networks.

It does not seem likely that there will be a BP-based "Internet" connecting many challenged networks that shares the same type of Metcalfe's law benefits as the Internet or whose adoption is driven by a "killer app", as the Internet was (Wood et al., 2009a), but, if this is to happen, it will require commonalizing on some specific profile of BP features and extensions. For DTN research in general, it seems that the community would benefit from clearer and more systematic descriptions of BP configurations used in specific experiments, or, ideally, specific baseline feature profiles, in order to increase the repeatability and understanding of results. For instance, many routing papers in the literature do not clearly describe fragmentation or custody transfer applicability, which creates confusion in comparisons.

In conclusion, while it is not without flaws or unfinished aspects (Wood and Eddy, 2009), after several years of research, DTN/BP remains the most widely applicable approach to challenged networking. Promising work on alternatives continues, especially work focused on individual applications or environments. Work on general alternatives will benefit from some of the lessons learned in the DTN community about flexibility and extensibility versus possible incompatibilities in deployment.

8.4 Future trends

The IRTF DTNRG that created the BP specifications and DTN architecture is still active and meets periodically. Many of the projects that originally spurred the group's activity are no longer funded (e.g., DARPA DTN, N4C, etc.), and the pace of work has slowed. Research community momentum has shifted recently towards ICN and outside DTN/BP, though many of the topics and issues being explored there are familiar. DTN/BP does not seem to have solid traction as a basis for ICN within the community at the moment, yet there is no clear reason why not, at this time.

Meanwhile, the international space exploration community is continuing with programs of demonstration and enabling slow infusion of DTN/BP into its operations. Deployments on the International Space Station have been active, communicating with corresponding ground systems. Projects in the US, Europe and Japan have been exploring DTN for space operations and demonstrating its capabilities. Agencies have investigated deployment of DTN relays as operational services to space missions, though none are known to have proceeded with funding or implementation plans for this yet.

8.5 Sources of further information and advice

DTN/Bundle Protocol Implementations:

- Interplanetary Overlay Network (ION) Source Code: Originally developed by NASA JPL, the ION code has since been distributed publically via a number of websites, including Open Channel Foundation, Ohio University, and recently Sourceforge. The code, which is primarily portable C, compiles for a number of different operating systems and platforms. The Sourceforge website for ION-DTN is: http://ion-dtn.sourceforge.net
- DTN2 Source Code: The DTN2 code, widely utilized by DTN research projects (including the NASA GRC experiments with the UK-DMC, the N4C project, and many others) is mostly written in C++, and compiles for a wide variety of operating systems and platforms. DTN2 source code is available via Sourceforge's website at: http://sourceforge.net/projects/dtn
- Several other implementations make source code and/or binaries available on the Internet; most of these can be located through links off the DTNRG's website: http://www.dtnrg.org

Other source code and websites:

- Haggle: Source code that builds for various operating systems, including some for popular mobile devices available through Google Code http://haggle.googlecode.com linked to from the Haggle website at http://haggleproject.org

References

Ali, S., Qadir, J., Baig, A., 2010. Routing protocols in delay tolerant networks—a survey. In: Proceedings of the 6th International Conference on Emerging Technologies (ICET) 2010, Islamabad, 18–19 October 2010.
Blake, R., 2012. Deployment of file based spacecraft communication protocols. In: Proceedings of Space Ops 2012, Stockholm, Sweden, 11–15 June 2012.

Boldrini, C., Le Boudec, J., Chaintreau, A., Conti, M., Crovella, M., et al., 2006. Final Specification of Forwarding Paradigms in Haggle. Haggle Deliverable 2.2, revision v1.0.

Cerf, V., Burleigh, S., Hooke, A., Torgerson, L., Durst, R., et al., 2007. Delay-Tolerant Networking Architecture. RFC 4838.

Chaintreau, A., Hui, P., Crowcroft, J., Diot, C., Gass, R., et al., 2006. Impact of human mobility on the design of opportunistic forwarding algorithms. In: Proceedings of IEEE Infocom 2006, Barcelona, Spain.

Consultative Committee for Space Data Systems, 2007. CCSDS File Delivery Protocol (CFDP) Part 1: Introduction and Overview. CCSDS 720.1-G-3.

Consultative Committee for Space Data Systems, 2012. CCSDS Bundle Protocol Specification. CCSDS 734.2-R-1.

Eddy, W., Wood, L., Ivancic, W., 2009. Reliability-only Ciphersuites for the Bundle Protocol. Internet-Draft (work in progress), draft-irtf-dtnrg-bundle-checksum-09 https://trac.tools.ietf.org/html/draft-irtf-dtnrg-bundle-checksum-09. (Accessed 10 July 2014).

Fall, K., 2003. A delay-tolerant network architecture for challenged internets, proceedings of ACM SIGCOMM 2003. Comput. Commun. Rev. 33 (4), 27–36.

Farrell, S., Weber, S., McMahon, A., Meehan, E., Hartnett, K., 2009. AnN4C DTN router node design. In: Proceedings of Extremecon 2009, Laponia, Sweden, 8–14 August 2009.

Flentge, F., 2011. Study on CFDP and DTN architectures for ESA space missions. In: Proceedings of the Third International Conference on Advances in Satellite and Space Communications (SPACOMM), Budapest, Hungary, 17–22 April 2011.

Haddow, C., Pecchioli, M., Montagnon, E., Flentge, F., 2012. File based operations—the way ahead? In: Proceedings of Space Ops 2012, Stockholm, Sweden, 11–15 June 2012.

Hui, P., Chaintreau, A., Gass, R., Scott, J., Crowcroft, J., et al., 2005. Pocket switched networking: challenges, feasibility, and implementation issues. In: Proceedings of WAC 2005, Vouliagmeni-Athens, Greece.

Hui, P., Leguay, J., Crowcroft, J., Scott, J., Friedman, T., et al., 2006. Osmosis in pocket switched networks. In: Proceedings of the First International Conference on Communications and Networking in China, Beijing, China.

Interagency Operations Advisory Group, Space Internetworking Strategy Group, 2010. Solar System Internetwork (SSI) Issue Investigation and Resolution. https://www.ioag.org/Public%20Documents/SSI%20Ops%20Concept%20Issue%20Resolution%20-%20final%20version.pdf. (Accessed 10 July 2014).

Ivancic, W., Eddy, W., Stewart, D., Wood, L., Northam, J., et al., 2010a. Experience with delay-tolerant networking from orbit. Int. J. Satell. Commun. Netw. 28 (5–6), 335–351.

Ivancic, W., Paulsen, P., Stewart, D., Taylor, J., Lynch, S., et al., 2010b. Large file transfers from space using multiple ground terminals and delay-tolerant networking. In: Proceedings of the IEEE Global Communications Conference (Globecom 2010), Miami, Florida, USA, 6–10 December 2010.

Jones, R., 2009. Disruption Tolerant Network Technology Flight Validation Report. JPL Publication 9-2.

Kapadia, S., Krishnamachari, B., Zhang, L., 2011. Data delivery in delay tolerant networks: a survey. In: Wang, X. (Ed.), Mobile Ad-Hoc Networks: Protocol Design. InTech.

Khabbaz, M., Assi, C., Fawaz, W., 2012. Disruption-tolerant networking: a comprehensive survey on recent developments and persisting challenges. IEEE Commun. Surv. Tutorials 14 (2).

Krishnan, R., Basu, P., Mikkelson, J., Small, C., Ramanathan, R., et al., 2007. The SPINDLE disruption-tolerant networking system. In: Proceedings of MILCOM 2007, Orlando, Florida, USA, 29–31 October 2007.

Lindgren, A., Doria, A., Scheien, O., 2004. Probabilistic routing in intermittently connected networks. In: Proceedings of the First International Workshop on Service Assurance with Partial and Intermittent Resources (SAPIR) 2004, Fortaieza, Brazil.

Magaia, N., Pereira, P., Casaca, A., Rodrigues, J., Dias, J., et al., 2011. Bundles fragmentation in vehicular delay-tolerant networks. In: Proceedings of the 7th Euro-NF Conference on Next Generation Networks (NGI'2011), Kaiserslautern, Germany.

N4C: Networking for Communications Challenged Communities, 2011. Test Results— Documentation of Test Results From Tests. N4C Deliverable D8.4, Version 2.0.

Neufield, M., 2010. DIP: Disruption-Tolerance for IP. arXiv:1003.3996 [cs.NI] http://arxiv.org/abs/1003.3996. (Accessed 10 July 2014).

Ott, J., Kutscher, D., 2005. Applying DTN to Mobile Internet Access: An Experiment with HTTP. Technical Report TR-TZI-050701, July 2005.

Rodrigues, J., Dias, J., Isento, J., Silva, B., Soares, V., et al., 2011. The vehicular delay-tolerant networks (VDTN) Euro-NF joint research project. In: Proceedings of the 7th EURO-NGI Conference on Next Generation Internet (NGI) 2011, Kaiserslautern, Germany, 27–29 June 2011.

Rogerio, P., Casaca, A., Rodrigues, J., Soares, V., Triay, J., et al., 2012. From delay-tolerant networks to vehicular delay-tolerant networks. IEEE Commun. Surv. Tutorials 14 (4), 1166–1182.

Scott, K., Burleigh, S., 2008. Bundle Protocol Specification. RFC 5050.

Scott, J., Hui, P., Crowcroft, J., Diot, C., 2006. Haggle: a networking architecture designed around mobile users. In: Proceedings of IFIP WONS 2006, Les Menuires, France.

Symington, S., Durst, R., Scott, K., 2009. Delay-Tolerant Networking Custodial Multicast Extensions. Internet-Draft (work in progress), draft-symington-dtnrg-bundle-multicast-custodial-06 https://tools.ietf.org/html/draft-symington-dtnrg-bundle-multicast-custodial-06. (Accessed 10 July 2014).

Symington, S., Farrell, S., Weiss, H., Lovell, P., 2011. Bundle Security Protocol Specification. RFC 6257.

Voyiatzis, A., 2012. A survey of delay- and disruption-tolerant networking applications. J. Internet Eng. 5 (1), 331–344.

Walker, B., Tsuru, M., Caro, A., Keranen, A., Ott, J., et al., 2010. Panel: the state of DTN evaluation. In: Proceedings of the 5th ACM Workshop on Challenged Networks (CHANTS), pp. 29–30.

Wood, L., Eddy, W., 2009. A bundle of problems. In: Proceedings ofthe 2009 IEEE Aerospace Conference, Big Sky, Montana.

Wood, L., Holliday, P., 2011. Using HTTP for Delivery in Delay/Disruption-Tolerant Networks. Internet-Draft (work in progress), draft-wood-dtnrg-http-dtn-delivery-07 https://tools.ietf.org/html/draft-wood-dtnrg-http-dtn-delivery-07. (Accessed 10 July 2014).

Wood, L., Eddy, W., Ivancic, W., McKim, J., Jackson, C., 2007. Saratoga: a delay-tolerant networking convergence layer with efficient link utilization. In: Proceedings of the IEEE International Workshop on Satellite and Space Communications 2007 (IWSSC'07), Salzburg, Austria, 13–14 September 2007, pp. 168–172.

Wood, L., Holliday, P., Floreani, D., Eddy, W., 2009a. Sharing the dream: the consensual hallucination offered by the bundle protocol. In: Proceedings of the Workshop on the Emergence of Delay-/Disruption-Tolerant Networks (e-DTN 2009), St Petersburg, Russia.

Wood, L., Holliday, P., Floreani, D., Psaras, I., 2009b. Moving data in DTNs with HTTP and MIME: making use of HTTP for delay- and disruption-tolerant networks with convergence layers. In: Proceedings of the Workshop on the Emergence of Delay −/Disruption-Tolerant Networks (e-DTN 2009), St Petersburg, Russia.

Zhu, Y., Xu, B., Shi, X., Wang, Y., 2013. A survey of social-based routing in delay tolerant networks: positive and negative social effects. IEEE Commun. Surv. Tutorials 15 (1), 387–401.

Opportunistic routing in mobile ad hoc delay-tolerant networks (DTNs)

9

Zhensheng Zhang[a], Shengbo Chen[b], and Ju Ren[c]
[a]University of California, Los Angeles, CA, United States, [b]Henan University, Kaifeng, China, [c]Central South University, Changsha, China

9.1 Introduction

In the last decade, much research effort has been devoted to mobile ad hoc networks (MANETs), and many routing protocols, such as ad hoc on-demand distance vector (AODV) (Clausen and Jacquet, 2003) and optimized link state routing protocol (OLSR) (Perkins et al., 2003), have been proposed and implemented. However, these routing protocols assume that the network is connected and an end-to-end path exists from source to destination. In reality, this is not the case; the network is often intermittently connected or partitioned. Such networks can be sparse mesh, in which case intermittent connectivity is due to lack of physical connections, or dense mesh, in which case intermittent connectivity may be due to high interference or shadowing.

An extensive review of the routing protocols in delay-tolerant networks (DTNs) available before 2006 and 2007, respectively, is given in Zhang (2006) and Zhang and Zhang (2007). Since then, many kinds of opportunistic routing in DTN have been proposed. The key ideas of opportunistic routing are to fully utilize the broadcast nature of the wireless media and determine the path to the destination on the fly (not determined a priori as with most conventional MANET routing protocols) by selecting a set of candidate nodes to forward. In a wireless ad hoc network, when even a packet is sent (unicast) to a next-hop node, all the neighboring nodes of the sender may be able to hear the packet. It is possible that some of the neighbors may have received the packet correctly while the intended next-hop node did not. Based on this observation, a new routing paradigm, opportunistic routing (OR), has recently emerged. Instead of selecting one node to forward a packet to, the OR-enabled node selects a set of candidate nodes to forward a packet to. OR takes advantage of the spatial diversity and broadcast nature of wireless communications and is an efficient mechanism to combat the lossy, time-varying links. To see how opportunistic routing improves routing reliability, consider the example in Fig. 9.1, where the source, S, wants to transfer a packet to the destination, D. The loss rate from the source to any of the intermediate nodes is 50%. MANET conventional routing protocols route packets along a fixed, "best" path from the source to the destination. However, since all paths in the figure are 50% lossy, the traditional approach only provides 50% reliability. In contrast, with OR the source broadcasts its packets. Any node that hears a

Advances in Delay-tolerant Networks (DTNs). https://doi.org/10.1016/B978-0-08-102793-6.00009-6

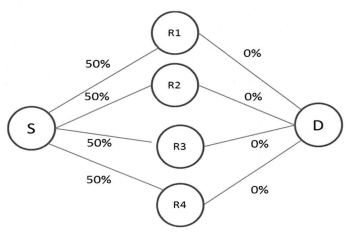

Fig. 9.1 Illustration of the benefit of opportunistic routing: links are 50% and 0% lossy, respectively.

packet can forward the information. Given four nearby nodes, the probability that none of them hears the packet is $0.5 \times 0.5 \times 0.5 \times 0.5 = 0.06$, that is, the reliability using OR can be as high as 94%.

OR looks at mobility, disconnections, partitions, etc., as features of the networks rather than exceptions. In this chapter, we discuss some of the key challenges in designing an efficient OR protocol, view some of the representative OR protocols, and point out a few open issues associated with OR. Recent developments in DTN using machine learning and artificial intelligence (AI), as well as cloud/fog computing will be discussed. Owing to space limitations, the overview of OR is not all-inclusive.

9.2 Challenges

There are several challenges in designing and implementing OR. The *first issue* is the selection of forwarding candidates. Selecting the size and the specific nodes of the set of forwarding nodes is critical to the efficiency of the OR. Selecting the most appropriate nodes to forward will increase the system throughput and reduce the latency, and choosing the right size of the forwarding nodes will increase the network utilization and decrease the network overhead. The selection of the set forwarding node(s) requires coordination among the candidate relays, which may require explicit exchange of state information among the neighboring nodes. The strategy is to keep the coordination overhead limited, as far as the number of participants is kept small. The *second issue* is the prioritization of the candidates. In general, it is better to forward the packet along the "shortest" path. The lower-priority forwarding candidates are essentially the backup to the node that is on the "shortest" path. However, due to the opportunistic nature, the "distance" from a certain node to its multihop-away destination will no longer be the same as that obtained by traditional shortest-path routing.

The path cost also depends on the spatial diversity opportunities along the path. How to quantify and incorporate the spatial diversity opportunities in OR has not been well understood. The *third issue* is when relays should forward packets. In DTNs, forwarding the received packets right away may not be the best solution, due to imperfect coordination among candidate relays and packet losses, which may cause multiple duplicate transmissions of the same packet. To avoid this unnecessary waste of network resources, it is better to establish a scheduling among the candidate relays, and to set different forwarding timers at the selected forwarders. The *fourth issue* is how to combine packets when network coding is used in OR.

In the following sections, we overview several key categories of OR in mobile ad hoc networks, such as scheduled-based, network coding-based, hybrid and social structure-based, machine learning, or AI or fog computing-based DTN.

9.3 Overview of multiple existing opportunistic routing protocols in mobile ad hoc networks

As mentioned, many routing protocols in DTNs have been proposed in the last decade, and a taxonomy of the routing protocols proposed before year 2006 is given in Zhang (2006), classifying protocols according to the information they require from the network as well as relevant routing strategies (e.g., pure forwarding, estimation of forwarding probability) into two main categories, namely, deterministic and stochastic routing. In the former case, node movement and future connections are known beforehand (i.e., nodes are completely aware about topology). In the case of stochastic approaches, the behavior of nodes and network is random and unknown; therefore, routing is opportunistic. The first and simplest OR protocol is epidemic routing, in which each intermediate node simply broadcasts its received packets to all its neighbors. It is simple, but less efficient, as many packets are unnecessarily transmitted many times in the network. More complex OR uses the history of encounters, node mobility patterns and message coding. In the following, we review some recent OR.

9.3.1 Extremely opportunistic routing protocol (ExOR)

One of the first schedule-based opportunistic routing protocols is presented in the paper by Biswas and Morris, who have proposed the extremely opportunistic routing protocol (ExOR) (Biswas and Morris, 2005). In ExOR the packets to be transmitted are grouped into batches according to their destination node. For each packet of the same batch, the source node selects a subset of optimal candidate forwarders, which are prioritized by closeness to the destination. The closeness property of a node is evaluated using the ETX metric, where ETX is defined as the average number of retransmissions needed to reach the destination. It is noted that ExOR only guarantees to transmit 90% of a batch using opportunistic forwarding, while the remaining 10% of packets are sent with legacy or conventional unicast routing. Thus, the implicit assumption underlying ExOR design is that a link-state routing protocol is also running in parallel to the OR to efficiently collect links' delivery probabilities.

The list of selected forwarders, ordered by node priority, is inserted into the header of each packet broadcast by the source. In this way, each node receiving a packet knows whether or not it has to participate in the forwarding process, and its position in the forwarding schedule. Owing to link loss, each candidate forwarder will successfully decode only portions of the packet batch it has received. In order to distribute information on which portion of the batch each forwarder has received and rebroadcast, each packet also contains a batch map. The batch map indicates the highest-priority node known to have received a copy of that packet and is used to update the local batch maps stored in the receiving nodes, which list the IDs of the node closest to the destination. A forwarder is allowed to broadcast only if the received packets have not been forwarded by any other higher-priority node. To avoid simultaneous transmissions by different nodes, whenever a forwarder receives a packet it sets a forwarding timer. This timer is an estimate of the time that higher-priority nodes need to transmit the remaining packets in the batch, and only when the timer elapses can the node rebroadcast the packets. The batch maps contained in the headers of transmitted packets play a crucial role in ExOR for the following reasons: (1) they are used to disseminate reception information from higher-priority nodes to lower priority nodes; (2) batch maps are used also for the acknowledgment process. That is, when the destination receives a new packet it sends its batch map back to the source, so that the source knows when the destination has received most of the current batch and when it can pass to transmit a new batch.

In ExOR, the size of a packet batch is needed before forwarding. To avoid using the packet batch, the simple opportunistic adaptive routing protocol (SOAR) is proposed in Rozner et al. (2006). Similarly to ExOR, SOAR employs a scheduling scheme relying on priority-based forwarding timers to avoid simultaneous transmissions by different nodes, also using ETX as the metric to estimate the node's closeness to the destination and its priority. Since SOAR does not use packet batches and operates on individual packets, the computation of forwarding timers is much simpler in SOAR than in ExOR, because SOAR can use constant timers proportional to the node priority, while ExOR employs variable timers whose duration depends on the number of packets. The method used by SOAR to establish the schedule among the candidate forwarders is different from ExOR. SOAR employs overhearing to coordinate forwarders' transmissions. Whenever a node overhears a transmission from higher-priority nodes, it will cancel its forwarding timer and remove that packet from its queue, thus avoiding duplicate transmissions. To ensure that the candidate forwarders are close enough to overhear each other with a high probability, SOAR uses only network paths in close proximity to the shortest route between source and destination. SOAR uses hop-by-hop retransmissions, which are different from ExOR and are driven by network-layer acknowledgments (ACKs) generated by the highest priority forwarder that received the packet. SOAR uses a combination of various ACK mechanisms, including selective ACK, piggyback ACK, and ACK compression to reduce ACK overhead. SOAR uses a classical sliding-window protocol to control the maximum number of outstanding data packets, suggesting a small window (only three packets) to limit the transmission delays.

9.3.2 Resilient opportunistic mesh routing (ROMER)

To avoid the complexity of strict scheduling among candidate relays, in Yuan et al. (2005) the resilient opportunistic mesh routing (ROMER) protocol is presented, which introduces a credit-based forwarding scheme. In ROMER, when a packet is generated by the source node, it also receives an amount of credits that it can spend during the forwarding process, and when a packet is forwarded its credit is reduced. The assigned credits are set as the sum of the minimum cost from the source to the destination (i.e., the shortest path cost), plus extra credits necessary to expand the path while being forwarded. According to the remaining credits of the packet and the cost of the shortest path from itself to the destination, whenever a node receives a packet, it decides whether it is an appropriate forwarder. The credit associated with each data packet is decremented at each forwarding step, which basically means that more credits are consumed as the packet moves away from the shortest path to the destination. The proposed approach allows a forwarding mesh to be created on-the-fly, centered around the minimum-cost path from the source node to the destination. With this approach, a key factor which affects the performance is how to distribute packet credits along multiple candidate forwarders. One strategy is to assign more credits to nodes closer to the source than to the destination. This has the effect of permitting a faster initial expansion of the forwarding mesh, and remaining closer to the shortest path when approaching the destination. Another strategy is to take into account interference or load distributions. To select the optimal forwarders, and to reduce the number of duplicate transmissions, ROMER employs a probabilistic strategy: the forwarding probability is set proportional to the link's current transmission rate and to the desired level of packet redundancy, so as to assign a higher forwarding probability to intermediate nodes that use higher transmission rates. This probability-based forwarding scheme exploits the best-rate links that are dynamically identified by the rate adaptation algorithms.

9.3.3 Multipath code casting

In contrast to prior work in OR, which required strong coordination across nodes to prevent information repetition, multipath code casting (MCC) (Gkantsidis et al., 2007) is based on network coding and does not require node coordination.

In recent years, network coding has attracted much research to improve the network capacity. In terms of the routing schemes in DTNs, network coding is an add-on to OR. It can be classified as a different group in the taxonomy, due to the qualitative conceptual changes that exploit storage of DTN nodes. Essentially, in OR based on network coding, each forwarding node combines the content of data packets and transmits in a broadcast manner the newly coded packet over the point-to-multipoint wireless medium. Every coded data packet received at an intended destination contains information about different original packets. Different received data packets contain information on some original packets, thus providing, in general, useful incremental information to the receiver.

MCC is designed as an extension on the forwarding path between the datalink layer and the network layer. The objectives of the MCC design are to (a) guarantee packet delivery from source to destination with low overhead, (b) efficiently use the system's resources to achieve maximum throughput, and (c) ensure fairness among competing flows. These goals lead to the following system components: multipath forwarding, packet coding and decoding, error control, and rate control. MCC proposes to integrate coding with multipath routing, which relies on a legacy routing protocol to find a set of multiple, not necessarily disjointed, paths between a source and a destination node. Forwarding nodes broadcast encoded packets generated using either native packets received by the source or other encoded packets received by neighbor nodes. Intermediate decoding is not allowed, and only the destination node collects encoded packets and reconstructs the original packets when it has received a sufficient number of linearly independent coded packets. In order to provide reliability, two error control mechanisms are defined: (1) a hop-by-hop local recovery is performed by each sending node, which overhears its neighbors' transmissions and retransmits some missed packets if necessary; (2) on a timeout, the destination sends a unicast request to the source for additional coded packets, which can also be intercepted and managed by any intermediate node holding the missed packets.

One critical part of the MCC is how to decide which next-hop a packet must be sent to, and how many encoded packets should be generated along each path. In MCC, a credit-based algorithm is proposed. Like SOAR, the source associates with each packet generation a given amount of credits, which represents the total number of packets (i.e., including coded packets) that should be used to transfer that block of native packets to the destination. The number of packets the source is allowed to send per time unit is specified as a function of the generated credits. Whenever a packet is successfully transmitted to a next-hop, a credit is also transferred to that node. Based on the credits associated with each node, forwarders are aware of the number of packets that are waiting to be sent over each link. To determine the best next hop to which a node should forward the packet it has received, the routing protocol applies a back-pressure algorithm to the total credits accumulated by each node. It is noted that one drawback of this approach is that the optimal scheduling is hard to implement, and the required node state grows exponentially with the number of neighbors.

9.3.4 Coding opportunistically (COPE)

In MCC, packets belonging to the same flow (per-flow or intra-flow) are encoded at intermediate nodes. To extend the concept to packets belonging to other flows, an inter-flow network coding in ad hoc networks, the coding opportunistically (COPE) protocol, is proposed in Katti et al. (2008). COPE relies on a legacy routing protocol to select a minimum-cost path (according to the ETX metric) between nodes and allows intermediate forwarders to mix packets from multiple unicast flows. Each node also implements opportunistic listening to overhear packets that are not intended for it but that can be used for efficient coding. The overheard packets are stored in a local buffer for a limited time period. Whenever the media access control (MAC) protocol grants a node the permission to transmit, this forward node selects from its local buffers the packets to code together in such a way that all next-hops of encoded packets will be able to reconstruct their corresponding native packets. More precisely, each node combines, using the XOR operator, n distinct packets headed for n different next-hop relays only if it is true that every intended next-hop already has all the $n-1$ packets required to decode the native packets encoded together. COPE has been engineered by way of sniffing the wireless channel and coding multiple packets so that a single transmission can benefit multiple receivers. One critical aspect of the COPE design is to ensure that each node can learn the state of neighbors' buffers. One approach is for each node to broadcast reception reports listing the packets it has stored. This is similar to the batch maps used in ExOR. One problem with the reception reports is that these messages may be lost, or arrive too late for coding purposes. For these reasons, each node may anticipate whether a particular packet has been received by a certain neighbor based on the delivery probability. In COPE, the packet-coding algorithm is based on the principle of never delaying packets whenever the wireless medium is available. Thus, the node transmits a combination of packets if a coding opportunity exists, giving preference to packets of the same length; otherwise, it simply forwards the native packet, if any, at the head of its transmission queue. COPE uses pseudo-broadcast transmissions instead of conventional broadcast. That is, the destination MAC address of the encoded packet is set to one of the intended next hops, while an additional COPE-header specifies all the next hops of the native packets mixed together. By setting radio interfaces in promiscuous mode, COPE enables each node to overhear multiple encoded packets, while, at the same time, unicast transmissions ensure a higher level of reliability. The advantage of wireless broadcasting is not fully achieved by COPE alone. COPE relies on traditional routing to find paths between sources and destinations. Since the routes are fixed and only one path is used at a time, the availability of coding-enabling topologies is limited, and thus the coding potential in the network is not fully exploited.

9.3.5 Hybrid opportunistic routing protocols

As discussed previously, EXOR, SOAR, and ROMER focus on opportunistic forwarding without coding intermediate packets while MCC and COPE focus on networking coding without utilizing the opportunistic feature of the DTNs. In this section, we will review two hybrid schemes, MAC-independent opportunistic routing and encoding (MORE) and optimized overlay-based opportunistic routing (O3), both protocols that integrate OR with network coding.

9.3.6 MORE

The proposed MORE protocol (Chachulski et al., 2007) is a practical system to combine intra-flow random linear coding with EXOR-style OR. As in MCC, packets are grouped into blocks, which are now called batches, and coding is restricted to linear combinations of packets of the same batch. Similarly to EXOR, OR is performed by letting the sender specify a prioritized list of candidate forwarders. Some OR protocol, however, requires tight node coordination. Different nodes in a network must have knowledge of which packets other nodes have received. Furthermore, the nodes have to agree on which nodes should transmit which packets. Such coordination becomes fragile in dense or large networks. MORE avoids node-coordination and is rooted in the theory of network coding. Routers code packets going to the same destination and forward the coded versions. The destination decodes and recovers the original packets. In contrast to ExOR, there is no structured transmission schedule among the forwarders; the coding permits random transmissions regulated through the 802.11 MAC protocol. The source node divides the file to be transmitted into batches of native packets, creates random linear combinations and broadcasts the resulting packets after adding a MORE header containing the forwarder list. A packet is said to be innovative if it is linearly independent of the other packets stored in the node's local buffer. Each receiving node discards a packet if it is not innovative or if the receiving node does not appear in the associated forwarder list. Otherwise, it linearly combines the received coded packets and rebroadcasts the newly encoded packets. Note that a linear combination of coded packets is also a linear combination of the corresponding native packets. As soon as the destination is able to decode the whole batch, it sends an ACK to the source using shortest-path routing, causing the sender to stop forwarding packets from that batch and start processing the next batch. The intermediate nodes stop coding/sending packets from a certain batch as soon as they intercept the ACK for that batch sent by the destination, or they receive a packet belonging to a new batch. This strategy leads to a faster synchronization among nodes without requiring complex coordination procedures.

Even though network coding reduces the number of required transmissions for successfully delivering a packet, the opportunistic paradigm allows any potential forwarder to send many coded packets, resulting in potential redundant transmissions. The trade-off is between transmitting a sufficient number of coded packets to guarantee that the destination has enough innovative packets to reconstruct the native packets, and avoiding injecting into the network unnecessary packets that may cause congestion. To address this issue, MORE uses a heuristic algorithm to estimate the maximum number of transmissions that each node can perform after receiving a packet from an upstream node, which is a node farther from the destination than itself. This limit is computed by each node considering the loss probability in sending a packet to its neighbors, and the probability that the packet to be transmitted has not yet been overheard by downstream nodes, which are nodes closer to the destination than itself. The MORE protocol has been evaluated in a 25-node experimental testbed. Experimental results show that MORE provides an average throughput increase of 60% and a maximum of 10-fold, demonstrating that the theoretical gains promised by network coding are realizable in practice.

9.3.7   O3: Optimized overlay-based opportunistic routing

When integrating OR and network coding, a key challenge to achieve the goal of optimizing end-to-end performance is a strong tension between OR and inter-flow network coding: to achieve high reliability, OR uses intra-flow coding to spread information across multiple nodes; this reduces the information reaching an individual node, which in turn reduces inter-flow coding opportunity. Therefore it is challenging to simultaneously leverage opportunistic forwarding to combat wireless losses and exploit inter-flow coding to compress traffic. To address this challenge, O3 (Han et al., 2011) decouples OR and inter-flow network coding by proposing a framework (making a wireless network consist of an overlay and underlay) where an overlay network performs overlay routing and inter-flow coding without worrying about packet losses, while an underlay network uses optimized OR and rate limiting to provide efficient and reliable overlay links for the overlay network to take advantage of. Each traffic demand is routed over one or more overlay paths in the overlay network. Nodes on the overlay path perform overlay forwarding. They may also use inter-flow network coding to reduce the amount of overlay traffic generated and use inter-flow network decoding to extract the original content. The underlay network provides efficient and reliable overlay links by using opportunistic routing to spread information across multiple forwarders and letting them cooperatively forward the traffic. To prevent fine-grained coordination, each forwarder independently generates random linear combinations of traffic from the same flow at an appropriate rate so that the destination can extract the original data after receiving enough linearly independent packets. An optimization algorithm to jointly optimize opportunistic routes, rate limits, inter-flow and intra-flow coding was developed. Using Qualnet simulation, the authors studied the individual and aggregate benefits of OR, inter-flow coding and rate limits. Simulation results show that (i) rate limiting significantly improves the performance of all routing protocols, (ii) OR is beneficial under high loss rates, whereas inter-flow coding is more effective under low loss rates, and (iii) O3 significantly out-performs state-of-the-art routing protocols by simultaneously leveraging optimized OR, inter-flow coding, and rate limits.

CORE (Yan et al., 2010) is also a coding-aware routing protocol, but the priority of the candidates in the forwarding set is dynamically established according to the coding opportunities. A node is said to have more coding opportunities than other nodes when it holds more packets to be coded. In order to be a candidate, the node must be a neighboring node of the source and it must be (geographically) closer to the destination than the source.

9.3.8 Opportunistic routing in multi-radio multichannel multihop wireless networks

Previous work indicated that OR achieves much higher throughput than legacy routing in multihop wireless ad hoc networks. With the spur of modern wireless technologies, another way to improve system throughput is to allow more concurrent transmissions by installing multiple radio interfaces on one node with each radio tuned to a different orthogonal channel. When merging these two techniques, an interesting question arises: "what is the end-to-end throughput bound or capacity of OR in multiradio multichannel systems?" Zeng et al., (2010) propose a methodology to answer this question. In order to maximize the end-to-end throughput of OR in multiradio multichannel multihop networks, they jointly address multiple issues: radio-channel assignment, transmission scheduling and opportunistic forwarding strategy. To provide a comprehensive study on these issues, the authors formulate the capacity of OR as a linear programming (LP) problem which jointly solves the radio-channel assignment and transmission scheduling. Leveraging their analytical model, they obtain the following insights into OR: (1) OR can achieve better performance than traditional routing under different radio/channel configurations; however, in particular scenarios, traditional routing can be preferable to OR; (2) OR can achieve comparable or even better performance than traditional routing by using less radio resource; (3) for OR, the throughput gained from increasing the number of potential forwarding candidates becomes marginal.

In the following, we consider a class of routing protocols based on social-aware mobility models, in which users' movements are driven by their social relationships and behavior.

9.3.9   Social-aware opportunistic routing protocol, dLife

dLife (daily life) (Moreira et al., 2012) is an opportunistic routing protocol that takes advantage of time-evolving social structures in intermittently connected networks. dLife operates based on a representation of the dynamics of social structures as a weighted contact graph, where the weights (i.e., social strengths) express how long a pair of nodes is in contact over different periods of time. It is a model-based solution, as mentioned in Zhang (2006). It considers two complementary utility functions: time-evolving contact duration (TECD), which captures the evolution of social interaction among pairs of users in the same daily period of time, over consecutive days; and TECD importance (TECDi), which captures the evolution of a user's importance, based on its node degree and the social strength toward its neighbors, in different periods of time. It is intended for use in wireless ad hoc networks where there is no guarantee that a fully connected path between any source–destination pair exists at any time, where traditional routing protocols are unable to deliver packets. The dLife protocol is expected to interact with the Bundle Protocol agent (RFC5050) for retrieving information about available bundles and for requesting bundles to be sent to another node. It is expected that the associated bundle agents are then able to establish a link, over the transmission control protocol (TCP) convergence layer or the user datagram protocol (UDP) convergence layer, to perform this bundle transfer. Scientific results (Moreira et al., 2012a,b) show that dLife is able to benefit from the predictability of human behavior in daily periods of time even in the presence of few contacts. However, the behavioral predictability can be estimated more accurately with a higher number of events.

9.3.10   Social group-based opportunistic routing

Zhao et al. (2012) propose a new group-based routing protocol for mobile opportunistic networks, in which the relay node is selected based on social group information obtained from historical encounters. A simple formation method to build multi-level social groups, which summarizes the wide range of social relationships among all mobile participants, is introduced. Notice that social relations and behaviors among mobile users are usually long-term characteristics and less volatile than node mobility. The group-based routing method forwards the packet greedily toward the destination's social groups.

9.3.11 Machine learning/artificial intelligence-based opportunistic routing

Given the recent rapid development and advances in AI and machine learning, it will be benefit to apply machine learning and AI in DTNs to improve routing efficiency and network performance.

In Ababou et al. (2018), the authors apply AI in DTNs to minimize the number of replications using fuzzy logic and ant colony optimization, and in Singh et al. (2017), using machine learning and AI to compute and learn the trust level to select the best candidate to relay messages.

In a Ph.D. thesis (Dudukovich, 2019) (and a conference paper (Dudukovich et al., 2017)), Dudukovich presents an architecture, problem formulation, and considerations for machine learning techniques relative to the environmental challenges in various sectors of the space and deep space network, specifically in situ landed networks, cislunar and similar networks and deep space networks. A software and network architecture for machine learning based techniques are developed. Several popular machine learning techniques such as Q-Routing, classification, multilabel classification, and clustering are implemented as potential improvements for DTN routing.

9.3.12 Fog computing and opportunistic routing

There are potential new applications for DTN in future internet architectures, such as the information centric network (ICN) (Monticelli et al., 2014). ICN shifts the attention from the host to data. Nodes can request data by mean of unique identifiers and provide data even though it is not the original source. The ICN approach may mitigate some of the problems in opportunistic networks, allowing for seamless content caching and replication. ICN seems a perfect match for the store-carry-forward paradigm used in DTNs.

DTN techniques can be applied to cloud and fog computing. In (Gao et al., 2017), the authors propose a hybrid data dissemination framework which applies software-defined network and DTN approaches in Fog computing. Fog computing moves Cloud services from remote Internet to the edge of networks and makes streaming content much closer to mobile users, which significantly decreases the streaming latency. Specifically, the authors decompose the Fog computing network with two planes, where the cloud is a control plane to process content update queries and organize data flows, and the geometrically distributed Fog servers form a data plane to disseminate data among Fog servers with a DTN technique. Extensive simulations show that the proposed framework is efficient in terms of data-dissemination success ratio and content convergence time among Fog servers.

In Manzoni et al. (2017), the authors present an architecture and a working proto-type of a fog computing "content island" which interconnects groups of "things" packed-up together to interchange data and processing among themselves and with other content islands. These islands are based on the integration of a publish/subscribe system with disruption tolerant networks (DTN) techniques to provide a higher flex-ibility with respect to data and computation sharing. Some preliminary results regard-ing the prototype performance are also given. To improve data delivery ratio, reduce end-to-end delay, and protect data privacy, designing an efficient data forwarding scheme iscrucial in guaranteeing the quality of data transmission. Fog computing can decreasethe amount of data transmission on the Internet and improvequality of services.

In Li et al. (2019), the authors present a fog computing assisted trustworthy forwarding (FCTF) scheme, which first selects the contact probability and the service degree as the basic trustworthy metrics between node pairs, andcombines high-performance optimization algorithms to design adynamic detection model of over-lapping trustworthy communities (DOTCs). Forwarding utilities for mobile devices and self-adaptive forwarding rules are defined to have better routing quality compared with some existing forwarding models in mobile IoT applications.

9.4 Combining on-demand opportunistic routing protocols

While current on-demand routing protocols are optimized to take into account unique features of MANETs, such as frequent topology changes and limited battery life, they often do not consider the possibility of intermittent connectivity that may lead to arbi-trarily long-lived partitions. The space-content-adaptive-time routing (SCaTR) frame-work, which enables data delivery in the face of both temporary and long-lived MANET connectivity disruptions, is proposed in Boice et al. (2006). SCaTR takes advantage of past connectivity information to effectively route traffic toward destina-tions when no direct route from the source exists. The SCaTR framework extends on-demand routing, taking action only when direct routes cannot be established by the underlying protocol. In the case of a route discovery failure, i.e., when the source and destination are in separate partitions, SCaTR tries to route data to the node or nodes in the source's partition that are likely to have a route to the destination in the near future. These nodes act as proxies for the destination and buffer messages until either the destination is discovered or another node is selected as a better proxy for those messages. Messages are replicated once at most, resulting in minimal data replication and duplicate message filtering overhead. Proxies are selected based on past connectivity information, which nodes keep in content-adaptive contact tables. As will become clear, contact tables, which are the equivalent of traditional routing tables, use time-dependent and space-dependent routing metrics. These metrics differ for different types of content or local constraints such as buffer size. For instance, if a proxy is running low on buffer space, it may decide to select as the next proxy for that destination the first node it hears from that has been in contact with the destination;

this is done even if the node's contact value is lower than its own. However, if the node has higher buffer availability, it can carry the data for a longer interval. Simulations results show that, when compared with traditional on-demand protocols, as well as opportunistic routing (e.g., epidemic), SCaTR increases delivery ratio with lower signaling overhead in a variety of intermittently connected network scenarios. It also shows that SCaTR performs, as well as on-demand routing in well-connected networks and in scenarios with no mobility predictability (e.g., random mobility). In the latter case, SCaTR delivers comparable reliability to epidemic routing with considerably lower overhead.

9.5 Open research topics and future trends

Although some of the OR protocols have shown their effectiveness in achieving higher throughput compared with traditional or legacy routing, there are still many important issues remaining unanswered or not well understood. In this section, we briefly discuss some of the open research topics.

1. Thorough understanding, through theoretic analysis, of the gain of OR combined with network coding. So far, none of the existing works provides a thorough understanding of how well OR can perform and how the selection of the forwarding candidate set will affect the routing efficiency. There is a lack of theoretical analysis on the throughput bounds achievable by the OR protocols. A more systematic study is necessary to quantify the impact of different design approaches on various network settings. For example, it is well-known that network coding can help to achieve the maximum flow capacity in arbitrary random networks, while it is not clear whether this is true when it is used together with OR.
2. One of the current trends in wireless communication is to enable devices to operate using multiple transmission rates and multiple devices. OR should explore this rate–distance trade-off and utilize the capability of multiple devices while avoiding interference.
3. Existing OR coordination schemes have some inherent inefficiencies, such as long delay and potential duplicate forwarding. Most OR protocols rely on link quality information to select and prioritize forwarding candidates. It is important to accurately measure the link quality in order to make the routing operate optimally.
4. Cross-layer optimization. There are many cross-layer designs in the MANETs to improve network efficiency. OR should apply the cross-layer concept and fully utilize information in the physical layer and application layers. Application-aware adaptation of queue management at the lower layers can improve the user experience.
5. Security is another major concern in wireless ad hoc networks. OR, by its indeterministic nature, is more robust to disruption or link quality time varying. However, it is vulnerable to adversary attack due to its randomness. It is valuable to investigate the security aspect of OR. How keys are distributed among users, especially when the connection is lost, must be addressed. One possible solution is to integrate available routing protocols and the existing security framework to provide a more robust and more secure information delivery service.
6. Integration of social network analysis and OR. There has been much activity in studying social networks in recent years. Social network analysis (Scott, 2000) has attracted significant attention in many research areas, such as computer science and engineering, biology, communication studies, and information science, and focuses on studying relationships among social entities and the patterns and implications of these relationships. Some of these

advances can be utilized in the social-based approaches for DTN routing: some social-based routing offers promising performance compared with OR, since these social-based approaches take advantage of relatively stable characteristics (social properties) efficiently to predict and deal with the dynamics of DTNs. To provide more accurate prediction, multiple social metrics could be applied together or by combining these social-based metrics with traditional opportunity-based metrics, which could improve the performance in different settings.

7. Opportunistic multicasting in cognitive radio networks. Multicasting is one important application, and most of the OR reviewed so far has focused on unicast. Another interest area is OR in cognitive radio networks, where users sense the channel condition and react appropriately.

8. As discussed earlier, there have been rapid development in machine learning and AI, new techniques and algorithms developed in the machine learning/AI should be taken into account in opportunistic routing.

9. Fog computing is another promising area to connect many Internet of Thing (IoT). There are certain relationships between fog computing and DTN routing. Techniques and recent development in fog computing should be incorporated into DTN routing design.

9.6   Sources of further information and advice

As mentioned previously, understanding the behavior of the social networks will benefit the design of efficient ad hoc networks. Two books on social network analysis have been published (Scott, 2000; Wasserman and Faust, 1994). More activities on social networking can be found at http://www.social-nets.eu/.

A recent book (Zeng et al., 2011) presents a comprehensive background to the technological challenges lying behind OR. The authors cover many fundamental research issues for this new concept, including the basic principles, performance limits, and performance improvement of OR compared with traditional routing, energy efficiency and distributed OR protocol design, geographic OR, opportunistic broadcasting, security issues associated with OR, etc. Furthermore, the authors discuss technologies such as multirate, multichannel, multiradio wireless communications, energy detection, channel measurement, etc. The book brings together all the new results on this topic in a systematic, coherent, and unified presentation and provides a much-needed comprehensive introduction to this topic.

The book *Mobile Opportunistic Networks* (Denko, 2011) covers many aspects of the opportunistic networks, including routing protocols, modeling, and quality of service guarantee in opportunistic networks. Some chapters cover vehicular ad hoc networks as well.

Vehicular ad hoc network (VANET) is a noninfrastructure-based network that does not rely on a central administration for communication between vehicles. The flexibility of VANETs opens the door to a myriad of applications; however, there are also a number of computer communication challenges that await researchers and engineers who are serious about their implementation and deployment. *Advances in Vehicular Ad-Hoc Networks: Developments and Challenges* (Watfa, 2010) tackles the prevalent research challenges that hinder a fully deployable vehicular network. This unique reference presents a unified treatment of the various aspects of VANETs and is essential not only for university professors, but also for researchers working in the automobile industry. An overview of the Vehicular DTN research is given in Pereira et al. (2012).

The delay-tolerant networking research group (DTNRG), www.dtnrg.org, is a research group chartered as part of the Internet research task force (IRTF) and is conducting research in DTN architectures and routing, including OR. Members of DTNRG are concerned with how to address the architectural and routing protocol design principles arising from the need to provide interoperable communications with and among extreme and performance-challenged environments where continuous end-to-end connectivity cannot be assumed.

References

Ababou, M., Bellafkih, M., El Kouch, R., 2018. Energy efficient routing protocol for delay tolerant network based on fuzzy logic and ant colony. Int. J. Intell. Syst. Appl. 10, 69–77.

Biswas, S., Morris, R., 2005. EXOR: opportunistic multi-hop routing for wireless networks. SIGCOMM Comp. Commun. Rev. 35 (4), 133–144.

Boice, J., Garcia-Luna-Aceves, J.J., Obraczka, K., October 2006. On-Demand Routing in Disrupted Environments, UCSC-CRL-06-16. USC, University of California, Santa Cruz.

Chachulski, S., Jennings, M., Katti, S., Katabi, D., August 2007. Trading structure for randomness in wireless opportunistic routing. In: ACM SIGCOMM 2007, Kyoto, Japan.

Clausen, T., Jacquet, P., October 2003. Optimized Link State Routing Protocol (OLSR). RFC 3626, IETF Network Working Group.

Denko, M., 2011. Mobile Opportunistic Networks: Architectures, Protocols and Applications. CRC Press. ISBN: 1466508124, Published 26 May 2011 by Auerbach Publications.

Dudukovich, R., January 2019. Application of machine learning techniques to delay tolerant routing. Ph.D. thesis, Case Western Reserve University.

Dudukovich, R., et al., 2017. A machine learning concept for DTN routing. In: IEEE International Conference on Wireless for Space and Extreme Environments (WiSEE).

Gao, L., Luan, T.H., Yu, S., Zhou, W., Liu, B., 2017. FogRoute: DTN-based data dissemination model in fog computing. IEEE Internet Things J. 4, 225–235.

Gkantsidis, C., Hu, W., Key, P., Radunovic, B., Rodriguez, P., et al., 2007. Multipath code casting for wireless mesh networks. In: Proc. ACM CoNEXT'07, 3rd Annual CoNEXT Conferences, 10–13 December, New York, NY, pp. 1–12.

Han, M.K., Bhartia, A., Qiu, L., Rozner, E., 2011. O3: Optimized overlay-based opportunistic routing. In: Proceedings of the Twelfth ACM International Symposium on Mobile Ad Hoc Networking and Computing (MobiHoc). ACM, New York, NY.

Katti, S., Rahul, H., Hu, W., Katabi, D., Médard, M., et al., 2008. XORs in the air: practical wireless network coding. IEEE/ACM Trans. Networking 16 (3), 497–510.

Li, J., et al., April 2019. Fog computing-assisted trustworthy forwarding scheme in mobile internet of things. IEEE Internet Things J. 6 (2).

Manzoni, P., et al., 2017. A proposal for a publish/subscribe, disruption tolerant content island for fog computing. In: SMARTOBJECTS'17, October 16, Snowbird, UT.

Monticelli, E., Arumaithurai, M., Psaras, I., Fu, X., Ramakrishnan, K.K., Combining Opportunistic and Information Centric Networks in RealWorld Applications. Available online. https://pdfs.semanticscholar.org/4b92/b74326d06c089438802ed0e56624a798dc54.pdf. (Accessed 28 July 2019).

Moreira, W., Mendes, P., Ferreira, R., Cequeira, E., October 2012. Opportunistic Routing Based on Users Daily Life Routine. draft-moreira-dlife-01, work-in-progress.

Moreira, W., Mendes, P., Sargento, S., 2012a. Opportunistic routing based on daily routines. In: Proceedings of the Sixth IEEE WoWMoM Workshop on Autonomic and Opportunistic Communications (AOC 2012), San Francisco, CA.

Moreira, W., de Souza, M., Mendes, P., Sargento, S., 2012b. Study on the effect of network dynamics on opportunistic routing. In: Proceedings of the Eleventh International Conference on Ad-Hoc Networks and Wireless (AdHoc Now 2012), Belgrade, Serbia.

Pereira, P., Casaca, A., Rodrigues, J., Soares, V., 2012. From delay-tolerant networks to vehicular delay-tolerant networks. IEEE Commun. Surv. Tutorials 14 (4), 1166–1182.

Perkins, C., Belding-Royer, E., Das, S., 2003. Ad hoc On-Demand Distance Vector (AODV) Routing. RFC 3561, IETF Network Working Group.

Rozner, E., Seshadri, J., Mehta, Y.A., Qiu, L., 2006. SOAR: Simple Opportunistic Adaptive Routing Protocol for Wireless Mesh Networks, WIMESH06.

Scott, J., 2000. Social Network Analysis: A Handbook. Sage Publications, London.

Singh, A.V., Juyal, V., Saggar, R., 2017. Trust based intelligent routing algorithm for delay tolerant network using artificial neural network. Wirel. Netw 23, 693–702.

Wasserman, S., Faust, K., 1994. Social Network Analysis: Methods and Applications. Cambridge University Press, Cambridge.

Watfa, M., May 2010. Advances in Vehicular Ad-Hoc Networks: Developments and Challenges. University of Wollongong, UAE. 384 pages, ISBN13: 9781615209132, ISBN10: 1615209131.

Yan, Y., Zhang, B., Zheng, J., Ma, J., 2010. CORE: a coding-aware opportunistic routing mechanism for wireless mesh networks. IEEE Wirel. Commun. 17 (3), 96–103.

Yuan, Y., Yang, H., Wong, S., Lu, S., Arbaugh, W., September 2005. ROMER: resilient opportunistic mesh routing for wireless mesh networks. In: Proc. IEEE Workshop Wireless Mesh Networks (WiMesh'05), Santa Clara, CA.

Zeng, K., Yang, Z., Lou, W., March 2010. Opportunistic routing in multi-radio multi-channel multi-hop wireless networks. In: IEEE Infocom, San Diego, CA.

Zeng, K., Lou, W., Li, M., 2011. Multihop Wireless Networks: Opportunistic Routing (Wireless Communications and Mobile Computing), first ed. Wiley.

Zhang, Z., 2006. Routing in intermittently connected mobile ad hoc networks and delay tolerant networks: overview and challenges. IEEE Commun. Surv. Tutorials 8 (1), 24–37.

Zhang, Z., Zhang, Q., 2007. Delay/disruption tolerant mobile ad hoc networks: latest developments. In: Wireless Communications and Mobile Computing. Wiley InterScience, pp. 1219–1232.

Zhao, L., Li, F., Zhang, C., Yu, W., December 2012. Routing with multi-level social groups in mobile opportunistic networks. In: Proceedings of the IEEE Global Communications Conference (IEEE Globecom 2012), Anaheim.

Reliable data streaming over delay-tolerant networks (DTNs)

Emmanuel Lochin[a], J. Lacan[a], P.-U. Tournoux[b], and J. Leguay[c]
[a]ENAC, University of Toulouse, Toulouse, France, [b]Université de la Réunion, Saint-Denis Réunion, France, [c]Thales Communications, Colombes, France

10.1 Introduction

Current delay-tolerant networks (DTNs) research has mainly addressed the problem of routing in various mobility contexts with the aim of improving bundle delay delivery and data delivery ratio, leaving aside the application requirements. However, transporting data in a reliable and delay-efficient manner over a DTN is a tough challenge, as most Internet applications assume a form of persistent end-to-end connection. The support of streaming-like applications over DTN is complex: first, due to the dynamic nature of the topology and second, because mobile nodes observe frequent connectivity disruptions. In other words, and from the application point of view, there is a very low probability that an end-to-end path exists between a given pair of nodes at a given time, and data transmission has unpredictable performance due to the network topology dynamic.

Due to these constraints, common ad hoc transport approaches cannot be used and, as a result, novel alternatives must be considered. This leads to the design of specific transport-layer mechanisms able to provide both full reliability and fast in-order delivery services needed by several kinds of applications. To cope with these intrinsic characteristics, the use of redundancy recovery mechanisms seems inevitable (Jain et al., 2005), and some schemes based on erasure coding have been proposed to improve the data delivery ratio. Among these proposals, two classes of transport protocols are commonly envisioned in DTN, depending on the existence of a feedback path. The first class does not use feedback messages, while the second considers both the use of feedbacks and various acknowledgment strategies, as explained in the following.

The first class is intended to improve the bundle delivery ratio and/or delivery delay with mechanisms such as replication, erasure codes, and multipath. The benefit of erasure coding with various path allocations has been studied by Jain et al. (2005). In particular, the authors show how to obtain the optimal successful delivery probability of a given source data thanks to erasure coding schemes. In another study (Liao et al., 2006), Liao et al. propose to combine an erasure coding-based scheme with a routing protocol based on the predictability of node mobility pattern. Later, Altman and De Pellegrini (2011) proposed an analytical approach to assess the effect of Fountain

---

☆ This chapter is a reprint of the chapter originally published in the first edition of Advances in Delay-Tolerant Networks (DTNs): Architecture and Enhanced Performance.

Advances in Delay-tolerant Networks (DTNs). https://doi.org/10.1016/B978-0-08-102793-6.00010-2

codes on the overall network performances. In the context of communication in sparsely populated areas, Kutscher et al. (2007) proposed the use of both forward error code (FEC) and rounds of source-based retransmission to achieve full reliability with a probability near to 1. Even if network coding techniques do not belong to the transport layer, Zhang et al. (2006) demonstrated that involving intermediate nodes in the coding process allows delay to be reduced in the case of a network with constrained resources.

In the second class, we find the works of Harras and Almeroth (2006), which investigate various acknowledgment strategies: the hop-by-hop partial reliability confirms that the bundle is transmitted to another relay; the active receipts are disseminated as new messages; passive receipts progress through the nodes of the network to acknowledge all nodes previously infected by a bundle message; finally, they propose the network-bridged receipt that transits through a cellular network. Their use depends on the trade-off between affordable complexity and delay. At last, and in the context of deep-space networking (DSN), transport protocols such as Licklider Transmission Protocol-Transport (LTP-T), proposed by Farrell and Cahill (2007) for large delay links with connectivity disruptions, or Saratoga, a simple file transfer protocol that can be used to transfer DTN bundles, rely on Automatic Repeat ReQuest (ARQ) and/or unequal error protection to reduce the amount of nonmandatory retransmissions.

All these algorithms and protocols are designed to carry bundle units with a reliable or unreliable service and, for some of them, only in a hop-by-hop fashion. For instance, Saratoga only enforces reliability between two DTN nodes in contact and not the whole transmission over a DTN path. Furthermore, they have been designed to carry stored data and, as a result, they can only enable a stored data streaming service because the whole piece of data to transfer must be available at the beginning of the transfer to effectively start the transmission. This fact simplifies and allows, for instance, the generation of erasure code blocks which can be spread or routed among the next set of contact opportunities, such as in Jain et al. (2005) and Altman and De Pellegrini (2011). However, live data streaming raises several other issues that these mechanisms cannot solve. Furthermore, and as shown by Jain et al. (2005), these algorithms require a challenging configuration closely linked to the network characteristics without providing any guarantee in terms of delay and reliability. Indeed, none of these solutions tackle how to carry streams of data bundles. Finally, in this context, if an application seeks to maintain a loss rate below the maximum threshold it can tolerate, the complexity of this configuration increases significantly and might even become impossible.

The so-called *streaming-like* applications which produce *continuous data* over time and require *to be consumed in sequence* (data must be ordered) at the receiver side require the use of adapted mechanisms to correctly perform over a DTN network. A possible solution has been designed in Tournoux et al. (2010), where an on-the-fly coding scheme is proposed to cope with the numerous issues raised by a live data streaming. This recent class of coding scheme (Tournoux et al., 2011; Sundararajan et al., 2008), which considers an intermittently connected feedback path, allows redundancy packets to be built on-the-fly. In other words, it means that the content does not need to be stored to send the data, making this kind of coding scheme a perfect candidate to enable reliable streaming over DTN. In the following, we illustrate why on-the-fly coding overcomes streaming support challenges without delay compromise while remaining independent of any mobility or routing schemes.

This chapter is organized as follows. We first detail in Section 10.2 the challenges raised for streaming support in DTNs. Then Section 10.3 presents how on-the-fly coding schemes allow a robust DTN streaming service to be enabled. An evaluation of several coding schemes is given in Section 10.4. Finally, Section 10.5 provides a discussion on the previously tested mechanisms, while Section 10.6 concludes the chapter.

10.2   Challenges for streaming support in DTNs

In a DTN, link and buffer capacities, node mobility and routing strategies impact on the bundle delivery performance in terms of delay, in-order delivery, and losses. As end-to-end paths used for bundle delivery depend on the mobility pattern of each peer, bundles sent between a given source and destination do not necessarily use the same path. As a result, despite a relatively stable average transmission delay, the standard deviation of each bundle's delay of a given flow can be very large. This could lead to a reordering ratio (following Morton et al.) often higher than 50%, with some peaks up to 90%, while in classical networks this value is usually below 5% (Zhou and Van Mieghem, 2004). The network structure also evolves as a function of time in terms of connexity, contact density, and node betweenness (Grossi and Pedersini, 2010) (the betweenness of a given node i defines the number of short paths between any nodes going through node i). In this case, the average delay might also drastically vary.

Each bundle remains in the network until it's time to live (TTL) reaches zero. This implies that the evolution of the network can lead to a high loss rate due to TTL expiration. Furthermore, the amount of data that can be transmitted during a given contact is limited. In the case of congestion, this can slow down the dissemination of the bundles in the network and lead to TTL expiration. Congestion can also lead to a buffer overflow, with a high probability of dropping all the copies of a given bundle, and, as a result, increase the loss rate.

All these intrinsic DTN characteristics limit the use of applications that need either in-order delivery or full reliability services. Indeed, when network reordering ratio is very low or when jitter is quite stable, ARQ mechanisms might be a possible solution. In the context of DSN, this is currently the choice enabled by the LTP-T Protocol (Farrell and Cahill, 2007). In this case, the number of false loss detections is small and it is possible to wait a fixed and reasonable amount of time before taking a retransmission decision. However, in the case of DTN, both reordering ratio and jitter prevent the use of ARQ mechanisms that would lead to spurious retransmissions. Concerning in-order delivery, the receiver needs to set a timer to allow waiting for missing bundles before transmission to the application. As for ARQ, it is necessary to determine whether a bundle is lost or reordered, and a practical solution is to wait for a fixed period of time. This implies both higher delay and higher perceived loss rate, as bundles that arrive after this waiting period might be considered as lost. The design of a mechanism that would determine whether a packet is lost or disordered within a reasonable delay bound is thus difficult, and possible solutions, if any, would be context-dependent (from both mobility and application points of view). Under certain conditions (i.e., when the redundancy ratio is higher than the loss rate), on-the-fly coding prevents the use of ARQ mechanisms while enabling full reliability without data retransmissions in certain conditions. Lost or delayed bundles are recovered in order, making it possible to accelerate data transmission to the upper layer and to solve the technical problem which arises when we have to decide whether a packet is lost or delayed.

10.3   Using on-the-fly coding to enable robust DTN streaming

This section presents the on-the-fly coding principle. The running example provided, named Tetrys, comes from Tournoux et al. (2011). This choice is generic enough, as the concept remains similar to other implementation, such as Sundararajan et al. (2008). We illustrate how this coding scheme can provide a robust streaming protocol when applied to DTNs.

The original goal of Tetrys was to propose an implicit acknowledgement strategy for long-delay networks (Lacan and Lochin, 2008), where recovery of missing packets is thus not possible with classical retransmission mechanisms (i.e., ARQ). Tetrys is based on an elastic encoding window updated dynamically according to the feedback information returned by the receiver. One important property of Tetrys is its capability to reduce recovery delay of lost data at the receiver side. In its most general form, this approach is compatible with any kind of traffic, and full reliability can be achieved as soon as the encoding ratio is higher than the average loss rate.

The main principle of this erasure coding scheme is to generate *repair bundles* every k source bundles, where k is an integer determined according to the expected bundle loss rate. The resulting coding rate used is then equal to $\frac{1}{k+1}$. These repair bundles are a random linear combination of the bundles included in the variable size encoding window, which contains the bundles sent but not yet acknowledged by the receiver. They are computed as shown in Eq. (10.1):

$$R_{(i,..,j)} = \sum_{u=i}^{j} \alpha_u^{(i,j)} \cdot B_u \tag{10.1}$$

with B_i to B_j the bundles that belong to the encoding window and $\alpha_u^{(i,j)}$ the coefficient randomly chosen in the finite field F_q used to encode the uth bundle in the repair bundle $R_{(i,...,j)}$. Only the seed of the random coefficient generator, which is specific to each repair bundle, has to be embedded in the latter. The source bundles are sent unmodified (i.e., the code is systematic: meaning that input data are embedded in the encoded output), which leads to reduced coding overhead.

For each repair bundle, the receiver subtracts all the available source bundles encoded. The resulting repair bundles, then, consist of linear combinations of the lost source bundles. The rest of the decoding simply solves this linear system. One should note that decoding can only be done as soon as the number of received repair bundles (the equations) is at least equal to the number of lost source bundles (the unknowns).

Source bundles are removed from the encoding window (at the sender side) with the reception of *acknowledgement bundles*, which assess their proper reception or decoding at the receiver side. Acknowledgements include SACK vectors, which follow the Transport Control Protocol RFC Selective ACKnowledgement (TCP RFC SACK) specifications. The purpose of this acknowledgement scheme is to reduce the number of source bundles involved in the encoding/decoding processes. Their frequency, availability or losses do not impact the performance in terms of reliability or decoding delay. They can be carried by one of the various transmission schemes investigated by Harras and Almeroth (2006). In particular, the authors propose the use of active receipts, which are disseminated as new messages; passive receipts, which evolve through the nodes to acknowledge all nodes previously infected by the bundle; and, finally, network-bridged receipts that transit through a cellular network. As the acknowledgement delay varies with these different methods, their use depends on the affordable encoding/decoding complexity for the nodes.

Fig. 10.1 illustrates the overall mechanism with a simple bundle exchange. For the sake of simplicity, small delay and jitter are considered. After a correct reception of bundle B1, B2 is lost. However, using the received repair bundle R(1,2) received

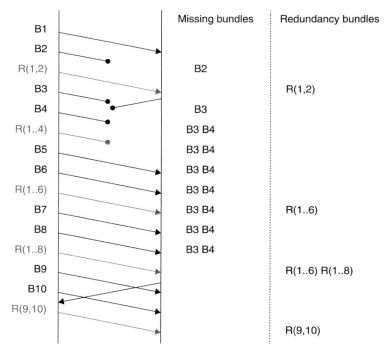

Fig. 10.1 A simple data exchange.

justafter, the receiver rebuilds B2. We assume that the receiver sends back an acknowledgement (according to one of the methods investigated by Harras and Almeroth (2006), for instance) to inform the sender that bundles older than B3 should be removed from the encoding window. However, in this example, we lost this acknowledgement during the transmission. We note that this loss does not compromise the following transmissions and the sender simply continues to compute repair bundles from B1. After this, we see that B3, B4 and R(1.0.4) bundles are also lost. None of these bundles needs to be retransmitted, as they are rebuilt thanks to the bundles received from B5 to R(1.0.8). Indeed, the receiver can rebuild B3, B4 by first "subtracting" all the received source bundles from the repair bundles in order to obtain $(R'(1.0.6) \ R'(1.0.8))$, which are linear combinations of B3 and B4. By solving this linear system, B3 and B4 can be recovered.

After receiving the acknowledgement, the encoding in the repair bundle R(9,10) starts from 9 and not from 1. The receiver was also able to acknowledge B_5 and B_6 with a SACK block, reducing more significantly the size of the window (not represented in the figure). If no feedback is received, the sender simply sends R(1.0.10). In this reliability scheme, the purpose of the acknowledgements is to reduce the number of source bundles handled by a repair bundle, as illustrated in Fig. 10.1, where, after receiving the acknowledgement, the repair bundle starts from 9 and not from 1. It is important to keep in mind that the reception of an acknowledgement is not mandatory. The loss of acknowledgements does not compromise either the reliability or the sending process (in Fig. 10.1, if no feedback is received, the sender simply sends R (1.0.10)). The loss of an acknowledgement impacts "only" on the coding window size increase and the encoding relative complexity. Indeed, only binary XOR operations are done.

In summary, this scheme allows a full reliability even if some source data bundles, repair bundles or acknowledgements are lost. More interestingly in the context of DTN, the decoding does not depend on the feedback received, and thus the loss recovery delay is totally independent of the Round Trip Time (RTT). As the decoding process is realized in order, we do not need to decide whether or not a given bundle is lost before we deliver the others to the application. The problem of retransmission with ARQ is also avoided, since Tetrys does not need to request retransmission.

10.4   Evaluation of existing streaming proposals over a DTN network

In this section we evaluate the performance of Tetrys in terms of delivery ratio and application delay against classical reliability schemes.

10.4.1   Relevant schemes

We now propose to compare the on-the-fly coding mechanism Tetrys with the other erasure coding schemes and with a standard uncoded scheme such as a DTN routing scheme.

10.4.1.1   Uncoded

In this scheme, messages are sent *as is* and are carried out by the routing scheme used, such as Spray and Wait (SW).

10.4.1.2   Erasure coding scheme (FEC (n,k))

We consider here a classical block erasure code (also called FEC) such as the one described by Rizzo (1997). The parameters n and k impact on the coding as follows: after the emission of k source bundles $B_1,\ldots, B_k$, FEC (n,k) adds $(n-k)$ repair bundles $(F_1\ldots F_{n-k})$ sent interleaved with the source bundles of the next block. This results in the following emitted sequence: $B_k, F_1, B_{k+1}, F_2..B_n, F_{n-k}, B_{n+1}, B_{n+2},(\ldots)$. We assume a maximum distance separable (MDS) code, which means that the decoding of the lost bundles from a given block is possible as soon as k bundles (B or F) are received.

Note that, concerning the parameter of correction capability, MDS codes are optimal. They are thus better than Fountain codes, studied in Altman and De Pellegrini (Altman and De Pellegrini, 2011), which require $(1 + \varepsilon) \cdot k$ bundles (with $\varepsilon > 0$) on average. Indeed, the main advantages of Fountain codes are the large number of potential redundant bundles and the decoding complexity. These two parameters are not critical in our context.

For both FEC (n,k) and Tetrys, we denote R: the redundancy ratio such that $R = \frac{k}{n-k}$.

10.4.1.3   Automatic repeat ReQuest

As FEC may not provide any full reliability guarantee, it would have been interesting to consider an ARQ-based protocol. Following Section 10.2, it is clear that this class of mechanism is not applicable. Even if the technical problems related to loss detection were solved, the recovery delay of the lost bundles would be increased by at least one RTT. The same remark also applies to Hybrid-ARQ, in which the basic principle is to code data with FEC and retransmit lost packets (ARQ) to rebuild the blocks that got more than $n-k$ losses.

In practice, either FEC or uncoded strategies implement a timer to determine whether bundles are lost or delayed before transmitting them to the application. This obviously introduces a delay increase. In our evaluation, FEC does not have such a delay overhead, as we use an Oracle to determine whether a bundle is lost. However, Tetrys does not benefit from this Oracle, as it provides a full reliable service. We use both in-order bundle delivery delay and bundle loss rate metrics (i.e., the ratio between the number of bundles received and the number of bundles sent) observed by the application.

10.4.2   Network settings

To evaluate these mechanisms, we are replaying real connectivity data from the RollerNet dataset (Tournoux et al., 2009). This dataset plots a high intensity of interactions between mobile nodes where the number of connectivity disruptions prevent the use of classical Mobile Area NETwork (MANET) protocols.

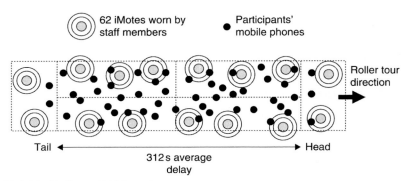

Fig. 10.2 The Rollernet DTN topology.

We consider data streams winch generate $\frac{1}{2}$ bundles per second between both extremities of the tour (i.e., specific head and tail nodes are deployed, as shown in Fig. 10.2). We use SW routing (Spyropoulos et al., 2005) with 16 copies to disseminate bundles across the 62 iMotes deployed. For the feedback of Tetrys, we use active receipts (Harras and Almeroth, 2006), which are sent using epidemic routing. The packet's lifetime is not limited and bundles are produced during 5000 s. Bundles can be received until the end of the experiment, set at time 7000 s.

The links capacities are limited according to the Bluetooth bit rate (≈ 700 KB/s) and each buffer is fixed to 300 MB. In order to generate a background traffic, 50 peers are randomly selected to generate $\frac{1}{2}$ bundle per second. All bundles (sources or repairs) have a fixed size of 200 KB. We have estimated a reordering ratio of 95% with an average extent of 348 bundles and a delivery ratio of 82% (see Morton et al., 2006, Section 4.2).

10.4.3 Results

Fig. 10.3 plots the distribution of the in-order delivery delay of Tetrys, FEC(600, 300), FEC(1000, 500) and the uncoded strategy. We can see that, under the same conditions, the probability for a bundle to be delivered to the application before a given delay is significantly higher with Tetrys than with FEC. As an example, a bundle is delivered before 1000 s with a probability of 0.48 for Tetrys, while we obtain only 0.16 for FEC

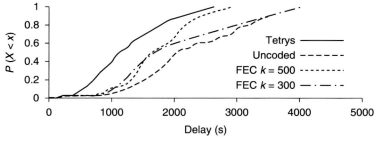

Fig. 10.3 Distribution of the in-order delivery delay with $R = 0.5$.

Reliable data streaming over delay-tolerant networks (DTNs)

Table 10.1 Average in-order delivery delay (seconds)—delivery ratio as a function of the redundancy ratio.

	Uncoded	Tetrys	FEC $k = 10$	FEC $k = 50$	FEC $k = 100$
$R = 1/2$	2172–0.82	1251–1.00	1870–0.89	1486–0.92	1430–0.92
$R = 1/3$	2172–0.82	2145–1.00	1829–0.87	1700–0.88	1564–0.89
$R = 1/4$	2172–0.82	2246–1.00	1912–0.85	1660–0.85	1775–0.85
	Uncoded	**Tetrys**	**FEC $k = 200$**	**FEC $k = 300$**	**FEC $k = 500$**
$R = 1/2$	2172–0.82	1251–1.00	1600–0.96	2035–1.00	1718–1.00
$R = 1/3$	2172–0.82	2145–1.00	1586–0.88	1689–0.92	1779–0.90
$R = 1/4$	2172–0.82	2246–1.00	1573–0.87	1662–0.88	2024–0.85

(1000, 500). It is interesting to observe that, in addition to recovery from the losses, the coding schemes allow reduction of the delay resulting from network reordering. In particular, Tetrys achieves a delay nearly half the uncoded scheme.

The first line of Table 10.1 shows that, for a redundancy ratio of 0.5, Tetrys has an average delivery delay of 1251 s, which is significantly smaller than the FEC's minimum average delay of 1430 s achieved with $k = 100$. We can also notice that FEC reaches full reliability when $k \geq 300$. However, compared with Tetrys, FEC-based schemes did not achieve the optimum delay and delivery ratio for the same values. This leads to an intricate trade-off between delay and reliability. On the contrary, Tetrys allows the smallest delivery delay on average without reliability compromise. Table 10.1 also shows that, if the redundancy ratio is reduced to $R = 1/3$ or $R = 1/4$, FEC does not rebuild all the lost bundles, even with large blocks. In this case, we cannot accurately compare in-order delivery delay of Tetrys and FEC, as the latter does not rebuild all the bundles and takes advantage of the Oracle. This issue is highlighted when considering only 85% of the Tetrys bundles, as the delay decreases to 1786 s. This corresponds to the same couple (delay, reliability) achieved by FEC with $k = 10, 50, 100, 500$.

Although we focused on the in-order delivery, we must notice that Tetrys remains a good candidate when we relax the in-order delivery constraint (e.g., without the de-jitter part of the delay). With $R = 1/2$, the average delay is 1125 s for Tetrys, 1295 s for FEC with $k = 300$, and 1243 s for FEC with $k = 500$. Eventually, we can notice that, as long as FEC rebuilds all lost bundles, Tetrys delay will be lower than or equal to FEC delay.

10.5 Implementation discussion

10.5.1 Choice of parameters

To benefit from the use of Tetrys, protocol engineers must ensure that the average loss rate will remain below the redundancy ratio. The main strength of Tetrys is to rebuild lost or delayed bundles within a delay tuneable according to the amount of redundancy and totally independent of the RTT.

Table 10.2 Average loss recovery delay as a function of the bundle loss rate and the redundancy ratio.

Bundle loss rate (%)	R = 1/2	R = 1/3	R = 1/4
15.0	217 s	792 s	1209 s
18.0	252 s	928 s	1395 s
19.3	446 s	1064 s	ϕ
22.2	419 s	1117 s	ϕ
24.6	625 s	1376 s	ϕ
35.8	854 s	ϕ	ϕ
40.0	937 s	ϕ	ϕ
47.5	1380 s	ϕ	ϕ

Table 10.2 shows the variation of the average loss recovery delay as a function of the loss rate (we changed the size of the forwarding node's buffer from 140 MB to 310 MB, to obtain different loss rates) and redundancy ratio. For each recovered bundle, it corresponds to the delay between the detection that a bundle is lost using the sequence numbers and its reconstruction. The table contains ϕ in the case where Tetrys did not decode all the lost bundles when the loss rate becomes higher than the redundancy ratio. As shown in this table, a redundancy slightly higher than the current loss rate estimated is mandatory to perform with Tetrys.

Concerning the decoding delay, it can be observed for a given loss rate that the higher the redundancy ratio, the less the decoding delay. As an example, for a bundle loss rate of 18%, Tetrys achieves an average decoding delay of 252 s for a redundancy ratio of 0.5, which increases to 928 s and 1395 s for a redundancy ratio of 0.33 and 0.25, respectively. Obviously, for a fixed redundancy ratio, the decoding delay increases with the loss rate as observed for $R = 1/2$.

According to the results presented, we observe that the estimation of the average loss rate is necessary to assess the performance of the system. As it depends on the routing and the mobility context, this information can be provided by the lower layer thanks to a cross-layer metric. This information can also be provided by the receivers' feedback in order to dynamically refine the redundancy ratio.

Finally, Table 10.2 shows that, when the redundancy ratio is reduced to 33% and to 25%, Tetrys achieved full-reliability comparison to FEC even with large block size. The decrease of the redundancy ratio directly impacts on the in-order delivery delay of Tetrys, as it increases to 1719 and 1928, respectively. We also note that, with a redundancy ratio of 33% and 25%, Tetrys achieves a delay in the same order of magnitude as that achieved by FEC with a redundancy ratio of 50% and block size correctly sized in order to get a full delivery ratio.

10.5.2 Complexity

We discuss here the time and space complexity of Tetrys.

10.5.2.1 Computation complexity

The per-bundle computation complexity at sender and receiver sides is mainly $O\left(\frac{RTT\lambda}{k}\right)$ with (n,k). The coding parameter, γ, is the average sending rate of the application in bundles per second, and RTT the average delay between the emission of a bundle and the reception of the acknowledgment feedback. For comparison purposes, the per-bundle complexity of the FEC mechanism considered in the previous section is O (k). If we look at the example in Fig. 10.3, the average size of the Tetrys encoding window is 521.6 bundles. Following Dairaine et al. (2005) and the encoding window size, we would have been able to achieve a maximum bit rate of 210 KB/s on a Pentium III 933 MHz, twice the bit rate assumed in our settings. Depending on the trade-off between the complexity and the cost of the feedback mechanism, we can adapt the latter (active/passive/bridged) (Harras and Almeroth, 2006) since the lower the feedback delay, the lower the complexity.

10.5.2.2 Buffer complexity

The complexity of Tetrys in space is also mainly $O(RTT^{*}\gamma)$ as the source bundles remain in the source node buffer until acknowledged. At the receiver side, the source bundles remain in the buffer as long as they are encoded in the repair bundles received. FEC needs a fixed number of k bundles at both the sender and receiver sides. In our evaluation, without congestion, the average buffer size required at the receiver side is 500 bundles for FEC with $k = 500$, 128 bundles with $k = 100$, and 131 if $k = 10$ (note that these values take into account the number of packets needed for reordering). Tetrys requires an average of 149 bundles and 141 bundles without any coding strategy.

In the studied scenarios, in both time and space, the complexity of Tetrys is in the same order of magnitude as other schemes.

10.6 Conclusion

This chapter has presented some potential solutions to enable stored and live streaming over a DTN network. Compared with stored data streaming, which has several solutions, live data streaming remains a challenging problem, and the use of appropriate recovery mechanisms is mandatory. After an overview of the challenges and existing solutions to enable a stored data streaming, we have presented how an on-the-fly coding scheme allows live data streaming to be performed over a DTN network. As an example of such a scheme, we focus on the Tetrys coding scheme, which is, to date, the first proposal that enables a robust live streaming service over a DTN. Tetrys achieves full reliability when the redundancy ratio is greater than the average loss rate. Furthermore, the loss recovery delay is independent of the RTT and tolerates bursty losses on either data or feedback paths. These characteristics allow such a scheme to fit both network and application requirements. Indeed, Tetrys accepts any reliability thresholds required by an application while recovering the bundles ordered. Tetrys prevents the use of a complex (and possibly inaccurate) algorithm to determine whether a bundle is lost or delayed, as implementing such a timer is complex and might involve spurious detection. Compared with other mechanisms that aim to improve the delivery ratio, Tetrys does not depend on any mobility context or routing mechanism, and the sole parameter required for its configuration is the average loss rate. This makes an on-the-fly coding scheme a perfect candidate to enable a reliable and live streaming service over a DTN network.

References

Altman, E., De Pellegrini, F., 2011. Forward correction and fountain codes in delay tolerant networks. IEEE/ACM Trans. Networking 19 (1), 1–13.

Dairaine, L., Lancérica, L., Lacan, J., Fimes, J., September 2005. Content-access QoS in peer-to-peer networks using a fast MDS erasure code. Comput. Commun. 28 (15), 1778–1790.

Farrell, S., Cahill, V., September 2007. Evaluating LTP-T: a DTN-friendly transport protocol. In: International Workshop on Satellite and Space Communications (IWSSC), Salzburg, Austria.

Grossi, G., Pedersini, F., 2010. Hub-betweenness analysis in delay tolerant networks inferred by real traces. In: Proceedings of the 8th International Symposium on Modeling and Optimization in Mobile, Ad Hoc and Wireless Networks (WiOpt), 2010, Avignon, France, pp. 318–323.

Harras, K.A., Almeroth, K.C., 2006. Transport layer issues in delay tolerant mobile networks. In IFIP NETWORKING 2006. Networking technologies, services, and protocols; performance of computer and communication networks; mobile and wireless communication systems. Lect. Notes Comput. Sci 3976, 463–475.

Jain, S., Demmer, M., Patra, R., Fall, K., 2005. Using redundancy to cope with failures in a delay tolerant network. SIGCOMM Comput. Commun. Rev. 35 (4), 109–120.

Kutscher, D., Greifenberg, J., Loos, K., 2007. Scalable DTN distribution over uni-directional links. In: NSDR'07: Proceedings of the 2007 Workshop on Networked Systems for Developing Regions, Kyoto, Japan. Association for Computing Machinery (ACM), New York, NY, pp. 1–6.

Lacan, J., Lochin, E., October 2008. Rethinking reliability for long-delay networks. In: IEEE International Workshop on Satellite and Space Communications, 2008. IWSSC 2008, Toulouse, France, pp. 90–94.

Liao, Y., Tan, K., Zhang, Z., Gao, L., 2006. Estimation based erasure-coding routing in delay tolerant networks. In: Proceedings of the 2006 International Conference on Wireless Communications and Mobile Computing (IWCMC), Vancouver, BC, Canada, pp. 557–562.

Morton, A., Ciavattone, L., Ramachandran, G., Shalunov, S., Perser, J., 2006. Packet Reordering Metrics. RFC 4737 (Proposed Standard).

Rizzo, L., April 1997. Effective erasure codes for reliable computer communication protocols. ACM Comp. Commun. Rev. 27, 24–36.

Spyropoulos, T., Psounis, K., Raghavendra, C., 2005. Spray and wait: An efficient routing scheme for intermittently connected mobile networks. In: Proceedings of the 2005 SIGCOMM Workshop on Delay-Tolerant Networking (WDTN), Philadelphia, PA, pp. 252–259.

Sundararajan, J.K., Shah, D., Medard, M., July 2008. ARQ for network coding. In: IEEE Int. Symp. on Information Theory, Toronto, Canada, pp. 1651–1655.

Tournoux, P.-U., Leguay, J., Benbadis, F., Conan, V., de Amorim, M.D., et al., April 2009. The accordion phenomenon: analysis, characterization, and impact on DTN routing. In: INFOCOM 2009. IEEE, Rio de Janeiro, Brazil, pp. 1116–1124.

Tournoux, P.U., Lochin, E., Leguay, J., Lacan, J., 2010. Robust streaming in delay tolerant networks. In: 2010 IEEE International Conference on Communications (ICC), Cape Town, South Africa, pp. 1–5.

Tournoux, P.U., Lochin, E., Lacan, J., Bouabdallah, A., Roca, V., 2011. On-the-fly erasure coding for real-time video applications. IEEE Trans. Multimedia 13 (4), 797–812.

Zhang, X., Neglia, G., Kurose, J., Towsley, D., April 2006. On the benefits of random linear coding for unicast applications in disruption tolerant networks. In: 4th International Symposium on Modeling and Optimization in Mobile, Ad Hoc and Wireless Networks, 2006, Boston, MA.

Zhou, X., Van Mieghem, P., 2004. Reordering of IP packets in internet. In: Proc. Passive and Active Measurement, Antibes, Juan-les-Pins, France. Springer Berlin Heidelberg, pp. 237–246.

Further reading

Ivancic, W.D., Paulsen, P., Stewart, D., Eddy, W., McKim, J., et al., 2010. Large file transfers from space using multiple ground terminals and delay-tolerant networking. In: Global Telecommunications Conference (GLOBECOM 2010), 2010. IEEE, Miami, FL, pp. 1–6.

Rapid selection and dissemination of urgent messages over delay-tolerant networks (DTNs)

M. Asplund and S. Nadjm-Tehrani
Linköping University, Linköping, Sweden

11.1   Introduction

The number of connected devices in the world is increasing at an enormous rate. A whole new range of products and services are starting to appear that take advantage of high-speed wireless communication being available anywhere at any time. Cellular technology has led the way, with very good coverage and reasonable bandwidth capabilities. However, these centralized systems are also vulnerable to overloads and unforeseen events (e.g., extreme weather), and there are still many rural areas with poor coverage. This motivates the need for alternative forms of communication whereby devices communicate directly with each other, through wireless mesh networks (Akyildiz et al., 2005) or other hybrid solutions. We are slowly starting to see the emergence of such systems with the WifiDirect standard for household devices and the 802.11p standard for vehicular communication.

Dissemination of information in such wireless ad hoc networks requires coping with intermittent connectivity and may at times need to cope with high network traffic load. Moreover, for many potential applications, such as vehicular communication with warnings about hazards on the road (e.g., ice, fallen trees, and flooded areas), it is imperative that messages are disseminated within a short and predictable time frame. While there is a rich body of work on how to best make use of the limited resources in delay-tolerant networks (DTNs) (especially for unicast routing), there are not as many works that focus on urgent one-to-many message dissemination in this context. By urgent, we mean that messages should be delivered within some seconds[a] rather than within minutes or even hours, which is not an uncommon assumption in research on DTNs.

In this chapter, we explore the provision of timely dissemination of information in intermittently connected networks. We provide a brief overview of the existing

---

[☆] This chapter is a reprint of the chapter originally published in the first edition of Advances in Delay-Tolerant Networks (DTNs): Architecture and Enhanced Performance.

[a] There are some applications that require messages to be delivered within milliseconds, but such systems can hardly afford to be intermittently connected at any time.

Advances in Delay-tolerant Networks (DTNs). https://doi.org/10.1016/B978-0-08-102793-6.00011-4

research in the area as well as a more in-depth study on how to achieve a higher level of predictability of the message dissemination latency. Specifically, we explore how different prioritization mechanisms can be used to increase the delivery of urgent messages and thereby also increase the global system performance. We focus on the case when the system is subject to a traffic overload, which would normally reduce the system performance significantly.

We build on Random Walk Gossip (RWG), which is a manycast protocol for partition-tolerant networks tailored for disaster area networks. In its original design, RWG treats all messages equally until the time when their time to live (TTL) counter expires. In this work, we extend the design of the protocol by differentiating messages based on their deadline and progress so far. Due to the unique way RWG works, we can categorize the level of the lateness of a message ranging from very early to very late, and use this categorization when setting dynamic priorities to provide the best possible use of the limited resources. In contrast to several other works, we do not need to collect system statistics to make differentiation decisions. Instead, the necessary information is immediately accessible from the message header.

We study four different prioritization policies, including randomly ordering messages, which we consider as our baseline. The other three policies are Earliest Deadline First (EDF), which is well known from task scheduling theory, Least Informed First (LIF), which prioritizes messages with little progress, and Least Slack First (LSF), which is also a standard task scheduling algorithm but which we have adapted to fit RWG.

To evaluate these prioritization policies, we have conducted a simulation-based experimental evaluation using highly realistic traces from a vehicular scenario. Starting from traces of a large-scale realistic simulation of car movements in the city of Cologne, we have extracted a subset of traces that correspond to one-fifth of all cars within 3 km from the city center and used this in our simulations. The results show that message differentiation (MD) can significantly increase the number of messages that the system can handle without being severely overloaded. Moreover, we found that LSF provided the best results overall, but that EDF would still outperform the other policies when the system is highly overloaded.

The remainder of this chapter is organized as follows. Section 11.2 gives an overview of the state of the art in one-to-many communication in resource-constrained environments with intermittent connectivity. Section 11.3 presents the RWG protocol, which we have extended with message prioritization mechanisms, as explained in Section 11.4. In Section 11.5 we present a simulation-based experimental evaluation of the differentiation policies. Section 11.6 concludes the chapter.

11.2 One-to-many communication in resource-constrained environments

Timely message dissemination in challenged networks with intermittent connectivity and varying load is difficult at best, and there is a rich body of research concerning different aspects of this problem. In this section, we present a brief overview of the state of the art in this area. We begin with a more general look at one-to-many message dissemination in intermittently connected networks, proceed with a summary of the most relevant differentiation mechanisms, and finally present a selection of DTN protocols in which MD is a key element.

11.2.1   DTN multicast/broadcast

As wireless networks were entering our everyday lives in the 1990s, researchers began to wonder whether and how it would be possible to deliver messages in networks with extremely poor connectivity. This gave rise to what is commonly referred to as disruption-tolerant networks or DTNs. In these networks messages that cannot be forwarded due to lack of reachable neighbors are stored, to be forwarded at future encounters, giving rise to the notion of a *custodian*. A significant portion of the research on DTNs has focused on unicast routing, where each message has one sender and one destination. Other forms of communication, such as one-to-many or many- to-one, have often been considered as just extensions of unicast routing. However, in many emerging application areas such as vehicular networks and disaster area networks, *information sharing* is a crucial mechanism, for which one-to-many communication is very suitable.

There are several variants of one-to-many communication, including *broadcast,* where a message should reach all nodes in the network, *multicast* (Zhao et al., 2005; Mongiovi et al., 2012), where a message should reach a designated set of nodes, *geocast* (Baldoni et al., 2007), where a message should reach all nodes within a given area, and *manycast* (Carter et al., 2003), where a message should reach a minimum *number* of nodes.

One of the early proposals for partition-tolerant multicasts is the Hyperflooding protocol by Obraczka and Tsudik (1998). The idea is fairly straightforward. Each message keeps a hop count and a TTL, and each node keeps track of its neighbors. When a node discovers a new neighbor, packets that have not expired in TTL or hop count are propagated. Later, Viswanath and Obrazcka (2002) improved this algorithm by adapting the flooding policy to the estimated network conditions. The worse the connectivity, the more aggressive the flooding policy.

Continuing along the same lines, Khelil et al. (2007) use a two-stage approach called hypergossiping to achieve efficient broadcasts in partitioned networks. A message is the first broadcast within the current partition (using gossip). Every node keeps track of its neighbor nodes using regular hello messages. When a new neighbor is detected, the hello message is appended with a record that lists the recently received messages. By sharing these lists, the nodes can conclude that the new neighbor comes from a different partition, which means that the nodes should exchange messages with each other. To find out which messages to share, a node sends the messages it already has so that the neighbor nodes can send the missing ones.

Vollset and Ezhilchelvan (2005) present a manycast algorithm called Scribble. It is designed to be partition-tolerant and uses a node signature to keep track of informed nodes (several different signature types are suggested). In Scribble, a subset of nodes (termed *responsible*) periodically sends messages until the message is considered to be delivered by a sufficient number of nodes (or another node takes over that responsibility).

Cooper et al. (2009) propose a gossip-based broadcast algorithm called encounter gossip. The basic idea is to let a node repeatedly broadcast a message a fixed number of times, but only when encountering a new neighbor. This means that both the overhead and the delivery ratio can be kept constant for varying encounter frequencies, without the need for a time-based expiry. That is, if the nodes take a long time to meet, then the counter will be slowly incremented so the packets will remain longer in the network. If, on the other hand, nodes meet frequently, the counter is updated more often, and the packet can be discarded earlier. This model seems to be most suitable for scenarios with homogeneous mobility, where nodes are likely to meet new nodes at more or less regular intervals.

Much of the more recent interest in the DTN has focused on social forwarding algorithms, in which the previous interaction patterns of nodes are used to make routing decisions. Such an approach is also taken by Gao et al. (2012), who use network centrality metrics to select message relay nodes.

11.2.2   Differentiation mechanisms

The notion that critical resources such as bandwidth and energy are constrained in wireless networks has always been at the core of research efforts in wireless and ad hoc networks. Naturally, for sparse networks with intermittent connectivity, the problem of limited resources becomes even more important. When faced with a situation where demand exceeds the available resources (i.e., an overload situation), it can make sense to prioritize which messages to disseminate and which ones to ignore. Such a prioritization approach can be based on application-defined priorities (sometimes called message utility), or based on system-defined priorities to improve overall system performance metrics.

If the message priorities are defined by the application layer, or if the network traffic can easily be divided into different quality-of-service classes, then the network protocols and buffer policies can be tailored to uphold the quality of service of the most important messages at the expense of the less important messages. This can be done in several different ways, including weighted fairness and class-based queuing (El-Gendy et al., 2003).

Message differentiation can also be useful even if there is no easy way to determine the relative importance of messages. By considering other message properties, such as their size, destination, and TTL, it is possible to optimize how network resources are used to provide better delivery rates and lower latencies. Two main strategies can be discerned for how to construct a message prioritization policy: *message-centered* and *system-centered*. In the message-centered approach, message properties are used to determine an ordering among messages, for example by giving higher priority to messages with a closer deadline. The resulting policies are then evaluated through analytical and experimental studies. In the system-centered approach, an analytical model of the system is created, which allows parameter values to be analytically derived that optimize some particular system performance metric (e.g., delivery ratio). However, it might be difficult to obtain a good analytical model of a complex network, and the resulting optimal policy might require global knowledge of the current network properties.

11.2.3 Message differentiation in intermittently connected networks

There are several ways in which MD can be used in DTNs. In *buffer management*, MD is used to determine which messages to keep and which ones to drop when the buffer is full. Gossip-style protocols can use MD to determine whether or not to forward a message, and in which order (Ramanathan and Singh, 2008; Cornejo et al., 2013). MD can also be used to determine the order in which to send messages when a node discovers a new neighbor; this is often called *message scheduling.*

Due to the overwhelming focus on unicast routing in the DTN community, most works on MD consider only unicast routing, even though the policies are often general enough to be applicable to one-to-many communication as well. In particular, there are several works (Ramanathan et al., 2007; Li et al., 2010; Shin and Kim, 2011) that apply MD to epidemic routing, which is basically a one-to-many message dissemination protocol used to reach a single destination.

For example, the PRioritised EPidemic (PREP) routing algorithm by Ramanathan et al. (2007) creates a topological map of the network, including a link estimation metric. This information is used to prioritize messages that are closer (shorter paths) to their destination. Moreover, the expiry and creation time of messages are considered when scheduling messages for transmission.

The RAPID protocol by Balasubramanian et al. (2007) is a utility-driven unicast protocol that tries to estimate the remaining time until a packet is delivered based on the historical meeting patterns in the network. This information is used by the protocol to greedily prioritize messages with, for example, the smallest expected delivery time.

Krifa et al. (2012) derive a globally optimal policy for buffer management and message drop in networks with exponentially (possibly with different pair-wise means) distributed intermeeting times. They also discuss a framework to collect network statistics as required to make decisions at a local node level. A similar approach is taken by Elwhishi et al. (2012), who also use a network estimation framework to collect information necessary to make scheduling decisions. However, their analytical framework is based on a fluid epidemic model rather than discrete.

Differences among nodes in their available energy and bandwidth can have a big impact on the success of message delivery. Sandulescu et al. (2013) focus on the estimation of these node-specific resources in the vicinity of a node. This allows custodian selection and message ordering to take into account the available bandwidth and energy of its neighbors and thereby make better use of these resources, especially when messages differ in size.

Common to many of the above works is that they try to estimate the current state of the network to improve the quality of MD mechanisms. In this chapter, we take a different approach and investigate how MD can be achieved using rich message metadata that itself captures the relevant information. The benefit of this approach is that it does not require any a priori knowledge of the network characteristics, does not cause a lot of signaling overhead, and is not affected by changes in the network.

11.3   Random walk gossip (RWG)

Random Walk Gossip (Asplund and Nadjm-Tehrani, 2009) is a message dissemination protocol designed to cope with the challenges faced in disaster area networks, including scarcity of bandwidth and energy, as well as unknown and unpredictable network topologies with partitions. RWG is a manycast protocol, which means that a message is intended to reach a given number k of nodes. When k nodes have been reached, the message is k-delivered and does not need to be propagated any more, thus not wasting bandwidth and energy. RWG uses a store-carry-forward mechanism to cope with partitions.

We present an overview of the protocol in three steps. First, we discuss the basic random walk mechanism in which messages are actively propagated in the network. Next, we discuss the message structure and how the protocol uses this to perform message dissemination efficiently. Finally, we discuss the message activation mechanism, which is of importance when adding MD to the protocol.

11.3.1   Random walk and handshake mechanism

When a message is sent in a connected part of the network, it performs a random walk over the nodes, until all the nodes in the partition are informed of this message. This is controlled by a three-way packet exchange, as shown in Fig. 11.1. First, a Request to Forward (REQF), which includes the message payload, is sent by the current custodian of the message (grey node in the picture). The neighboring nodes that hear the REQF reply with an acknowledgment packet (ACK), unless at least L nodes have already sent an acknowledgment (L being a parameter for which 3 has been found to give the best results). The custodian randomly chooses one of the nodes that acknowledged and sends an OK to Forward (OKTF) to this node, indicating that it will be the next custodian. The other nodes retain the message without actively disseminating it. In addition to delivering the message to the application layer, they keep the message as *inactive* until it expires. Partitions can be overcome by the movement of nodes. Thus, new uninformed nodes will be informed by some node that keeps the

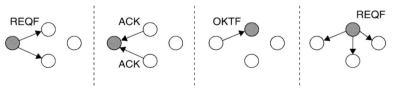

Fig. 11.1 Random Walk Gossip handshake.

message as *inactive* and restarts to disseminate. This process will continue as long as no more uninformed nodes remain in the network or the message is k-delivered.

Finally, when a node realizes that a message is k-delivered, it sends a Be Silent (BS) packet to its vicinity. This packet will cause all receiving nodes to also realize that the message is k-delivered and thus remove it from their buffers. No new BS packets are sent upon the reception of a BS packet.

11.3.2 Message metadata

All the packet types share the same header structure. Since there are some complex features of the protocol, which we do not discuss here, we concentrate on the most important header fields:

- Packet ID, a 64-bit identifier
- Group size (i.e., the k parameter), 16-bit integer
- TTL, remaining time in ms until the packet expires, 32-bit integer
- Starting TTL, the lifetime of the message when first sent (stays constant), 32-bit integer
- Informed, bit vector that represents informed nodes, 256 bits.

Note that the starting TTL field is not part of the original RWG protocol, but has been added in this study since this information is necessary for some of the prioritization policies, which we investigate. This adds 4 bytes to the header, which slightly increases the overhead to the protocol. The full header size is 92 bytes, and in our simulations, we have used a payload of 500 bytes, so the increase in packet size is less than 1 per cent.

The *informed* bit vector (implemented as a Bloom filter) is used to keep track of which nodes have seen a given message (see Fig. 11.2). When a node receives the message it produces a hash of its own address and puts a "1" in the bit vector in the field corresponding to the hash. This allows the protocol to know when a message is k-delivered, and to tell the potential future recipients of the message how far the message has reached towards its dissemination goal (summing the number of "1"s indicates the current known local knowledge of this).

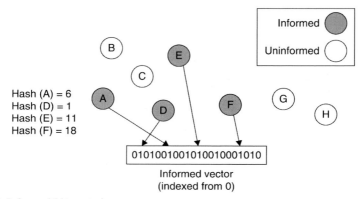

Fig. 11.2 Informed bit vector.

11.3.3 Message activation mechanism

The *informed* vector is also used for deciding whether or not to activate an inactive message. When a node A hears a new neighbor B, A will go through the messages stored in its buffer (i.e., the set of inactive messages) to see whether B has not yet been informed of any of these messages, in which case those messages will be reactivated and broadcast to node B (and other uninformed nodes in the vicinity). If the hash of node B is marked with a "0" in the informed vector, then it is likely that B has not heard of this particular message. Naturally, the information in the informed vector represents a limited and potentially stale view of the network, meaning that some messages will be unnecessarily activated. Moreover, since the informed vector uses a hash function to index nodes, hash collisions can happen, which will cause the message not to wake up even if that would have caused it to spread to a new node.

11.4 RWG and message differentiation

The original version of RWG does not make any distinction or prioritization between different packets. The protocol is designed so that each message should always produce a minimal amount of overhead independently of whether there are other packets in the system. Even in cases of high load, the protocol will not adapt, but will simply send packets to the MAC layer, where they are put in a queue. To make RWG better suited to deal with overloads, we added the possibility to differentiate messages based on their properties. This message differentiation is used in two cases: (1) when activating messages and (2) in the acknowledgment step of the handshake mechanism.

Message activation, which mainly happens when a node hears another (potentially new) neighbor and activates messages, was explained in Section 11.3.3. With message prioritization enabled, messages are activated in priority order. Thus, high- priority messages are always activated first, allowing them to spread faster in the network compared with other messages.

In the acknowledgment step, as explained in Section 11.3.1, nodes will not send an acknowledgment if enough nodes have already replied. With message prioritization enabled, this acknowledgment decision will also consider whether the message has a higher priority than the other currently active messages, in which case an acknowledgment will be sent in any case. Again, this speeds up the delivery of high-priority messages at the expense of lower-priority messages.

We now proceed to present the four different prioritization policies, which we have studied as possible options for how to decide which messages should be given higher priority.

11.4.1 Random order

As a baseline, we consider a random ordering of the messages, which is basically the same as having no differentiation mechanism at all. This policy has the benefit of being simple and straightforward as well as providing fairness among messages.

Other policies, such as first-in-first-out, could also have been used as a baseline. However, due to the way RWG frequently removes and inserts elements in the list of inactive messages, there is no obvious notion of "first in queue", which is why we opted for the random order as the most logical baseline.

11.4.2 Least informed first

When the system is overloaded, so that some of the messages might miss their deadline, but most are delivered in time, then it makes sense to help the messages that have not yet reached many nodes. This is the rationale for the LIF policy, which gives higher priority to messages with fewer informed nodes.

11.4.3 Earliest deadline first

EDF is one of the classical scheduling algorithms. EDF is optimal among all scheduling algorithms for uniprocessor systems in the sense that for any task set, if there exists an algorithm that can schedule the tasks without any deadline overruns, then EDF will also find such a schedule (under some assumptions such as task independence). EDF gives the highest priority to the task with the earliest deadline. This means that task priorities can change dynamically over time and that it is possible to dynamically schedule arriving tasks.

11.4.4 Least slack first

The LSF scheduling can be seen as a combination of the EDF and LIF policies in the sense that it considers both the deadline of a message and how much progress the message has made. It is based on the notion of slack, which in uniprocessor task scheduling is defined as $s(t, i) = (D(i)-t)-r(i,t)$, where t is the current time, $D(i)$ is the deadline of task i, and $r(i,t)$ is the remaining computation time of task i at time t. We use the same notion for messages by defining the remaining time as the number of remaining nodes to inform times and information spreading rate. Mathematically, we write this as $r(i,t) = (k(i)-I(i,t))R(i)$, where $k(i)$ is the group size of message i, $I(i,t)$ is the number of nodes that have been informed by i at time t, and $R(i)$ is the expected rate at which new nodes are informed. LSF gives higher priority to messages with a smaller slack value.

11.5 Evaluation with vehicular mobility models

In this section, we present our simulation-based evaluation of RWG when combined with the above four different prioritization policies. The main objective of the evaluation is to investigate how these policies behave in a realistic large-scale scenario where the system load exceeds the network capacity.

We consider a vehicular scenario where cars use direct communication and local forwarding to exchange information about the current traffic scenario. Such information could be that a queue is building up in a certain location, that an emergency vehicle is approaching, or that there are obstacles in the road. Common to these examples is that they are relevant only for a short period of time and that they need to be delivered within a short time frame to be of any use for the receiver. Moreover, the information is mostly local and messages need not reach all nodes in the network.

11.5.1   Experimental setup

We used the Ns3 network simulator version 3.16, which provides a packet-level simulation environment with a well-documented structure. In accordance with the evaluation scenario, we consider messages with a relatively short TTL. One-third of the messages have a TTL of 5 s, one-third have a TTL of 10 s, and the rest can live for 20 s. We assume that, even if some events can be of relevance for a longer duration than 20 s, such information will be continuously resent as new messages. All messages have a group size of 30 nodes (unless otherwise stated), meaning that they should be delivered to at least 30 nodes in order to be k-delivered. Messages are generated at a constant rate but from randomly selected senders at a rate ranging from 5 to 25 messages per second. We have used the 802.11p Wi-Fi standard for car2car communication with a transmission power of 28.8 dBm, which is the maximum allowed power according to the standard and corresponds to a transmission range of approximately 300 m in our simulation.

We logged a total of 300 messages for each point on the illustrated charts. In order to remove any boundary effects caused by the simulation, we ran the simulation for 20 s (same as the longest TTL) before starting to log messages, and we let the simulation continue 20 s after creating the last of the logged messages, which resulted in a simulation time ranging from 52 s to 100 s. The simulation parameters are summarized in Table 11.1.

11.5.1.1   Mobility

To get as realistic a vehicular mobility model as possible, we used a large-scale mobility trace generated by the microscopic (i.e., each car is simulated with a driver model) traffic simulator Sumo. The trace is made available by the TapasCologne project (Uppoor et al., 2013) and is based on the German city of Cologne. The trace file we used ranges from 6 am to 8 am and contains a total of 121,140 vehicles.

We assume that only 20% of all vehicles are equipped with communication capabilities, which reduces the number of nodes to 23,492. In order to reduce the scale of the simulation, we then decided to concentrate on the vehicles whose initial position was within 3 km of the center and which had some movement in a 20-min period from 6:50 to 7:10. This reduced the number of cars to 1092, which is still a large number of nodes to handle with a packet-level simulator like Ns3.

We further take advantage of the fact that most journeys are relatively short, meaning that most of the time the car is turned off (and thus not taking part in the network). It turns out that, out of the 1092 vehicles in total, only 366 are driving around at any point in time, while the rest are parked. Therefore, in the simulator, we use 366

Table 11.1 Simulation parameters.

Time to live	5, 10, 20 s
Group size (k)	30
System load	5–25 messages/s
Wi-Fi standard	802.11p SCH
Transmission power	28.8dbm
Number of logged messages	300 per measurement point
Simulation time	52–100 s
Mobility model	Vehicular
Number of nodes	1092 (at most 366 active)
Area	28.3 km^2

different node entities that behave like the 1092 original vehicles. The simulator allows nodes to move instantaneously from one position to another (when a node changes identity) as well as completely removing any information about which messages a removed node has encountered or stored. The end result is that we get the same system behavior as the original 1092 nodes, which corresponds to 20% of all the vehicles within 3 km from the city center.

11.5.1.2 Metrics

We measure the performance of the differentiation mechanisms using the k-delivery ratio and the k-delivery latency. The k-delivery ratio is defined as $\frac{1}{N}\sum_{n=1}^{N} d_i$. where N denotes the number of logged messages, and $d = 1$ if message 1 was delivered to at least k nodes, and $d = 0$ otherwise. The latency of a k-delivered message is defined as the time from when the message was created to when it became k-delivered.

11.5.2 Comparison of message differentiation mechanisms

MD makes the most sense when some messages cannot be given the required service due to restricted resources, which is what happens in an overload situation. Thus, to assess the effectiveness of the different differentiation mechanisms, it makes sense to consider a scenario in which the load negatively affects the system performance. In the case of our experimental setup, this happens when the total number of messages introduced in the system (we denote this *system load*) exceeds 10 messages per second.

Fig. 11.3 shows the k-delivery ratio of the different mechanisms as the load varies from 5 messages/s (system not overloaded) to 25 messages/s (severe overload). As expected, the performance when the system is not overloaded (5–10 messages/s) is basically the same independently of which differentiation mechanism is used. When the load exceeds 10 messages/s, the performance with random message ordering (which basically means no message MD) drops drastically, whereas prioritizing messages with a short deadline or slack keeps an acceptable performance for considerably higher loads. Specifically, if the LSF mechanism is used, the system is capable of

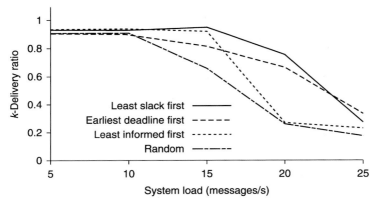

Fig. 11.3 k-Delivery ratio versus system load ($k = 50$).

handling at least a 50% higher load compared with random message ordering in our scenario.

EDF seems to behave similarly to LSF, but, as we will soon see, there are some differences that cannot be seen in this graph alone. The LIF mechanism, which prioritizes messages that have not yet reached as many nodes, performs well at a moderate overload but quickly degrades in performance as the load increases further. This makes sense since LIF prioritizes messages that have not yet reached a lot of nodes. If the overload is moderate, this will help the messages that would have otherwise missed their deadline. As the overload gets worse, and there is no chance of delivering all messages, LIF is counterproductive, since it gives a higher priority to messages that have no chance of being delivered in any case.

Since RWG is a manycast protocol, it is also interesting to consider how the group size affects the protocol performance under different differentiation mechanisms. Fig. 11.4 that shows the k-delivery ratio as a function of group size, tells a similar story to Fig. 11.3. The LSF and LIF mechanisms allow the k-parameter setting to be 25% higher (from 40 to 50) compared with the baseline without degrading the system performance.

As the load increases, and the more transmissions that will happen in the system, the longer each message will need to wait in each step of the dissemination process. Fig. 11.5 shows how this affects the average latency. As expected, the latency increases significantly as the system becomes overloaded, and the random order baseline shows the worst performance. More interestingly, as the overload becomes severe with 25 messages/s, the EDF approach seems to perform much better compared with the other mechanisms.

11.5.3 Detailed analysis of resulting latency distributions

In the previous section, we discussed how the delivery ratio and average latency were affected by increasing load and group size under four different scheduling policies. To gain a deeper understanding of these results, we also analyze the latency distributions,

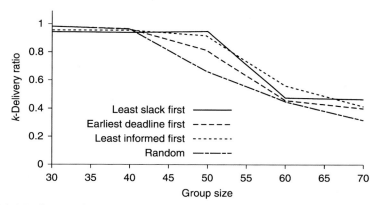

Fig. 11.4 k-Delivery ratio versus group size.

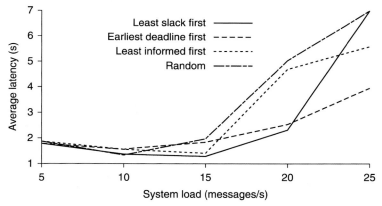

Fig. 11.5 Average latency versus system load.

which are more useful to illustrate how quickly messages are disseminated in the network.

Fig. 11.6 shows the latency distribution as a Cumulative Distribution Function (CDF) for a system load of 15 messages/s. Here we consider undelivered messages as having an infinite latency, which is the reason why the distributions never reach 1. Interestingly, while the final delivery ratio differs among the different policies, they all have a steep rise before 5 s, after which not many messages are delivered.

Fig. 11.7 shows the corresponding results for a system load of 25 messages/s, meaning that the system is severely overloaded. There are some interesting differences. First of all, EDF seems to perform much better than the other policies. We could see this also in the average latency, which was lower for EDF at 25 messages/s. Now we can see how this is reflected in the shape of the latency distributions. For EDF, the shape is similar to those in Fig. 11.6, in the sense that most of the k-delivered messages have a latency of less than 5 s. For the other policies, the shape is closer to a straight line, with message latencies being spread evenly in the range of 0–20 s. Note that after

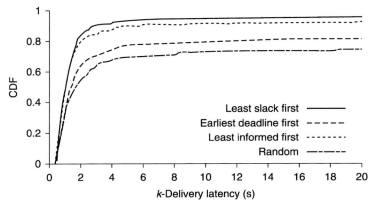

Fig. 11.6 Cumulative distribution function (CDF) at 15 messages/s.

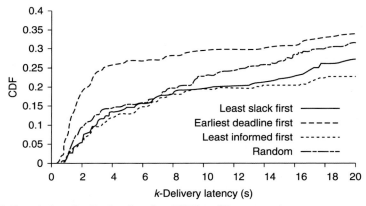

Fig. 11.7 Cumulative distribution function (CDF) at 25 messages/s.

20 s all curves stay at a constant level since no messages are delivered after more than 20 s, which is the longest TTL of any message in the system.

Finally, Figs. 11.8 and 11.9 show the latency distribution for the most urgent messages (those with a TTL of 5 s) for load profiles of 15 and 25 messages/s, respectively.

At 15 messages/s, the latency distribution of the urgent messages is similar to when considering all messages. The main difference is that EDF performs significantly better for the urgent messages (compared with EDF for the total messages in Fig. 11.6). At 25 messages/s (Fig. 11.9), the difference is more significant. Here EDF clearly outperforms the other policies.

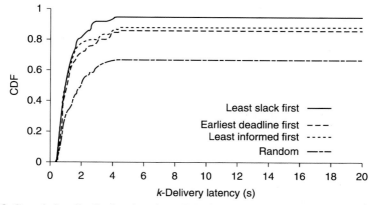

Fig. 11.8 Cumulative distribution function (CDF) for urgent messages at 15 messages/s.

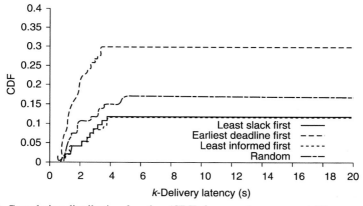

Fig. 11.9 Cumulative distribution function (CDF) for urgent messages at 25 messages/s.

11.6   Discussion

In this chapter we discussed MD protocols for intermittently connected networks, and how MD has been used to increase message delivery and reduce latency in such networks. A lot of progress has been made in understanding intermittently connected networks in the last few years. Most general basic phenomena are fairly well understood by the community, and deeper insights are often inhibited by the lack of proper domain-specific data that can be analyzed. MD has been discussed more or less actively since the early paper on RAPID by Balasubramanian et al. (2007), but the main focus has been on unicast routing for networks with very long delays.

We have tried to complement the existing body of work by studying how MD mechanisms can be effectively used in an overloaded network with very short message deadlines. The study was based on four different prioritization policies: random order, EDF, LIF, and LSF. Using realistic vehicular traces, we simulated the performance of an overloaded network running the RWG manycast protocol in conjunction with these four policies. Due to the way RWG keeps track of informed nodes, it is possible to implement these prioritization policies without collecting network statistics over time to estimate some locally unknown parameters.

LSF, which is basically an adaptation of a well-known uniprocessor scheduling algorithm, seems to provide the best overall results. For more severe overloads, the simpler EDF policy outperforms the other policies, especially in keeping down the latency of the delivered messages.

The bottom line is that relatively simple time-based prioritization mechanisms can be a powerful tool to increase the performance in wireless networks with intermittent connectivity and constrained resources. In particular, for vehicular networks with urgent MD requirements, the system load can be increased by 50% by using the LSF policy.

References

Akyildiz, I.F., Wang, X., Wang, W., 2005. Wireless mesh networks: a survey. Comput. Netw. 47 (4), 445–487.

Asplund, M., Nadjm-Tehrani, S., 2009. A partition-tolerant Manycast algorithm for disaster area networks. In: 28th International Symposium on Reliable Distributed Systems (SRDS). IEEE, pp. 156–165.

Balasubramanian, A., Levine, B., Venkataramani, A., 2007. DTN routing as a resource allocation problem. ACMSIGCOMM Comp. Commun. Rev. 37 (4), 373.

Baldoni, R., Ioannidou, K., Milani, A., 2007. Mobility versus the cost of geocasting in mobile ad-hoc networks. In: Pelc, A. (Ed.), 21st International Symposium on Distributed Computing (DISC). Springer-Verlag, pp. 48–62.

Carter, C., Yi, S., Ratanchandani, P., Kravets, R., 2003. Manycast: exploring the space between anycast and multicast in ad hoc networks. In: 9th Annual International Conference on Mobile Computing and Networking (MobiCom). ACM, pp. 273–285.

Cooper, D.E., Ezhilchelvan, P., Mitrani, I., 2009. Encounter-based message propagation in mobile ad-hoc networks. Ad Hoc Netw. 7 (7), 1271–1284.

Cornejo, A., Newport, C., Gollakota, S., Rao, J., Giuli, T., 2013. Prioritized gossip in vehicular networks. Ad Hoc Netw. 11 (1), 397–409.

El-Gendy, M.A., Bose, A., Shin, K.G., 2003. Evolution of the internet QoS and support for soft real-time applications. Proc. IEEE 91 (7), 1086–1104.

Elwhishi, A., Ho, P.-H., Naik, K., Shihada, B., 2012. A novel message scheduling framework for delay tolerant networks routing. IEEE Trans. Parallel Distrib. Syst., 871–880.

Gao, W., Li, Q., Zhao, B., Cao, G., 2012. Social-aware multicast in disruption-tolerant networks. IEEE/ACM Trans. Networking 20 (5), 1553–1566.

Khelil, A., Marron, P.J., Becker, C., Rothermelns, K., 2007. Hypergossiping: a generalized broadcast strategy for mobile ad hoc networks. Ad Hoc Netw. 5 (5), 531–546.

Krifa, A., Barakat, C., Spyropoulos, T., 2012. Message drop and scheduling in DTNs: theory and practice. IEEE Trans. Mob. Comput. 11 (9), 1470–1483.

Li, Y., Jiang, D., Jin, D., Su, L., Zeng, L., Wu, D.O., 2010. Energy-efficient optimal opportunistic forwarding for delay-tolerant networks. IEEE Trans. Veh. Technol. 59 (9), 4500–
4512.

Mongiovi, M., Singh, A.K., Yan, X., Zong, B., Psounis, K., 2012. Efficient multicasting for
delay tolerant networks using graph indexing. In: 2012 Proceedings IEEE TNFOCOM.
IEEE, pp. 1386–1394.

Obraczka, K., Tsudik, G., 1998. Multicast routing issues in ad hoc networks. In: Universal Personal Communications, 1998. ICUPC'98. IEEE 1998 International Conference on,
pp. 751–756.

Ramanathan, P., Singh, A., 2008. Delay-differentiated gossiping in delay tolerant networks. In:
2008 IEEE International Conference on Communications. IEEE, pp. 3291–3295.

Ramanathan, R., Hansen, R., Basu, P., Rosales-Hain, R., Krishnan, R., 2007. Prioritized epidemic routing for opportunistic networks. In: Proceedings of the 1st International MobiSys
Workshop on Mobile Opportunistic Networking – MobiOpp'07. ACM Press, New York,
pp. 62–66.

Sandulescu, G., Schaffer, P., Nadjm-Tehrani, S., 2013. Exploiting resource heterogeneity in
delay-tolerant networks. Wirel. Commun. Mob. Comput. 13 (3), 230–243.

Shin, K., Kim, S., 2011. Enhanced buffer management policy that utilises message properties
for delay-tolerant networks. IET Commun. 5 (6), 753.

Uppoor, S., Trullols-Cruces, O., Fiore, M., Barcelo-Ordinas, J., 2013. Generation and analysis
of a large-scale urban vehicular mobility dataset. In: IEEE Transactions on Mobile Computing. PP(99), p. 1. To appear.

Viswanath, K., Obrazcka, K., 2002. An adaptive approach to group communications in multi
hop ad hoc networks. In: Seventh IEEE Symposium on Computers and Communications
(ISCC). IEEE Computer Society, pp. 559–566.

Vollset, E.W., Ezhilchelvan, P.D., 2005. Design and performance study of crash tolerant protocols for broadcasting and reaching consensus in MANETs. In: 24th IEEE Symposium on
Reliable Distributed Systems (SRDS), pp. 166–175.

Zhao, W., Ammar, M., Zegura, E., 2005. Multicasting in delay tolerant networks: semantic
models and routing algorithms. In: WDTN'05: Proceedings of the 2005 ACM SIGCOMM
Workshop on Delay-Tolerant Networking. ACM, pp. 268–275.

Using social network analysis (SNA) to design socially aware network solutions in delay-tolerant networks (DTNs)☆

B. Jedari, F. Xia, A. Mohammed Ahmed, P. Pirozmand, and Y. Najaflou
Dalian University of Technology, Dalian, China

12.1 Introduction

In the last few years, several networking solutions have been proposed which take advantage of social network analysis (SNA) techniques (Wasserman and Faust, 1994) to promote human interaction and improve the performance of data forwarding services in delay-tolerant networks (DTNs) (Fall, 2003; Khabbaz et al., 2012). In this paradigm, mobile devices such as smartphones and tablets are carried or controlled by humans. Hence, users' social characteristics and relationships as well as their movement and contact patterns can be exploited to devise efficient protocols in DTNs. For example, people with the same social similarities and interests may contact each other with higher probability. Therefore, they could form a community and share data efficiently through the community via their mobile phones. Consequently, it is of paramount importance to study the influence of users' various social behaviors on the performance of networking protocols in DTNs.

Communication in human-centric DTNs is mainly based on opportunistic contacts between mobile devices based on low-cost and short-distance wireless technologies such as Wi-Fi and Bluetooth. Through the pair-wise contacts, mobile users exchange various kinds of data such as photos, advertisements, software updates, etc. SOCIALNETS (http://www.social-nets.eu) is a sample prominent project in this area which attempts to exploit the underlying social network structure to develop adaptive data delivery protocols in intermittently connected networks. This project exploits the social interactions and habits of users to drive the design of protocols for online social networks and opportunistic wireless networks.

Recently, SNA techniques and methods have been introduced into existing protocols in DTNs (Musolesi and Mascolo, 2007; Hossmann et al., 2010). For instance, socially aware networking (SAN) (Xia et al., 2013) is a new trend for DTNs which mainly encompasses the exploration of mobility patterns as well as similarities

---

☆ This chapter is a reprint of the chapter originally published in the first edition of Advances in Delay-tolerant Networks (DTNs): Architecture and Enhanced Performance.

Advances in Delay-tolerant Networks (DTNs). https://doi.org/10.1016/B978-0-08-102793-6.00012-6

between mobile users in pervasive environments in order to uncover useful interaction and evolutionary social patterns among mobile users.

Schurgot et al. (2012) explored the social structure of mobile users and revealed that, beyond physical and contact graphs, social graphs (formed by interpersonal relationships) can be utilized to enhance the efficiency of the networking protocols in DTNs considerably. In addition, innovative SNA metrics such as centrality, similarity, and tie strength were proposed to extract the social features of mobile users and quantify their important properties in DTNs.

Human mobility is a challenging issue, which significantly affects the simulation and evaluation of social-based protocols in DTNs. Mobility data for DTNs are mainly produced in two different ways: real-world traces and simulation-based models. Real-world mobility traces are acquired using different kinds of communication systems and devices (Aschenbruck et al., 2011) (most real-world mobility datasets are accessible on the CRAWDAD archive; crawdad.cs.dartmouth.edu). However, most of the real traces have been captured in bounded environments such as universities and they are not scalable. Furthermore, they are not controllable and flexible for changing system parameters such as node density and node velocity. To this end, simulation-based models have been proposed which aim to mimic the movement of humans in real life and simulate their mobility synthetically using parametric methods (Karamshuk et al., 2011). Some literature has also analyzed mobility data in order to characterize the spatial, temporal, and connectivity properties of human movement. For example, experimental analysis by Phithakkitnukoon et al. (2012) demonstrated that users frequently visit the locations with which they have strong social ties.

There exists no stable end-to-end route between sender and receiver devices in DTNs, and they commonly use a *store-carry-and-forward* scheme to exchange data with each other. Hence, data routing and forwarding become a very challenging issue in this paradigm. Recently, several social-based data dissemination protocols for DTNs have been proposed, since users' social properties tend to be stable over time (Jedari and Xia, 2014; Zhu et al., 2013). The protocols in this class mainly utilize characteristics such as community concept or other features like contact duration, node centrality, similarity, and interest to enhance the efficiency of the forwarding algorithms in DTNs. Generally, the main objective is to select the most appropriate relay nodes with the highest probability of meeting the destination node(s) in order to improve the data delivery ratio and delay. In some cases, some users may not be perfectly willing to store or relay messages on behalf of the others. Therefore, the impact of selfish behaviors on the performance of existing algorithms should also be considered and appropriate incentive mechanisms should be applied to stimulate selfish users to cooperate in data delivery.

This chapter explores the social characteristics of DTNs in three main categories. The remainder of the chapter is organized as follows. The application of SNA to enhance the efficiency of DTN protocols is elaborated in Section 12.2. Some pioneering social-based mobility models for DTNs are presented in Section 12.3. The prominent social-based routing and data distribution algorithms in DTNs are explored in Section 12.4. Section 12.5 concludes the chapter.

12.2   Social characteristics of delay-tolerant networks (DTNs)

SNA has recently gained much attention in many fields, such as anthropology, communication studies, economics, computer science, and engineering. Contemporary research in this area mainly explores relationships and ties among social entities in order to reveal useful patterns and structures of interaction among them. With the increasing proliferation of new information technologies such as smart sensing and mobile networking, SNA plays an important role in analyzing and designing new policies, protocols, and applications for ubiquitous environments.

In the DTN context, SNA techniques can be greatly utilized to analyze state information of the network nodes in order to devise effective and efficient protocols and evaluation metrics. Exploring mobile data, SNA can retrieve inherent relationships and structures of mobile nodes and deduce important properties such as node centrality, similarity, tie strength, community structure, etc. In the rest of this section, the most important issues regarding SNA and its applications in DTNs with respect to social graphs, measurement metrics, and community structure are studied.

12.2.1   The social graph

A social graph (sometimes called a sociogram) is a useful structure to represent the connectivity of ties among social actors and analyze various relationships among them, such as community structure or friendship. Formally, it is a graph-based representation of a social network, where nodes correspond to social entities (e.g., individuals) and edges represent social ties between them. A social graph in a DTN can be formed by interpersonal and friendship relations, which can be defined between network nodes based on their contact information (i.e., contact frequency and duration). In this setting, the graph can be demonstrated with some nodes as the individuals and an edge between two nodes if they had at least a contact opportunity with each other during the period of interest. Alternatively, a weighted edge can be inserted, with a weight representing the number or duration of contacts between the two nodes during the period of interest. However, the structure of such a graph can change dynamically, since relationships between users evolve in social life.

12.2.2   Measurement metrics

Several social-based metrics have been formalized to characterize and evaluate different properties of mobile individuals and groups and use them to make better decisions in DTN protocols. The available metrics can be discussed in three main classes: node centrality, social similarity, and tie strength. According to Daly and Haahr (2007), centrality in network analysis is a measure of the relative importance of a node within the graph. There are several ways to measure centrality. Three widely used centrality measures are degree centrality, closeness centrality, and betweenness centrality. Degree centrality is measured as the number of direct links that involve a given

Degree and closeness centrality node

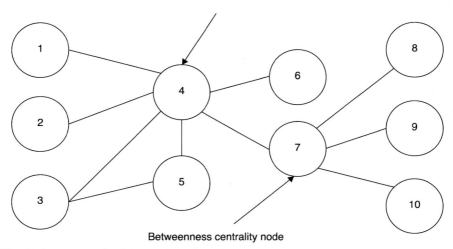

Betweenness centrality node

Fig. 12.1 An example of degree centrality, closeness centrality, and betweenness centrality nodes.

node. Closeness centrality is the shortest path between a node and all other reachable nodes. Closeness centrality of a node is a measure of how long it will take information to spread from a given node to other nodes in the network. Betweenness centrality is a centrality measure of a node that acts as a bridge along the shortest path between two other nodes. An example of three kinds of centrality metrics is shown in Fig. 12.1.

Considering the above definitions, node centrality measures do not take into account the strength of the links between network nodes. Tie strength (Granovetter, 1973) is a quantifiable property that characterizes the link between two nodes and is defined as "a combination of the amount of time, the emotional intensity, the intimacy, and the reciprocal services, which characterize a tie." In most cases, ties are represented by weighted edges connecting two nodes of a graph. The social similarity is another important metric, which indicates the group of nodes depending upon common contacts or interests, which can be measured by the ratio of common links between individuals. It has been believed by sociologists, such as Milgram (1967), that there is a higher probability of two people being acquainted if they have one or more other acquaintances in common (small world theory). In a social network, the probability of two nodes being connected by a link is higher when they have a common neighbor.

12.2.3 Community structure

A community is a structural subunit of individuals in a network with stronger ties to members within the community than to members outside the community. The connections between the nodes inside a community could be family, friends, or common locations. In general, individuals in the same community meet each other more frequently. Consequently, their contact patterns can be utilized to choose the appropriate forwarding paths between nodes. However, the social characteristics and behaviors of a mobile user can change dynamically, which makes community discovery a very challenging issue in mobile networks.

Broadly, two main steps are carried out to form a community structure in a dynamic network. As a first step, an evolving network is converted into static graphs at different time steps. Then, a community detection algorithm is used to infer relationships between partitions at different periods. A sample of evolving networks containing different communities is shown in Fig. 12.2. There has been a considerable amount of work done to identity common parts between two partitions in static graphs, which is known as a matching problem (Hopcroft et al., 2004; Spiliopoulou et al., 2006).

A number of researchers have striven to identify critical events that characterize the evolution of communities in DTNs. For example, a community may merge with another one, split or disappear. Palla et al. (2007) identified events by applying the clique percolation method on a graph formed by the communities discovered at two consecutive snapshots. Similarly, Bubble Rap (Hui et al., 2011) was proposed. This uses weighted network analysis and k-clique algorithms for community detection, and the contact threshold for extracting the k-clique community. Another event-based framework to capture all the transitions between communities at two consecutive snapshots was provided by Takaffoli et al. (2011). In this method, a series of significant events and transitions are defined to characterize the evolution of networks in terms of their communities and individuals. The event definition formula aims to track the transitions of communities over the entire observation time, not only between two consecutive snapshots. In addition, to capture the changes that are likely to occur for a community, five events (form, dissolve, survive, split, and merge) and four transitions (size transition, compactness transition, leader transition, and persistence transition) are defined.

12.3   Social-based human mobility models

Movement trajectories of mobile carriers in their daily life routines are strongly affected by their personal and social characteristics as well as environmental parameters. For example, interests and friendship relations are the two main features that considerably distinguish mobility patterns of different users. The users might also be attracted to some specific locations, such as historical places and shopping malls, and/or some people, like tour guides or close friends. Various aspects of human

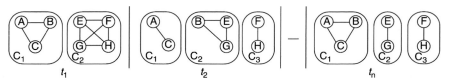

Fig. 12.2 An example of community evolution at different time steps. *C*, community.

mobility patterns have been characterized in some recent studies. For example, the authors (Phithakkitnukoon et al., 2012) analyzed real-life human mobility data and revealed that users frequently visit the locations with which they have strong social ties. Furthermore, it is concluded that users visit just a few locations, where they spend the majority of their time (Song et al., 2010). In most cases, they often travel over short distances and rarely migrate long distances (Gonzalez et al., 2008).

Human mobility models have been extensively utilized to simulate and evaluate the performance of protocols in DTNs. In some researchers, real-world traces that have been collected using wireless LAN, Bluetooth, and Wi-Fi technologies on campuses, conferences, and entertainment environments are used to set up mobility in DTNs. An anonymous version of most of these data are now available to other research groups in the CRAWDAD archive. Interested readers are also referred to Aschenbruck et al. (2011) for more details about the trace-based mobility models. However, most of the real traces are environment-specific and they are not collected on a wide scale. Furthermore, parameters of network nodes, such as the number of nodes, transition range of the devices, or bandwidth, cannot be modified. These problems forced researchers to use synthetic mobility models to generate mobility traces which are flexible, and parameters of the mobility models can be changed according to problem specifications. Simulation-based mobility models are the most widely used models, which attempt to mimic the mobility behavior of nodes without the support of an existing real trace dataset. In the rest of this section, some pioneering social-based mobility models are introduced. For further study about simulation-based mobility models, interested readers are referred to Karamshuk et al. (2011).

The community-based mobility model (CMM) (Musolesi and Mascolo, 2007) was one of the first mobility models to be directly founded on social network theory. Network nodes in this model are divided into several groups (called communities), based on their social relationships, and each group is assigned a geographical place. Various kinds of social relationships can be defined between the nodes in different time periods (i.e., a day or a week). In CMM, when a node leaves its home community toward another community, nodes that have social relationships with this node are attracted to the same community. The authors in Boldrini et al. (2007) showed that the CMM model is not able to capture the attraction exerted on users by physical locations. This problem is called the gregarious behavior of nodes. To address this issue, the home-cell community-based mobility model (HCMM) (Boldrini and Passarella, 2010) was proposed, which considers both social communities and location attractions by defining preferential locations.

Ironically, sociological interaction mobility for population simulation (SIMPS) (Borrel et al., 2009) and general social mobility (GeSoMo) (Fischer et al., 2010) have been proposed, considering the fact that social relations can not only attract nodes toward each other but also separate them. SIMPS is a model of human crowds with a pedestrian motion that is based on a process model called "sociostation." In this model, node movement is based on two behavioral states: socialize and isolate. In the socialize state, node movement is based on users' social relationships, whereas in the isolation state they try to escape from neighbors with which they do not have social relations. GeSoMo separated the core mobility model from the structural description of the social network. This model receives a social network as input and creates a mobility trace, which is a schedule for the movement of each individual node in the input social network such that this trace creates meetings between the nodes according to their social relations. Considering social relationship and location dependencies, different kinds of attraction and repulsion between nodes are defined in this model.

Some mobility models integrate spatial and temporal features of human movement with its connectivity properties. Geo-CoMM (Zignani, 2012) was a sample model in this class that is able to properly reproduce in a model the three aspects of human mobility that can be observed in real traces. In this model, the simulation area is divided into some spatial regions to model different urban conditions and people move within a set of geo-communities. In addition, temporal ranges (e.g., between 8:00 am and 8:00 pm) can be defined to determine the time durations nodes are active. Inside the communities, nodes move according to a random model, while a list of remaining communities (sorted by their geographical distances) is attached to each community, which helps nodes to choose the next destination and travel to another community. Similarly, the arrival-based framework (Karamshuk et al., 2012) is a flexible and controllable mobility model which is based on the three dimensions of human movement. In this model, a social graph is used to represent the social relationships between users. Then, based on the input social graph, communities are detected and assigned different locations. Thus, users belonging to the same community share a common location where the members of the community meet and visit these locations over time based on a configurable stochastic process. Short descriptions of the explored mobility models are summarized in Table 12.1.

12.4 Socially aware data forwarding in DTNs

Data delivery is a nontrivial issue in DTNs since there is a lack of end-to-end paths between nodes at the majority of times. Hence, most existing forwarding protocols in this disconnected setting employ the store-carry-and-forward method to exchange data between the nodes. If there is no connection available at a particular time, a node can carry the data until contact is made with other network nodes. When the carrier has a forwarding opportunity, all encountered nodes could be the candidates to relay the data. Consequently, an efficient forwarding strategy should be applied to select optimal relaying nodes. Considering network resources, mobile device criteria, and user characters, certain mechanisms could be utilized at different stages of a routing process.

A preliminary method for a mobile node to forward data in DTNs would be to relay all the generated data in its buffer to all the encounter nodes, like epidemic routing (Vahdat and Becker, 2000). However, unlimited replication of data in DTNs saturates network resources, such as storage and bandwidth, which results in a data congestion problem. To tackle this issue, several single-copy and multicopy protocols have been proposed which typically aim to limit the number of replicas of data and leverage a trade-off between resource usage and the probability of data delivery. As an example,

Table 12.1 Some prominent social-based human mobility models for delay-tolerant networks (DTNs).

Model	Description
CMM (Musolesi and Mascolo, 2007)	Nodes are assigned to a number of subareas using preferential attachment. The attractiveness of one area is determined by the current number of nodes assigned to that area.
HCMM (Boldrini and Passarella, 2010)	Combines notions about the sociality of users with spatial properties observed in real users' movement patterns.
SIMPS (Borrel et al., 2009)	Derives the motion of users in such a way that individuals' movements are governed by both their social relationships and geographically surrounding individuals.
GeSoMo (Fischer et al., 2010)	Separates the core mobility model from the structural description of the social network underlying the simulation.
Geo-CoMM (Zignani, 2012)	Each node selects a subset of predefined sets of locations and moves between them based on a customizable behavior.
Arrival-based (Karamshuk et al., 2012)	A mobility framework that takes a social graph as input. Then, the spatial and temporal dimensions of mobility are added.

PROPHET (Lindgren et al., 2003) was an extended version of the epidemic routing that makes use of a delivery predictability metric to calculate how likely it is that a node will be able to deliver a message to the destination. Contact frequency is considered to calculate delivery predictability between nodes in this method.

Social characteristics of mobile users have been widely utilized to streamline routing protocols in DTNs. The main reason is that users' social attributes and relationships have long-term characteristics and they are less volatile. As a result, they can be significantly utilized to make effective routing decisions. Broadly, socially aware forwarding protocols in DTNs mostly attempt to group mobile nodes into communities and/or take advantage of other social features such as contact duration, node centrality, similarity, and user interest in data delivery. In the rest of this section, prominent socially aware forwarding proposals in DTNs with respect to community structure, context-awareness, multicasting, user selfishness, and incentive schemes are overviewed.

12.4.1 Community-based forwarding

Several data forwarding algorithms have been proposed which take advantage of community structure to improve the performance of data forwarding in DTNs. Two main steps are carried out to set up a community-based forwarding in DTNs. First, mobile nodes are grouped into several communities by a certain community detection algorithm. Then, some special nodes called hubs or brokers are selected to construct network overlays and make a community structure. However, different nodes can play the role of a hub or a broker inside a community to forward data between the communities. In this phase, if the relay nodes are out of the destination community, an intercommunity forwarding strategy relays data to the destination community. When data reach a node in the destination community, an intracommunity forwarding strategy is used to pass the data to the destination node.

LABEL (Hui and Crowcroft, 2007) is an initiative protocol in this class in which every node possesses a label informing other nodes about its affiliation. To transfer data between communities, it compares the labels of the potential relay nodes and the label of the destination node and consequently forwards the data to nodes that belong to the same community as the destinations. The performance of this protocol can be degraded significantly if nodes belonging to destination communities do not mix well in the network. Bubble Rap (Hui et al., 2011) is a well-known community-based protocol in which nodes are structured into communities based on node centrality. In this method, network nodes have global and local rankings. The global ranking denotes the popularity of the node in the entire network, whereas the local ranking shows its popularity within its own community. A message is forwarded between the communities based on the global ranking until a node in the destination community receives the message. Inside the destination community, the message is transferred to nodes with higher local rankings to be finally delivered to the destination node.

LocalCom (Li and Wu, 2009) is a community-based epidemic forwarding scheme that utilizes the social network properties to improve the forwarding efficiency in DTNs. In this method, similarity metrics, according to the nodes' encounter history, are considered to construct the neighboring relationship between each pair of nodes. Then, a distributed algorithm is proposed for community detection which only utilizes the nodes' local information. Furthermore, two strategies are utilized to first select and then prune gateways that connect communities to control redundancy and facilitate efficient intercommunity forwarding. Friendship-based routing (Bulut and Szymanski, 2010) made use of contact history by introducing three behavior features: high frequency; longevity; and regularity, to define friendship strength between nodes and form social communities. In this method, two metrics, namely social pressure metric (SPM) and conditional SPM, are defined for direct and indirect friendship. Finally, friendship-based routing is proposed, in which temporally differentiated friendships are used to make the forwarding decisions of messages.

12.4.2   Community-independent forwarding

Community-based data forwarding in DTNs suffers from the overhead of community formation and management. Hence, some data delivery protocols have been presented which take advantage of other social characteristics such as node centrality, similarity, and user interest to make forwarding decisions. SimBet (Daly and Haahr, 2009) is a leading algorithm in this category, which uses an ego network analysis technique to identify betweenness centrality and social similarity for network nodes. The concept of ego network is exploited, where only local information and structure of network nodes are analyzed. In this method, an encountered node with higher centrality is selected for data forwarding in order to increase the possibility of finding the potential carrier to the final destination. SimBet has good overall performance regarding message delivery, close to the epidemic routing but with a highly improved delivery cost.

Some community-independent forwarding protocols suffer from high delay as well as congested traffic around central nodes. For instance, in the SimBet algorithm, the top 10% of nodes carry out 54% of all the forwards and 85% of all the handover. To tackle this problem, some strategies have been proposed in order to keep low delays, high success ratios, and high availability of nodes. Congestion-aware forwarding (Randkovic and Grundy, 2011) defined three aspects of a node's ego network congestion—delay, buffer availability, and congestion rate of the ego networks—to find alternative routes, offload the traffic from congested parts of the network, and spread it over less congested parts. Similarly, FairRoute (Pujol et al., 2009) was inspired by the social process of perceived interaction strength, where messages are preferably forwarded to users that have a stronger social relationship with the target of the message. It also uses an assortatively based queue control mechanism to limit the exchange of messages to those users with similar social status and makes the traffic load more balanced.

Social ranking and mobility prediction-based routing have also been considered in some recent DTN forwarding protocols. PeopleRank (Mtibaa et al., 2010) assign ranks to nodes using tunable weighted social interaction information to identify the most popular nodes. Then, this ranking is used to decide on the next hop for data exchange, as it is known that socially connected nodes become the best forwarders for message delivery. CiPRO (Nguyen and Giordano, 2012) is a mobility prediction-based routing algorithm which considers both spatial and temporal dimensions of contact so that the source device knows when and where to start the routing process to minimize the network delay and overhead. Similarly, predict and relay (PER) (Yuan et al., 2012) considers the time of contact, determines the probability distribution of future contact times, and chooses a proper next hop in order to improve the end-to-end delivery probability.

Hypercube-based social feature routing (HSFR) (Wu and Wang, 2012) uses the internal social features of each user to convert a multipath routing problem in an unstructured contact space to a static and structured feature space where nodes with the same social features are grouped together. The groups in the static feature space can be mapped into an m-dimensional hypercube, in which two groups are connected if and only if they differ in only one feature. HSFR included two unique processes: social feature extraction and multipath routing. In the social feature extraction, entropy is used to extract the m most informative social features. The routing method then becomes a hypercube-based feature-matching process, where the routing process is a step-by-step feature difference-resolving process. Table **12.2** summarizes the most important features of social-based routing algorithms in DTNs.

12.4.3   Social-based multicasting

Multicasting is an effective method for efficient data dissemination and multiparty communication in DTNs. For example, in sparse vehicular ad hoc networks, a vehicle may disseminate live traffic information to other following vehicles (Pereira et al., 2012).

Table 12.2 Description and important features of social-based forwarding protocols in DTNs.

	Protocol	Description	Centrality	Tie strength	Similarity
Community-based forwarding	LABEL (Hui and Crowcroft, 2007)	A benchmark scheme in which every node compares labels of relay and destination nodes and forwards the data only to encounter nodes that belong to the destination community.	×	×	×
	Bubble Rap (Hui et al., 2011)	In this algorithm, nodes have global and local rankings. A message is forwarded to nodes having higher global ranking until the message is delivered to a node in the destination community. Inside the destination community, local ranking of nodes is used to deliver a message to destination node.	√	×	×
	LocalCom (Li and Wu, 2009)	LocalCom defines some similarity metrics to establish relationship between nodes and form communities. Then, two selection and gateway pruning schemes connect communities and facilitate intercommunity forwarding.	√	√	√
	Friendship-based routing (Bulut and Szymanski, 2010)	This algorithm utilizes a SPM metric to define friendship community between the nodes. Then, temporally differentiated friendships are used to make the forwarding decisions.	×	√	×
Community-independent forwarding	SimBet (Daly and Haahr, 2009)	A centrality-based algorithm which uses *SimBet* utility function to forward messages to destination nodes.	√	√	√
	Congestion aware (Randkovic and Grundy, 2011)	A multipath algorithm which uses a number of social, buffer and delay utility metrics to offload the traffic from congested parts of the network and spread it over less congested parts.	√	√	√
	FairRoute (Pujol et al., 2009)	FairRoute exploits the social process of perceived interaction strength based on the interaction strength between nodes in a short-term and long-time scale to make fair the load distribution.	×	√	×

Continued

Table 12.2 Continued

Protocol	Description	Centrality	Tie strength	Similarity
PeopleRank (Mtibaa et al., 2010)	A distributed algorithm that ranks nodes in a social graph and makes use of stable social information between nodes to decide on forwarding.	×	×	√
CiPRO (Nguyen and Giordano, 2012)	A context prediction-based algorithm which applies the Back Propagation Neural Network model to predict the mobility behavior of device carriers and select optimal intermediate nodes.	×	×	√
PER (Yuan et al., 2012)	A mobility prediction-based method to determine the probability distribution of future contact times for network nodes which aims to choose a proper next-hop relay node in order to improve the end-to-end delivery probability.	×	×	×
HSFR (Wu and Wang, 2012)	A social feature-based scheme that consists of two parts: social feature extraction using entropy to extract the most informative social features and multipath routing. The feature difference between the source and the destination is resolved during the routing process.	√#	√	×

Note: √ if the protocol satisfies the property, × if not.

According to Ye et al. (2009), some problems can arise if Internet-based multicast routing methods are applied in DTNs. First, it is difficult to maintain the connectivity of a multicast structure during the lifetime of a multicast session. Second, data transmissions would suffer from many failures and large end-to-end delays because of the disruptions caused by repeatedly broken multicast branches. Third, the traditional approaches are designed with the assumption that the underlying networks are basically connected.

The state-of-the-art multicasting protocols for DTNs have utilized community and centrality concepts to improve multicasting performance. For instance, the social network-aided multicast delivery (Chuah, 2009) scheme made use of the community concept to deliver multicast messages efficiently. The scheme is motivated by the following observations: a node should forward a copy to an encountered node if that encountered node either is in the same community as any of the destination nodes or has a higher chance of reaching any of the destination nodes. Quite recently, a probabilistic socially aware multicast (Gao et al., 2012) was proposed which uses two key concepts, centrality, and communities, to improve the cost-effectiveness of multicast. In this work, the authors propose a community-based approach that only requires nodes to maintain the probabilities of forwarding each data item to other nodes in the same community. When the destinations are in other communities, data forwarding is conducted through some gateway nodes which belong to multiple communities. However, this model increases computational complexity considerably.

12.4.4   User selfishness and incentive schemes

Almost all the routing protocols considered in this chapter assume that users are willing to help each other in data delivery. However, network and device resources such as energy, cache, and bandwidth in real applications are limited, and a user might not be interested in relaying messages for the other nodes. In other words, some nodes may have selfish behaviors in relaying data, which could significantly degrade the performance of a DTN routing algorithm. According to Li et al. (2012), two kinds of selfish behavior can be defined for a noncooperative node. In some cases, a selfish user wants to help other nodes with whom he has social relationships (e.g., friends, classmates, colleagues) because he received help from them in the past or will probably get help in the future. This kind of selfish behavior is called individual selfishness. A selfish user may show a different degree of selfishness (or cooperation) to other nodes based on social tie strength. This behavior is called social selfishness. An example of data delivery in DTNs with unselfish, individually selfish, and socially selfish nodes is shown in Fig. 12.3.

In the past few years, several incentive schemes have been proposed to stimulate selfish users to more cooperation, in spite of the fact that this issue is extraordinarily challenging in DTNs due to intermittent connections between mobile carriers. The existing incentive strategies can be categorized into three classes: tit-for-tat (TFT), credit-based, and reputation-based schemes. In the following, some pioneering incentive schemes in each category are highlighted.

In TFT mechanisms, two encounter nodes exchange the same number of packets with each other in a "give one to get one" manner. Consequently, message selection is

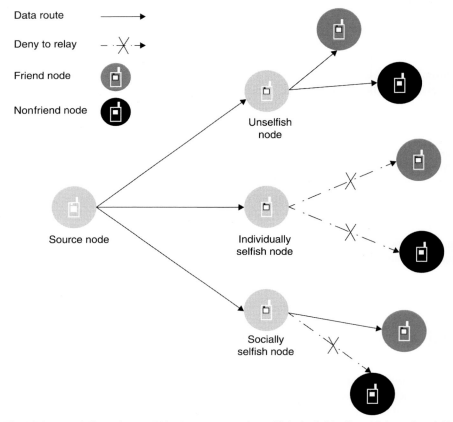

Fig. 12.3 Data delivery in a DTN in the presence of unselfish, individually selfish, and socially selfish users.

a very important issue in this method, which can significantly affect the performance of a data delivery protocol in terms of data delivery and delay. MobiTrade (Krifa et al., 2011) is a prominent scheme in this class that uses an optimal buffer allocation policy to split the buffer of a node to each channel. In reality, the amount of data for exchange between two encounter nodes is not equal. This issue can lead to some problems such as fairness issues or deadlocks when some interesting content cannot be disseminated between the relay nodes. To resolve this problem, a trading mechanism was utilized in MobiTrade that allows a node to buy, store, and carry content for other nodes so that it can later trade it for content it is personally interested in. Similarly, barter trade (Buttyan et al., 2010) discouraged selfish behavior based on the principles of barter. In this method, messages are classified into two types, namely, primary messages and secondary messages, which can be traded between the users. First, encounter nodes send a description of the messages that they want to exchange. Then, a message selection process is applied in such a way that the nodes agree to download from each other one by one. The message selection process is considered as a two-person game to increase the message delivery ratio in this method.

Credit-based incentive approaches utilize the concept of virtual credit to resolve the unfairness problem of TFT strategies. Practical incentive (Pi) (Lu et al., 2010) was one of the first proposals in this class, aiming to improve the performance of data forwarding in DTNs in the presence of individually selfish nodes. In this method, some incentives are attached to each message, which is not only attractive for relaying nodes but also fair to all network nodes. To achieve fairness in Pi, the intermediate relay nodes get credit from the source node if the messages are delivered to destination nodes successfully. Otherwise, the intermediate nodes will get a reputation from a trusted authority that aims to guarantee fairness. SMART (Zhu et al., 2009) is a secure multilayer credit-based scheme which is based on the notion of a layered coin to encourage selfish users to cooperate in data delivery. The first layer of the coin, called the base layer, is generated by the source to indicate the payment rate (credit value) and other rewarding policies. Furthermore, each intermediate node adds a new layer, named an endorsed layer, which implies that the forwarding node agrees to provide forwarding service under the predefined forwarding policies. In summary, in credit-based strategies, source and destination nodes are required to have access to a trusted third party to manage their payment policies.

Reputation-based schemes have also been considered in the literature to stimulate selfish users to cooperate in DTNs. In MobiID (Wei et al., 2011), a node is allowed to manage its reputation evidence and show this to demonstrate its reputation whenever necessary. Furthermore, the concepts of self-check and community-check are defined to speed up reputation dissemination between nodes and allow them to form consensus views toward targets in the same community based on a social metric. IRONMAN (Bigwood and Henderson, 2011) utilized self-reported social network information to establish a trust mechanism in order to detect and punish selfish nodes. In addition, a reputation method is used to allow nodes that have been deemed selfish to improve their trust score. Short descriptions and important features of the discussed incentive schemes are outlined in Table **12.3**.

Table 12.3 Comparison of incentive schemes in DTNs.

	Algorithm	Characteristic	Properties		
			Single-copy	Multicopy	Fairness
Tit-for-tat	MobiTrade (Krifa et al., 2011)	A trading mechanism that allows a node to buy, store, and carry content for other nodes so that it can later trade it for content it is personally interested in.	×	√	–

Continued

Table 12.3 Continued

	Algorithm	Characteristic	Properties		
			Single-copy	Multicopy	Fairness
Credit-based	Barter trade (Buttyan et al., 2010)	A game-theoretic model based on the Nash Equilibrium strategy to discourage selfish behavior based on the principles of barter.	√	√	√
	Pi (Lu et al., 2010)	An incentive model in which selfish nodes are stimulated to help forward bundles.	√	×	√
	SMART (Zhu et al., 2009)	An incentive scheme which stimulates bundle-forwarding cooperation with thwarting various attacks.	√	√	√
Reputation-based	MobiID (Wei et al., 2011)	A user-centric incentive scheme which allows a node to manage its reputation evidence.	×	√	–
	IRONMAN (Bigwood and Henderson, 2011)	An incentive mechanism that uses preexisting social network information to bootstrap the detection and discouragement of selfishness.	×	√	–

Note: √ if the model satisfies the property, × if not, and – for ambiguous cases.

12.5   Conclusion

In this chapter, we investigated the social characteristics of mobile users which have been extensively utilized to improve the performance of protocols in DTNs. Specifically, three social aspects of DTNs were explored in this chapter. First, some important SNA concepts and metrics such as a social graph, node centrality, tie strength, community structure, etc. were investigated. Then, well-known social-based mobility models were introduced, since simulation and evaluation of existing forwarding protocols in DTNs rely heavily on the reality of mobility models. Finally, the state of the art of social-based data forwarding in DTNs with respect to multicasting, user selfishness, and incentive schemes was presented and the most important features were outlined.

References

Aschenbruck, N., Munjal, A., Camp, T., 2011. Trace-based mobility modeling for multi-hop wireless networks. Comput. Commun. 34 (6), 704–714.

Bigwood, G., Henderson, T., 2011. IRONMAN: using social networks to add incentives and reputation to opportunistic networks. In: SocialCom2011, Boston, USA, pp. 65–72.

Boldrini, C., Passarella, A., 2010. HCMM: modelling spatial and temporal properties of human mobility driven by users' social relationships. Comput. Commun. 33 (9), 1056–1074.

Boldrini, C., Conti, M., Passarella, A., 2007. Users mobility models for opportunistic networks: the role of physical locations. In: IEEE WRECOM, Rome, Italy, pp. 1–6.

Borrel, V., Legendre, F., Amorim, M.D., Fdida, S., 2009. SIMPS: using sociology for personal mobility. IEEE/ACM Trans. Networking 17 (3), 831–842.

Bulut, E., Szymanski, B.K., 2010. Friendship based routing in delay tolerant mobile social networks. In: IEEE GLOBECOM, Miami, FL, USA, pp. 1–5.

Buttyan, L., Dora, L., Felegyhazi, M., Vajda, I., 2010. Barter trade improves message delivery in opportunistic networks. Ad Hoc Netw. 8 (1), 1–14.

Chuah, M.C., 2009. Social network aided multicast delivery scheme for human contact-based networks. In: Simplex 2009, 1st Annual Workshop on Simplifying Complex Network for Practitioners, Venice, Italy.

Daly, E.M., Haahr, M., 2007. Social network analysis for routing in disconnected delay-tolerant MANETs. In: ACM MobiHoc'07, Montreal, Canada, pp. 32–40.

Daly, E.M., Haahr, M., 2009. Social network analysis for information flow in disconnected delay-tolerant MANETs. IEEE Trans. Mob. Comput. 8 (5), 606–621.

Fall, K., 2003. A delay-tolerant network architecture for challenged internets. In: ACM SIGCOMM'03, Karlsruhe, Germany, pp. 27–34.

Fischer, D., Herrmann, K., Rothermel, K., 2010. GeSoMo—a general social mobility model for delay tolerant networks. In: IEEE MASS, Wuhan, China, pp. 99–108.

Gao, W., Li, Q., Zhao, B., Cao, G., 2012. Social-aware multicast in disruption-tolerant networks. IEEE/ACM Trans. Networking 20 (5), 1553–1566.

Gonzalez, M.-C., Hidalgo, C.-A., Barabasi, A.-L., 2008. Understanding individual human mobility patterns. Nature 453 (7196), 779–782.

Granovetter, M.S., 1973. The strength of weak ties. Am. J. Sociol. 78 (6), 1360–1380.

Hopcroft, J., Khan, O., Kulis, B., Selman, B., 2004. Tracking evolving communities in large linked networks. Proc. Natl. Acad. Sci. U. S. A. 101, 5249–5253.

Hossmann, T., Spyropoulos, T., Legendre, F., 2010. Social network analysis of human mobility and implications for DTN performance analysis and mobility modeling (Technical Report—TIK Report Nr. 323).

Hui, P., Crowcroft, J., 2007. How small labels create big improvements. In: IEEE PERCOMW'07, White Plains, NY, USA, pp. 65–70.

Hui, P., Crowcroft, J., Yoneki, E., 2011. Bubble rap: social-based forwarding in delay-tolerant networks. IEEE Trans. Mob. Comput. 10 (11), 1576–1589.

Jedari, B., Xia, F., 2014. A survey on routing and data dissemination in opportunistic mobile social networks. IEEE Commun. Surv. Tutorials. https://doi.org/10.1109/SURV.2014.022714.00153 (in press).

Karamshuk, D., Boldrini, C., Conti, M., Passarella, A., 2011. Human mobility models for opportunistic networks. IEEE Commun. Mag. 46 (12), 157–165.

Karamshuk, D., Boldrini, C., Conti, M., Passarella, A., 2012. An arrival-based framework for
human mobility modeling. In: IEEE WOWMOM, San Francisco, CA, pp. 1–9.

Khabbaz, M., Assi, C.M., Fawaz, W., 2012. Disruption-tolerant networking: a comprehensive
survey on recent developments and persisting challenges. IEEE Commun. Surv. Tutorials
14 (2), 607–640.

Krifa, A., Barakat, C., Spyropoulos, T., 2011. Mobitrade: trading content in disruption tolerant
networks. In: ACM CHANTS, Las Vegas, NV, USA, pp. 31–36.

Li, F., Wu, J., 2009. LocalCom: a community-based epidemic forwarding scheme in disruption-
tolerant networks. In: IEEE SECON, Rome, Italy, pp. 1–9.

Li, Q., Gao, W., Zhu, S., Cao, G., 2012. A routing protocol for socially selfish delay tolerant
networks. Ad Hoc Netw. 10 (8), 1619–1632.

Lindgren, A., Doria, A., Schelen, O., 2003. Probabilistic routing in intermittently connected networks. SIGMOBILE Mobile Comput. Commun. Rev. 7 (3), 19–20.

Lu, R., Lin, X., Zhu, H., Shen, X., Preiss, B., 2010. Pi: a practical incentive protocol for delay
tolerant networks. IEEE Trans. Wirel. Commun. 9 (4), 1483–1493.

Milgram, S., 1967. GPSR: greedy perimeter stateless routing for wireless networks. Psychol.
Today 1, 61–67.

Mtibaa, A., May, M., Ammar, M., Diot, C., 2010. PeopleRank: combining social and contact information for opportunistic forwarding. In: IEEE INFOCOM, San Diego, CA, USA, pp. 1–5.

Musolesi, M., Mascolo, C., 2007. Designing mobility models based on social network theory.
ACM Mobile Comput. Commun. Rev. 11 (3), 59–70.

Nguyen, H.-A., Giordano, S., 2012. Context information prediction for social-based routing in
opportunistic networks. Ad Hoc Netw. 10 (8), 1557–1569.

Palla, G., Barabasi, A.-L., Vicsek, T., 2007. Quantifying social group evolution. Nature 446
(7136), 664–667.

Pereira, P., Casaca, A., Rodrigues, J., Soares, V., Triay, J., Cervello-Pastor, C., 2012. From
delay-tolerant networks to vehicular delay-tolerant networks. IEEE Commun. Surv. Tutorials 14 (4), 1166–1182.

Phithakkitnukoon, S., Smoreda, Z., Olivier, P., 2012. Socio-geography of human mobility: a
study using longitudinal mobile phone data. PLoS One 7 (6), e39253.

Pujol, J., Toledo, A., Rodriguez, P., 2009. Fair routing in delay tolerant networks. In: IEEE
INFOCOM, Rio de Janeiro, Brazil, pp. 837–845.

Randkovic, M., Grundy, A., 2011. Congestion aware forwarding in delay tolerant and social
opportunistic networks. In: WONS, Bardonecchia, Italy, pp. 60–67.

Schurgot, M.R., Comaniciu, C., Jaffres-Runser, K., 2012. Beyond traditional DTN routing:
social networks for opportunistic communication. IEEE Commun. Mag. 50 (7), 155–162.

Song, C., Koren, T., Wang, P., Barabasi, A.-L., 2010. Modelling the scaling properties of human
mobility. Nat. Phys. 6 (10), 818–823.

Spiliopoulou, M., Ntoutsi, I., Theodoridis, Y., Schult, R., 2006. Monic: modeling and monitoring cluster transitions. In: ACM SIGKDD, Philadelphia, USA, pp. 706–711.

Takaffoli, M., Sangi, F., Fagnan, J., Zaiane, O.R., 2011. Community evolution mining in
dynamic social networks. Soc. Behav. Sci. 22, 49–58.

Vahdat, A., Becker, D., 2000. Epidemic routing for partially connected ad hoc networks
(Technical Report, CS-2000-06). Department of Computer Science, Duke University,
Durham, NC.

Wasserman, S., Faust, K., 1994. Social Network Analysis: Methods and Applications, Structural Analysis in the Social Sciences Series. Cambridge University Press.

Wei, L., Zhu, H., Cao, Z., Shen, X., 2011. MobiID: a user-centric and social-aware reputation based incentive scheme for delay/disruption tolerant networks. In: Frey, H., Li, X., Ruehrup, S. (Eds.), Ad-Hoc, Mobile, and Wireless Networks, Ser. Lecture Notes in Computer Science. vol. 6811. Springer, Berlin/Heidelberg, pp. 177–190.

Wu, J., Wang, Y., 2012. Hypercube-based multi-path social feature routing in human contact networks. IEEE Trans. Comput. 63 (2), 383–396.

Xia, F., Liu, L., Li, J., Ma, J., Athanasios, V., 2013. Socially-aware networking: a survey. IEEE Syst. J. https://doi.org/10.1109/JSYST.2013.2281262 (in press).

Ye, Q., Cheng, L., Chuah, M.C., Davison, B.D., 2009. Performance comparison of different multicast routing strategies in disruption tolerant networks. Comput. Commun. 32 (16), 1731–1741.

Yuan, Q., Cardei, I., Wu, J., 2012. An efficient prediction-based routing in disruption-tolerant networks. IEEE Trans. Parallel Distrib. Syst. 23 (1), 19–31.

Zhu, H., Lin, X., Lu, R., Fan, Y., Shen, X., 2009. SMART: a secure multilayer credit-based incentive scheme for delay-tolerant networks. IEEE Trans. Veh. Technol. 58 (8), 4628–4639.

Zhu, Y., Xu, B., Shi, X., Wang, Y., 2013. A survey of social-based routing in delay tolerant networks positive and negative social effects. IEEE Commun. Surv. Tutorials 15 (1), 387–401.

Zignani, M., 2012. Geo-CoMM: a geo-community based mobility model. In: IEEE WONS, Courmayeur, Italy, pp. 143–150.

Performance issues and design choices in delay-tolerant network (DTN) algorithms and protocols ☆

J. Morgenroth, W.-B. Pöttner, S. Schildt, and L. Wolf
Technische Universität Braunschweig, Braunschweig, Germany

13.1   Introduction

As with other protocols and technologies, performance is important for delay-tolerant networks (DTNs). In this chapter, we study several areas relevant to DTN performance and take a look at the main factors impacting the performance of a DTN system. In addition to generic discussions of DTN performance aspects, we give examples from real-world experiments and implementations. Furthermore, we give suggestions for implementers based on our experiences with several DTN systems and protocol implementations. However, the points examined here are not only applicable to specific implementations or even the Bundle Protocol. The discussed issues are general and relevant to all implementations and deployments of DTN-like systems.

We begin by introducing the metrics that are relevant to the performance of a DTN system in Section 13.1. In Section 13.2, we discuss the issue of processing overhead in a DTN Engine, which is strongly connected to the protocol data unit (PDU) transmission frequency. In a DTN, PDUs are usually persistently stored over an extended period of time. Therefore, DTN Engines may suffer performance degradation when copying PDUs too often, as elaborated in Section 13.3. The throughput in a DTN reflects the network capacity, and we provide some insight into influencing factors in Section 13.4. Although DTNs are delay-tolerant, for some applications the PDU delay may still be important. We discuss issues connected with the delay of PDUs in Section 13.5. Opportunistic contacts tend to be short and not well predictable and have to be used as efficiently as possible, as shown in Section 13.6. Finally, Section 13.7 concludes the chapter.

To substantiate the discussion regarding DTN performance, we present a number of measurement results throughout this chapter. Those results have been obtained and published earlier (Pöttner et al., 2011a,b) and are reproduced here to illustrate certain points. In those measurements, we used two high-performance PCs (Athlon II X4 IV 2.8 GHz, 4 GiB RAM) on a wired Gigabit Ethernet connection with Ubuntu Linux. Using iperf (http://iperf.sourceforge.net/l, we measured the Ethernet connection to

---

☆ This chapter is a reprint of the chapter originally published in the first edition of Advances in Delay-Tolerant Networks (DTNs): Architecture and Enhanced Performance.

Advances in Delay-tolerant Networks (DTNs). https://doi.org/10.1016/B978-0-08-102793-6.00013-8

allow a maximum transmission control protocol (TCP) throughput of approximately 940 MBit/s. We used DTN2 (Demmer et al., 2004) version 2.7, IBR-DTN (Schildt et al., 2011) version 0.6.3, and ION (Burleigh, 2007) version 2.4.0. All implementations have been measured using their respective default configuration unless otherwise noted.

13.2 Performance metrics

There are various performance metrics that will influence the overall performance of a DTN system. Some metrics may influence other metrics. The main performance metrics and their interdependence are depicted in Fig. 13.1.

13.2.1 Protocol design

While protocol design is not a performance metric, the chosen protocol fundamentally limits the performance a given implementation can achieve theoretically. There are three main types of protocols in a DTN. First, the basic DTN protocol itself defines how data is encoded and sent over the wire and might include mechanisms for reliability. A prominent example of a DTN protocol is the Bundle Protocol (Scott and Burleigh, 2007). As many DTN applications deal with opportunistic contacts, often a Discovery Protocol is needed to make neighboring nodes aware of each other. Finally, the choice of a routing mechanism influences the performance of the system. While we do not focus on specific routing protocols in this chapter, we want to point out the obvious relations between routing and performance: whenever a routing protocol makes the "wrong" forwarding decision, due to insufficient knowledge, metrics such as delivery delay will suffer. As a general rule, it can be said that most DTN routing protocols offset missing knowledge with increased replication: the system will transfer many identical copies of data on different paths. This can increase the chance

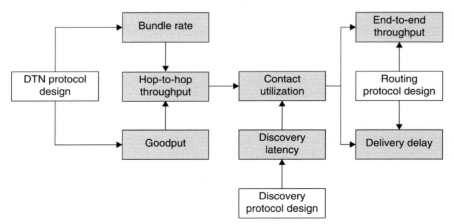

Fig. 13.1 Performance metrics and dependencies.

for optimal decisions, but it will put more load on the network and thus on the DTN implementations itself, influencing basically every metric discussed in this chapter.

13.2.2 Throughput and goodput

Throughput is the amount of data that can be transferred in a specific time period. This is limited by the efficiency of the employed protocols: Taking the overhead introduced by a protocol into account yields the *goodput*. This is the throughput that an application experiences. Protocol limitations aside, the throughput is largely determined by the available computing or network resources and the quality of the protocol implementation. *Hop-to-hop throughput* influences the *contact utilization*, which in turn impacts *end-to-end throughput*.

13.2.3 Delivery delay

Even though DTN stands for *delay*-tolerant, in most applications lower *end-to-end delays* are still preferred if possible. End-to-end delays are influenced by the path choices of a routing algorithm. Congestion affects *delivery delay*. Low contact utilization can lead to congestion. In a congested system, where the total network capacity available to the DTN is smaller than the capacity needed for yielding shortest-possible delivery delay for all bundles, scheduling mechanisms can influence the delay of individual bundles.

13.2.4 Discovery latency

In nondeterministic dynamic scenarios, common in DTN applications, the system performance is heavily influenced by *discovery latency:* after two nodes come into communication range, they can only exchange data after they become aware of each other's presence. The time it takes before detecting a new contact is the discovery latency. High discovery latencies negatively impact contact utilization. In a DTN, long discovery latencies can also prevent the system from using short communication opportunities at all.

13.2.5 Contact utilization

The achieved hop-to-hop throughput and discovery latency results in the contact utilization ratio, which defines how much of the capacity theoretically available during contact is actually used by a DTN implementation.

$$\text{contact utilization} = \frac{\text{transferred data}}{\text{link bandwidth} \times \text{contact time}}$$

13.2.6 Bundle rate

Some processing is needed to parse, generate, or forward bundles. The protocol design determines the processing overhead necessary to parse PDUs. As DTN systems implement the store-carry-and-forward approach, a (semi-)persistent storage is part of the system, leading to added input/output (I/O) overhead for storing and retrieving bundles. A DTN implementation needs to store a possibly large number of bundles, which leads to increased computing overhead for storage management. Thus, while throughput for very large bundles might be good, it might be limited by the per-bundle overhead, which ultimately limits the achievable bundle rate. A small bundle rate will decrease contact utilization.

13.2.7 Memory consumption

While memory consumption does not directly influence the network performance metrics of a DTN system, it is a crucial metric when choosing suitable hardware and software platforms for a DTN application. Besides storing bundles itself, which might happen on a mass storage system, for performance reasons it is to be expected that an implementation will keep data structures to store routing-specific information, storage indices, or neighbor information in memory. This is a challenge when deploying high-traffic DTN solutions on resource-constrained embedded hardware.

In this chapter, we will first take a detailed look at the processing necessary in DTN implementations. To do this we will introduce and define a basic set of processing steps, which every DTN implementation must perform. Next, we take a closer look at the most crucial component and performance metric of a DTN system: the storage system and I/O performance. This will lead us to the discussion of achievable overall throughput, i.e., the metric most relevant to many applications. We will also look at queueing approaches to selectively improve performance for specific data. Finally, we discuss the impact of discovery latencies and the balance between fast discovery and energy usage, which is the major performance challenge in opportunistic scenarios with short contacts.

13.3 Processing overhead

A DTN Engine is considered as performant if the throughput of data is high and may utilize all of the available bandwidth. If an interruption occurs very often and the nodes are not constantly connected to each other, we have to consider the contact utilization instead of the bandwidth only. In this section, we will discuss the reasons for processing overhead, which results in a low contact utilization.

13.3.1 Data flow

The datagram in Fig. 13.2 shows the typical flow of a PDU passing a DTN Engine. There are only two ways a PDU enters the scope of the DTN Engine: it is either generated by an *application* or received through a *convergence layer* (CL). First, we

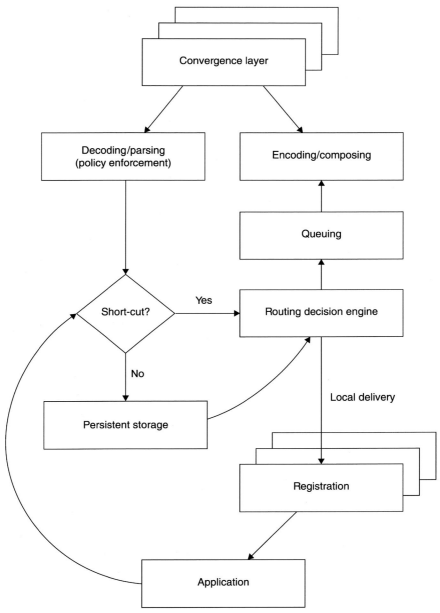

Fig. 13.2 Flow of a PDU passing a DTN Engine.

consider the CL as an entry point. A received PDU has to be decoded or parsed. During this process, it is possible to enforce some policies, which might restrict large PDUs, deny invalid security signatures or simple checks if the current PDU has already been received.

Next, an optional *short-cut* decision may be made. The DTN Engine can decide to forward the data directly to another peer instead of storing it first in the *persistent storage*. This is a very complicated decision and is very uncommon in DTN environments.

After the whole PDU has been stored, the *Routing Decision Engine* tries to determine a route. Further, it has to decide whether or not the PDU should be delivered locally to an *application* using its *registration*. If the PDU is not local, a route has been found and the next peer is reachable, the PDU will be queued for forwarding. Then the PDU has to be encoded or composed for the transmission and passed to the outbound CL.

The second way a PDU appears within the DTN Engine is if it is generated by an application. In this case, the processing starts at the short-cut decision, and then the PDUs are handled like the others.

13.3.2   Storage

One issue with DTNs, and especially with the Bundle Protocol specification (Cerf et al., 2007), is the huge amount of data a single PDU can carry. This leads to complex processing of PDUs and makes a persistent storage system sufficient. Since each PDU has to pass the storage, it takes place as the central and most significant performance factor in a classic DTN architecture. Typically, the storage fulfills several simple but frequently used procedures.

• *Duplicate checking*

Every time a PDU should be stored, the storage should check whether this PDU is already available in the storage. This is necessary because the different routing approaches in DTNs might generate a lot of copies which could be received over several different paths. In the worst case, all PDUs are stored using a linked list. In this case, the whole list must be iterated to find out that a specific PDU is not in the storage.

• *Insertion*

If a PDU is not already stored and needs to be added to the amount of stored PDUs, the PDU is inserted into some data structure, e.g., a linked list. Using a linked list, the insertion complexity is $O(1)$, which is quite good, but, on the other hand, duplicate checking or other operations like searching a specific PDU will cost a lot. Thus, more performance-oriented storage will implement some sort of index, which could be arbitrarily complex. A quite simple index data structure, which shows better characteristics, would be a sorted set. The insertion complexity is $O(\log n)$, while double-checking and insertion are combined as one operation.

• *Removal*

The storage has to provide a way to delete a specific PDU. The deletion might have several flavors depending on the specific implementation and the specific platform. In case of a file-oriented storage, it might be necessary to purge the corresponding files. In other systems, it would be sufficient to mark the amount of space as free. Implementations with one or more index data structures have to modify them according to the specific removal mechanisms, which also costs some resources.

- *Expiration*

A storage should take care of expired PDUs to regain space. Thus, it has to monitor when a PDU will expire. Additionally to the costs of the removal process, the storage needs a way to find the expired PDUs. One way is to iterate the whole storage and search for them, but this approach would be expensive in processing costs. Another approach is to manage an additional set of references sorted by the expiration time. The PDU which expires next is always the first PDU of such a set. Thus, checking whether there are PDUs to expire just requires checking the head of the set ($O(1)$). If one PDU is found to expire, the removal process is executed.

- *Iteration*

One very frequently used operation on storage is iteration. Every time the routing searches for PDUs to forward or deliver, it has to iterate through all PDUs to queue a subset of them. Typically, only a subset of the metadata is required to decide which PDU has to be selected. The application data unit (ADU) is not necessary in this case. A way to improve the performance here would be to maintain a routing-specific index, which lowers the complexity of the search process and avoids the complete iteration of all PDUs.

The only way to avoid such an exhaustive search completely is to queue PDUs in specific contact queues right after receiving/storing them. But this does not work with opportunistic routing schemes, which react to discovered neighbors and cost a lot of resources.

As seen, storage for a DTN might be a complex thing. In particular, if performance matters, it is hard to balance memory usage for index structures with performance. One way to bypass the storage is shown in Fig. 13.2. The junction short-cut tries to determine whether a direct forwarding to a connected neighbor is possible. This is a very special case and is only possible if strict conditions are met. One of these is that the next chosen hop has to be connected to the DTN Engine. However, no implementation using such a mechanism has yet been seen in publicly available DTN implementations.

13.3.3 PDU encoding

As shown in Fig. 13.2, each PDU has to be encoded for transmission. Due to complex encoding schemes, this may take some time and may involve specific encoding policies.

Basically, the encoding can be done during the transmission or a priori. A priori encoding is useful if the PDU is stored encoded anyway or if the DTN Engine has enough time to prepare the next transmission because it knows the future contacts. If the encoding of the PDUs has to be done during the contact time, it takes too much time to encode the PDU first, which means copying the whole data and then sending the encoded data to the foreign peer. A better approach is to encode the PDU during the transmission and write each encoded data-part directly to the CL.

On the receiver side, the PDU must be decoded or at least parsed to determine the next hop for it. This process might also be very expensive and should be done without any copying during the transmission. Only in this way is it possible to apply policy checking directly to the received data.

13.4   The curse of copying—I/O performance matters

Traditional Internet Protocol (IP) stacks adopt the notion of "streaming," in which a limited amount of data may be buffered but most of the data is sent out right away. If the outgoing link is currently unavailable or overloaded, packets are discarded. End-to-end data loss is usually prevented by end-to-end retransmissions of higher-level protocols. In a DTN the outgoing link may be unavailable over an extended period of time and data has to be stored on nodes. The DTN architecture (Cerf et al., 2007) further requires such storage to be persistent so that stored data survives system restarts.

Commercial Ethernet switches employ the "store-and-forward" paradigm in which frames are received, buffered (usually in RAM), and subsequently forwarded. While this allows switches to do error checking, it also requires enough temporary storage for at least a single frame. While Ethernet frames are limited in size, ADUs (which are transformed into PDUs by the DTN Engine) in a DTN are "possibly long" (Cerf et al., 2007) and normally not limited in size. When talking about the BP, PDUs have, in fact, limited size of 1.8×10^{19} bytes (because a self-delimiting numeric value (SDNV) can only hold $2^{64} - 1$ values).

So, a DTN Engine has to *persistently store* PDUs of *significant size*. While the performance of IP stacks is usually limited by the processing capabilities, DTN Engines will likely be limited by the storage bandwidth. Since PDUs have to be stored and retrieved during forwarding, the attainable throughput cannot exceed 50% of the storage bandwidth. Since persistent storage usually involves a hard disk drive (HDD) or flash memory, the storage bandwidth is significantly lower than RAM used in the switches. This makes it clear that DTNs are not a good match for streaming applications (many small ADUs) because the overhead per ADU is comparably high. Furthermore, supporting ADUs of arbitrary size causes certain handling problems, which will be discussed in this section.

13.4.1   Problem statement

On conventional DTN nodes, volatile memory is usually in the form of RAM and persistent memory in the form of flash memory or a hard disk. While RAM cannot be used as persistent storage and is also more expensive than flash or hard disk, it offers significant performance benefits. In Fig. 13.3, we show a network throughput measurement of the DTN2 reference implementation with PDUs stored in RAM or on HDD.

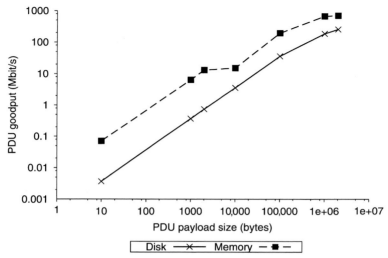

Fig. 13.3 DTN2 network throughput (Pöttner et al., 2011a) (log y-axis).

The attained throughput when using RAM is between 2.7 and 19.1 times faster compared with storing bundles on HDD. This clearly shows that storing or buffering ADUs in RAM can offer significant performance benefits. However, since RAM is volatile and will be lost on node restarts, not all PDUs can be stored in it. Those requiring special reliability (custody) have to be stored persistently before custody is accepted.

Even when the performance of the storage back end is sufficient to support high throughput, copying data can also drastically impact performance. In Fig. 13.4 we show a traditional DTN Engine in which data arrives at a CL and is immediately handed over to a storage module. This storage is likely not in RAM because the PDU has to be persistently stored and can be of arbitrary size that can easily exceed the RAM. When the routing module then takes care of the PDU, data is copied to the next storage module. When the PDU is forwarded, data is again copied from the storage module to the CL (and the respective storage module) to allow sending the PDU. Even when keeping PDUs in RAM, copying is expensive and impacts performance.

When keeping PDUs in persistent memory, copying has to be avoided as much as possible because the performance impact is even more significant.

13.4.2   Design advice: Central block storage mechanism

To allow the DTN Engine to achieve high performance, copying of block data has to be avoided as much as possible. The ideal case is shown in Fig. 13.5, in which a central storage component takes care of the PDU. The PDU enters the DTN Engine on the left side and is directly stored in the central component. Subsequently, references to the PDU are passed along until the PDU is forwarded to the next hop. In Fig. 13.6, we

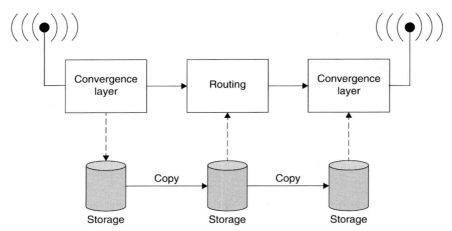

Fig. 13.4 DTN Engine PDU handling with (slow) copying of blocks.

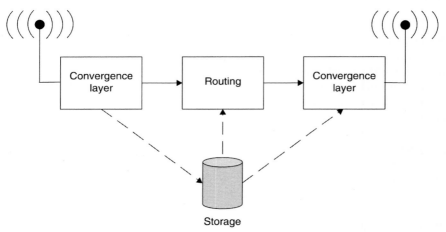

Fig. 13.5 DTN Engine PDU handling with central storage.

show a measurement with IBR-DTN, with and without a central block storage module. The performance increase of central storage that avoids copying is between 5.6% and 80.4%, depending on the size of the ADU.

Another issue with performance is the application programming interface (API). Sending and receiving applications have to be able to create and retrieve ADUs as fast as possible. In most implementations, copying the data at this point cannot be avoided. However, in an implementation that is ideal from the performance perspective, this copying would also be avoided by letting the application directly access the central storage component.

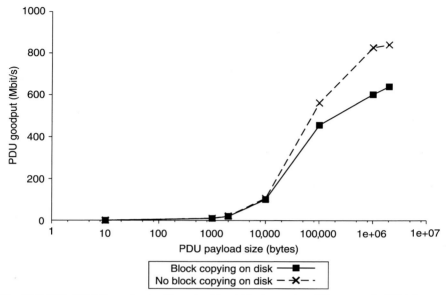

Fig. 13.6 IBR-DTN throughput with and without block copying (Pöttner et al., 2011a).

13.4.3 Design advice: Hybrid storage

As argued earlier, fast storage such as RAM is usually expensive and volatile. Persistent storage such as HDD is slow and cheap, while solid-state drives (SSDs) are in between because they are faster than HDDs but also more expensive. It is a characteristic of DTNs that the traffic patterns are bursty. During a contact, data has to be transferred as fast as possible because, especially for short contacts, time is precious. When no other node is in range, IO performance is of minor importance. A hybrid storage approach that combines the benefits of fast, expensive and slow, cheap storage is a good match for this kind of traffic pattern.

Fig. 13.7 shows the concept, in which two layers of storage are combined (Patterson and Hennessy, 2005). On write accesses, data is first of all written to volatile memory. Custody PDUs need to be written directly into persistent storage before accepting custody (write-through). However, conventional PDUs may be forwarded when residing on volatile memory. These PDUs can be written to persistent storage whenever there is time (write back). For write accesses, hybrid storages allow a certain amount of data to be stored with the native speed of the volatile storage. When the volatile storage is exceeded, the storage performance goes down to the performance of the persistent storage. This pattern is a good match for the bursty traffic pattern of typical DTNs.

For read accesses, it is desirable to use the performance of the volatile memory. However, the DTN Engine or the storage component would have to preload PDUs into volatile memory. In a network with predicted or scheduled contacts (see Section 13.6), this is very possible. Since the DTN Engine knows which neighbor

Write ────────┐   ┌──────► Read

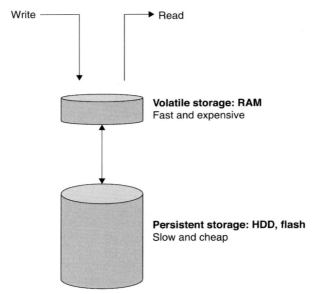

Volatile storage: RAM
Fast and expensive

Persistent storage: HDD, flash
Slow and cheap

Fig. 13.7 Hybrid storage architecture.

is going to show up next, ADUs for this neighbor can be preloaded and transferred at the bandwidth of the volatile storage. However, the prediction of opportunistic contacts is outside the scope of this chapter. In any case, ADUs that have not been preloaded into the volatile buffer have to read out of the persistent memory. Fortunately, flash as well as HDDs have the property that read access is faster (in terms of data rate) than write access. Therefore, preloading ADUs produces a smaller performance advantage than buffering write accesses.

The volatile buffer of the hybrid storage should be able to handle all data that is transferred during one contact. This ensures that data transfer can happen at maximum speed. For networks with a maximum contact duration of $t_{contactmax}$ and a networking link with a data rate r, the amount of volatile buffer that is necessary can be calculated as $t_{contactmax} \times r$. Furthermore, the intercontact time should be long enough to flush the volatile buffer into persistent storage.

13.5 Throughput

The throughput in a DTN can be measured from two different perspectives. On the one hand, we have the end-to-end throughput that can be attained between a sender and receiver pair, possibly over multiple intermediate nodes. In literature, the end-to-end throughput is often also referred to as the capacity of the network because it specifies the amount of data that can be transported per time interval. The end-to-end throughput depends on a number of factors, including routing and the hop-by-hop throughput. The hop-by-hop throughput, on the other hand, depends on the usable time per contact (see Section 13.6) and the throughput that can be attained during the contact. We have discussed the influence of I/O performance on the throughput in the previous section and discuss additional influences such as protocol overhead in this section. Influence of processing on the nodes can be found in Section 13.2.

13.5.1 Hop-by-hop throughput

13.5.1.1 DTN protocol stack overhead

DTNs are designed as overlay networks and reside on the application layer of the protocol stack. This means that the DTN protocol (e.g., the BP) is wrapped in multiple "exterior" protocols before the actual data is sent on the wire. When using the TCP convergence layer (TCPCL) of the BP, ADUs are wrapped in bundles, TCPCL segments, TCP segments, and IP packets. Depending on the lower layers, this plethora is then further wrapped into Ethernet frames. This seemingly long chain of protocols that are wrapped into each other exposes a significant protocol overhead that reduces the attainable throughput. While this may not be noticeable when using a Gigabit Ethernet connection, it becomes very real when using low-bandwidth wireless links that are found in many DTN applications. Taking the example of the BP, we calculate the overhead when using TCPCL over Ethernet in the following. We furthermore compare with the BP when using Compressed Bundle Header Encoding (CBHE) (Burleigh, 2011) and with cross-layer approaches, which reduce overhead significantly.

In Table 13.1, we show the minimal and typical overhead for the above-mentioned protocols. The minimal overhead is calculated by assuming realistic minimal values for all variable-length header fields. The typical case is an example of PDU in which all variable-length header fields use values that we see in our lab networks. While not all DTNs use IP or Ethernet as underlying layers, this example clearly shows that especially small PDUs suffer from the overhead. With a payload size of 10 bytes on a Gigabit Ethernet connection under ideal conditions, the goodput cannot exceed

Table 13.1 Overhead of a typical DTN protocol stack.

Protocol			Minimum size	Typical size
BP	Standard		28 Bytes	86 Bytes
	CBHE		22 Bytes	25 Bytes
TCPCL			2 Bytes	3 Bytes
TCP			20 Bytes	20 Bytes
IP	V4		20 Bytes	20 Bytes
	V6		40 Bytes	40 Bytes
Ethernet			18 Bytes	18 Bytes
Total			82–108 Bytes	86–167 Bytes

63.7 MBit/s when using IP Version 4 (IPv4). CBHE reduces the overhead of the BP so that the goodput cannot exceed 102.0 MBit/s for the same bundles.

The significant overhead of the BP shows that it is primarily designed for "larger" chunks of data for which the overhead ratio is small.

13.5.1.2   Cross-layer approaches for reduced overhead

DTNs are designed as overlay networks (Cerf et al., 2007) and make use of potentially heterogeneous underlying protocols between nodes. However, research (Pöttner et al., 2012) has shown that this does not have to be the case. The concepts of DTNs can also be applied to networks that avoid using an underlay network, thereby drastically reducing the overhead of the full protocol stack. In fact, a number of CLs for different underlying networking technologies exist, which all avoid using a full-blown underlay network. Those approaches can also be seen as cross-layer.

The DTN2 reference implementation includes an Ethernet-based CL (http://dtn. sourceforge.net/DTN2/doc/manual/cl-eth.html), which enables wired communication between neighboring nodes without using IP and/or TCP. Node discovery is supported as well. Further details about this CL are not published, but it appears that bundle segmentation up to 65,507 bytes is supported. DTN2 also includes a Bluetooth-based CL (http://dtn.sourceforge.net/DTN2/doc/manual/cl-bt.html) and an AX.25-based CL (Ronan et al., 2010). AX.25 is a link layer for packet service over HF, VHF, and UHF radio amateur links. Node discovery is not supported at the moment, but reactive fragmentation allows large bundles to be transmitted over intermittently available contacts.

Licklider transmission protocol (LTP) is a protocol for reliable data transmissions over links with extremely long round-trip times and is primarily intended for communication in space. A CL (Burleigh, 2012) allows bundles to be transmitted over LTP. μDTN (Pöttner et al., 2012) is a BP implementation for wireless sensor networks (WSNs) running on 8-bit microcontrollers. PDUs are wrapped in IEEE 802.15.4 (IEEE, 2007) radio data frames and transported between nodes.

13.5.1.3   Implementation details

Putting all previous considerations aside, hop-by-hop throughput is also dominated by implementation details. Implementing a DTN Engine is a complex task that also depends on the requirements of the application and the target platform. Furthermore, programming paradigms may also influence the achievable performance. We discuss certain performance-related implementation details in Section 13.2. However, to provide an impression of what difference the particular implementation can make, we show the attainable goodput over a Gigabit Ethernet connection with IPv4 in Fig. 13.8. The figure also shows the theoretical maximum throughput based on the overhead considerations stated above. The results in the figure clearly outline that, even when using the same protocol with the same overhead, the implementation makes a significant impact on the achievable throughput.

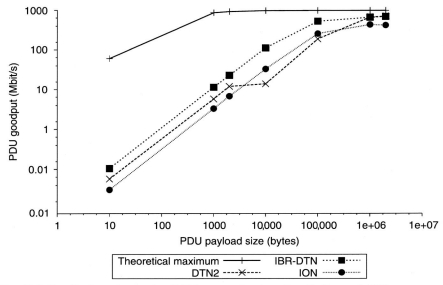

Fig. 13.8 Bundle throughput using RAM storage for three Bundle Protocol (BP) implementations (Pöttner et al., 2011a) (log y-axis).

13.5.2   End-to-end throughput

The throughput that can be achieved from source to destination node of a DTN is primarily a concatenation of multiple links in between. When PDUs are transferred on a single path, modeling the capacity is straightforward (Pöttner et al., 2012): the minimum capacity of all links in the path dominates how much data can be transported end-to-end.

13.5.2.1   DTN capacity model

The capacity of a link in a DTN is a function of the duration and frequency of contacts between nodes and the throughput. Since the amount of data that can be exchanged between nodes is usually fixed per time interval, longer contacts mean more exchanged data. Furthermore, a higher number of contacts also increases the capacity, because, with more contacts, data can be exchanged more often. In effect, the data path between individual hops of a DTN can be modeled as a bottleneck that can only transmit a certain amount of bundles of given size per time. Our model consists of a sender and a receiver node with limited storage capacities and certain link capacities and is expressed in the following paragraph. We assume The capacity $C_{i,j}$ of a link j in Eq. (13.1) is the total number of bundles that can be transferred in time interval i. $C_{i,j}$ can be calculated based on the number and duration of contacts as well as the rate at which bundles can be exchanged. The *BundleRate$_i$* is determined by the size of bundles and the data rate of the physical link over which the bundles shall be transported. The amount of bundles that can be transferred (T_i in Eq. 13.3) in time interval i is the minimum of the link capacity and the number of bundles waiting in the storage of the sender $S_{\text{Send},i}$. $S_{\text{Send},i}$ in Eq. (13.4) can be calculated using $S_{\text{Send},i-1}$, the bundles that have been transmitted during the last interval T_{i-1} and the bundles that have been created in the current interval N_i. Since the storage of the sender is limited, $S_{\text{Send},i}$ is the minimum of the theoretical amount of bundles in storage and the real storage capacity $S_{\text{Cap,Send}}$. The bundles in the storage of the receiver $S_{\text{Recv},i}$ in Eq. (13.5) depend on $S_{\text{Recv},i-1}$ and the number of transmitted bundles T_i, but cannot be higher than the storage capacity of the receiver $S_{\text{Cap,Recv}}$.

$$C_{i,j} = Contacts_i \cdot Duration_i \cdot BundleRate_i \tag{13.1}$$

$$Bundlerate_i = \frac{Bundlesize}{Linkdatarate} \tag{13.2}$$

$$T_i = \min\left(S_{\text{Send},i}, C_{i,j}\right) \tag{13.3}$$

$$S_{\text{Send},i} = \min\left(S_{\text{Send},i-1} - T_{i-1} + N_i, \ S_{\text{Cap,Send}}\right) \tag{13.4}$$

$$S_{\text{Recv},i} = \min\left(S_{\text{Recv},i-1} + T_i, \ S_{\text{Cap}}, \ S_{\text{Recv}}\right) \tag{13.5}$$

13.5.2.2 DTN capacity

The overall capacity of a DTN depends on a number of factors, as published in various papers. In practice, the throughput of a sender-receiver pair does not depend only on the least capable intermediate link but also on the load of the network induced by other nodes. Furthermore, DTN routing protocols are often based on replication, which generally helps to reduce message delays and may also increase reliability. However, with typically limited buffer spaces and slow (wireless) links, replication increases the load of the network and hence reduces end-to-end throughput.

In general, there is a trade-off between delay and throughput (Lee et al., 2007, 2009). However, at a certain point, the storage is the limiting factor, and then the throughput in the network decreases. Going back to the DTN capacity model above, replication leads to a higher number of PDUs. This in turn may overload the storages ($S_{\text{Cap,[Send,Recv]}}$) and lead to lower overall throughput. Furthermore, the mobility of nodes (Doering et al., 2010a) affects the contact time ($Duration_i$) and may also impact the network throughput. Increasing the radio range (e.g., when using a different radio technology or better antennas) leads to longer contacts (Doering et al., 2011a) and also increases the number of contacts.

13.6 Latency and queueing

Even in DTN structures, it is a reasonable goal to achieve a low end-to-end delay. Thus, in this section, we will discuss the additional latency introduced by processing overhead during the contact and common scheduling approaches.

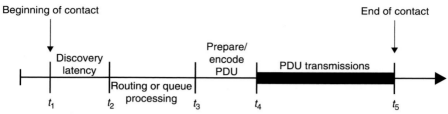

Fig. 13.9 Timeline scheme for contact utilization.

13.6.1 Contact utilization

As already considered in Section 13.2, the DTN Engine claims some processing capacity for each PDU it has to handle. The different factors are broken down in the timeline scheme for contact utilization (Fig. 13.9).

The first mark (t_1) denotes the beginning of a contact. During the interval $[t_1, t_2]$ the DTN Engine needs some time to discover the foreign peer. Under ideal conditions or when using hard-scheduled contacts, this interval would be zero. Once a peer has discovered the other, it can select the PDUs to transfer. Depending on the implementation, this involves the Routing Decision Engine. The cheapest approach would be to just select the corresponding queue for the peer and select the first PDU in it. Also, this would cost a certain amount of time. During the interval between t_3 and t_4, the first PDU has to be encoded for the transmission. Even if it is possible to encode the PDU partially during the transmission as described in Section 13.2.3, the encoding of the first segment consumes some time here.

As already discussed in Section 13.6, the interval $[t_1, t_2]$ is possibly huge due to energy constraints and can significantly limit the utilizable contact duration. Similarly, in the interval $[t_2, t_3]$, the process to find the right PDU to forward could be arbitrary complex and might involve some sort of handshake to exchange routing data (e.g., summary vectors), which implies several round trips. One way to keep that interval small is to prepare routing queues for each contact. Then the only costs would be in selecting the right queue. However, that approach would involve many resources if each node needs to store additional data for every node in the network. Moreover, many routing approaches depend on data exchange during a contact. That makes an a priori queuing system impossible. For that reason, it is necessary to search for PDUs to forward during the contact opportunity in typical DTN environments. In turn, the selection of PDUs depends on storage performance, as explained in Section 13.2.2. In the worst-case scenario, all this processing takes too long for a contact. More precisely, if the interval between t_1 and t_4 is greater than or equal to the contact itself, there will be no chance to transmit any data.

13.6.2 Throughput

If the previously discussed mechanisms are sufficiently fast for the contact duration, the DTN Engine will try to forward as many queued PDUs as possible. While the maximum amount of data to transfer is naturally limited through the contact capacity, the

PDU frequency (PDUs per time) depends on the processing overhead during the transmission.

As shown in Fig. 13.2, there are several tasks to do. The sender has to load and encode each PDU for the transmission. This process involves the persistent storage, and, as mentioned before, this is the bottle-neck in the whole system. Further, the routing decision engine of the sender has to select the PDUs to transfer. If the queue runs short or new PDUs have been received, the selection of PDUs starts all over again.

But the sender does not have to be the slowest part of the transmission in all cases. Assuming that the routing algorithm is not very complex and the storage can read PDUs very fast, the receiver has to decode and store the received PDUs. Since writing is more expensive than reading in general, the transmission is dominated by the receiver side. One way to improve the contact utilization would be to cache received PDUs in memory and delay the store. But this would only help if the contacts are very short and the amount of data is not very big, because memory is limited on most platforms.

Fig. 13.10 illustrates performance dependency. It shows the result of a throughput measurement which has been done by Pöttner et al. (2011a) to estimate the performance of different BP implementations. In this case, IBR-DTN (Schildt et al., 2011) is transferring several PDUs between two stations. The shared y-axis compares the PDU frequency (on the left side) and the achieved goodput (on the right side) dependent on the payload size of each PDU. Other tested implementations of the BP show similar behavior to IBR-DTN.

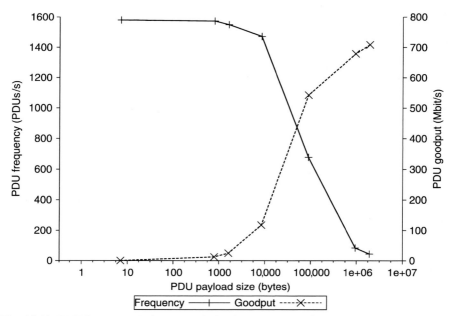

Fig. 13.10 PDU frequency during transmission in IBR-DTN (Pöttner et al., 2011a).

As a result, the DTN Engine seems to be limited to a maximum bound while sending PDUs with a very small payload. Here, the processing overhead is the limiting factor and consumes all available resources to encode, decode, and process the PDUs. Since the processing of the payload itself is extremely slight, the processing overhead of each PDU is dominating here. This starts changing with larger payload sizes (≥ 2000 bytes), where the processing of the payload grows and becomes significant.

With increasing payload size the achieved throughput is also raised until an upper bound is met. This bound is defined through the available bandwidth minus the protocol overhead. Additional to this, there might exist gaps in the transmission due to processing overhead. These gaps further lower the maximum achievable throughput.

13.6.3 Scheduling

Besides the obvious performance limitations caused by processing overhead and bandwidth bounds, there exists an additional factor, which might have an impact on the delivery delay. As explained before, each DTN Engine uses CLs to send and receive PDUs. Typically these CLs provide only one channel to other peers. If the channel is used for transmission, it is occupied and other PDUs are delayed until they are at the front of the queue. The result is that large PDUs might block the transmission opportunity and smaller PDUs are delayed because larger ones are queued in front of them. Also when using scheduling with priorities this would be an issue, because PDUs cannot be queued before an enduring transmission as long as it is impossible to abort them. It is even possible that small PDUs starve because they expire before they can be transmitted.

One approach to mitigate this issue would be to fragment all PDUs using proactive fragmentation if they exceed a maximum size. This would limit the payload volume of all the PDUs in the network and introduce a predictable upper bound for transmissions. Another way would be to multiplex several transmissions using one channel, but it is hard to decide how many of the concurrent transmissions are useful. Transmissions of several PDUs in parallel imply more overhead and prolong the transmission time of an individual PDU.

At this point, an example might help to clarify the mentioned issue. Assume a radio channel with 4 MBit/s and there are three PDUs to transfer. In encoded format, the PDUs have sizes of (1): 100 Kbytes, (2): 10 Kbytes, (3): 1 Kbyte. PDU (1) takes at least 200 ms to transmit, the second 20 ms and the third 2 ms. If all three PDUs are queued to the same destination, the third one will be delayed by at least 222 ms. If we introduce scheduling with priorities, it is possible to prioritize (3). That would change the queueing order to (3), (1), (2), because (3) will be inserted into the queue in front of (1). But if we queue (2) and (3) after the transmission of (1) has been started, the other PDUs have to wait until the transmission is done and will be delayed by at least 200 ms. Fragmentation would mitigate this issue by splitting (1) into several smaller PDUs, e.g., parts of 10 Kbytes. In such a case, the maximum additional delay for a prioritized PDU would be only 20 ms instead of 200 ms.

Basically, there exist many scheduling approaches with different goals. One might prioritize PDUs in some way using special flags, and another might prefer PDUs with a short remaining lifetime to increase the probability that they will reach the destination before they expire. An evaluation (Doering, 2012) of different common scheduling schemes for DTNs shows that they do not increase the global throughput of the network. But at least, if the individual latency of each PDU matters, a specific scheduling approach should be applied.

13.7 Discovery latency and energy issues

In opportunistic DTN scenarios, when there is a communication opportunity, nodes aim to utilize the contact as much as possible. Before using a contact, both nodes must be aware of the contact. How easy this depends on the type of contact. We extend and modify the terminology introduced in RFC4838 (Cerf et al., 2007) to define different types of DTN contacts relevant to discovery:

1. *Persistent contacts*: Contacts that are always available. An example is a DTN node connected to the Internet hosted inside some data centers. This is the equivalent of a server in classical Internet applications. As these nodes are considered to be always online, there is no need for any discovery mechanism.
2. *Opportunistic contacts*: Contacts that are neither scheduled nor predicted. From the system's point of view, these contacts are random. Examples are DTN Peer-to-Peer (P2P) applications using mobile devices. DTN applications of this kind are sometimes termed pocket switched networks (Hui et al., 2006).
3. *Scheduled contacts:* Scheduled contacts are established at a particular time and for a particular duration. Time and duration are known beforehand. There are two types of scheduled contacts:
   (a) *Hard-scheduled contacts*: Time and duration of contact are known with a high degree of precision. Examples are classical IPN scenarios. Just as with persistent contacts, a discovery mechanism is not necessary.
   (b) *Soft-scheduled contacts*: Time and duration of contacts are known beforehand; however, both time of establishment and duration have significant amounts of uncertainty attributed to them. Examples are DTNs in public transportation systems (Doering et al., 2010b), where a timetable provides some degree of determinism.
4. *Predicted contacts*: This class is somewhat less sharply defined. This can encompass contacts between scheduled and opportunistic. Depending on the history and some domain-specific information, future contacts will be predicted. This includes a wide range of contact situations: soft-scheduled contacts from timetables can be seen as predicted, as well as some statistical properties from opportunistic contacts, such as used by PRoPHET routing (Anders et al., 2004). While in the first case, contact can be predicted with a reasonable degree of certainty within a couple of minutes, PRoPHET, on the other hand, only "predicts" that some node might have a certain chance to be able to forward a packet to a destination sometime in the future.

13.7.1 The need for discovery

Fig. 13.11 shows the relation between the different types of DTN contacts. No discovery mechanism is needed for *persistent contacts* or *hard-scheduled contacts*: in the first case a node is assumed to be available all the time, and in the second case it

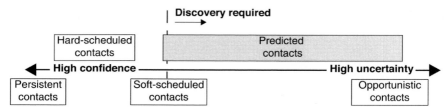

Fig. 13.11 Contact types in a delay-tolerant network (DTN).

is known exactly when a node will become available; thus a sender can just defer transmitting data until such a time as it knows the node will be ready for the reception. For *soft-scheduled contacts* the situation becomes a little bit more unclear: imagine a bus-based DTN, where the timetable is known, but due to traffic conditions the actual schedule will differ slightly from the timetable. Without a discovery mechanism, one needs to include a safety margin, i.e. if according to the expected schedule communication is possible between times t_0 and t_1, a system needs to allow communication only between $t_0 + t_{\text{safety}}$ and $t_1 - t_{\text{safety}}$. Thus, the utilized time of contact would be decreased by up to $2 \times t_{\text{safety}}$ from the optimum. Contact utilization is equal to a discovery mechanism that needs $2 \times t_{\text{safety}}$ time to detect another contact. However, no energy will be spent for discovery purposes. It is to be expected that for most soft-scheduled contacts a discovery mechanism will lead to better utilization, as t_{safety} needs to be chosen very conservatively to obtain a reliable system. For all contacts further on the right in Fig. 13.11, some form of discovery mechanism is required.

13.7.2   Contact utilization

Contact utilization in a DTN is fundamentally limited by the speed of the discovery mechanism. A lower contact utilization can influence end-to-end throughput and latencies. More specifically with

- bandwidth bw in bytes/s
- length of contact t in seconds
- transferred volume v
- discovery time d

when two nodes meet the amount $v = bw \times (t - d)$ $(t > d)$ of data can be moved. Dividing by the maximum possible utilization that could be leveraged in the case of persistent contacts or hard-scheduled contacts, we get the contact utilization U

$$U = \frac{bw \cdot (t-d)}{bw \cdot t} = \frac{t-d}{t} \quad \text{with } t > d$$

The relation between contact utilization and discovery time is shown in Fig. 13.12. Obviously, when $t < d$, a contact is useless and no data can be transferred. Therefore, especially for scenarios with short contacts due to movement or radio range, the

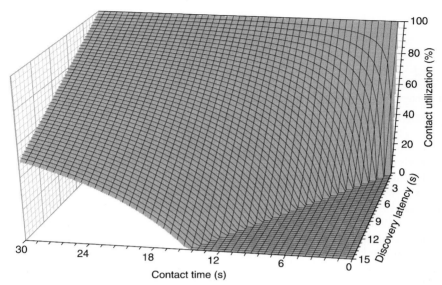

Fig. 13.12 Contact utilization with discovery.

discovery time d should be as short as possible. However, a common type of these scenarios is, for example, pocket switched networks, where battery-powered devices are used. This leads to the problem of energy usage for discovery.

13.7.3 Energy usage for discovery

The basic principle of discovery in wireless networks is always the same: nodes broadcast presence information that can be received by other nodes. The frequency of beaconing is directly related to the discovery time d. If a discovery beacon is broadcast with a frequency f Hz, on average, $d = (2f)^{-1}$ seconds cannot be used for each contact. For example, sending beacons with 0.1 Hz will lead to an average of 5 s lost on each contact. While transmitting beacons with a higher frequency uses energy, the real problem is the listening power consumption. Potential receivers must keep RF hardware in a high power state at all times to be able to receive beacons. This is the reason why the IBR-DTN BP implementation on Android (Morgenroth et al., 2012) is not performing continuous IP Neighbor Discovery (IPND).

One potential solution is simple duty-cycling methods, where the receivers know at what time to expect the beacon. Now nodes can keep their RF hardware in a low-power state and only go active when a discovery slot arises. In many practical scenarios, this is unfeasible, as it requires strict time synchronization between all nodes. Special schedules have been proposed to overcome this limitation and allow asynchronous discovery even if nodes are duty-cycling their radios. In the *DISCO* (Dutta and Culler, 2008) protocol, nodes choose two prime numbers and enter a discovery phase consisting of either beaconing, listening, or both in every time slot that is divisible by one of their primes. It is easy to see that the schedules of two nodes will overlap at some time, and, taking the chosen primes into account, an upper bound for the time needed before two nodes synchronize can be derived. Another scheduling approach called *Searchlight* with improved discovery latency over *DISCO* is presented in Bakht et al. (2012). *Searchlight* has also been implemented and tested on real smartphones. These approaches allow discovery to be combined with duty-cycling for energy-constrained nodes even when clocks are not synchronized perfectly.

In the common case of using Wi-Fi-enabled devices such as smartphones, a practical implementation hurdle is a long and device-dependent time to bring the Wi-Fi interface up and down (Bakht et al., 2012). This implies longer slot length, resulting in a practically achievable energy efficiency that is not yet as good as it theoretically could be. While a higher duty cycle is preferable from a contact utilization point of view, the Wi-Fi implementations in current mobile devices are limited by the long transition times, and devices may keep the Wi-Fi hardware in a high-power state for some time after the last transmission, which negatively affects energy efficiency. As Wi-Fi uses a serious amount of power, most mobile phones or tablets switch off the Wi-Fi hardware when they are on standby and no application is actively transmitting data. Therefore, mobile DTN applications, which do not require intermediate interaction can save a lot of energy when they opt to just use the Wi-Fi interface when it is active anyway due to the user using the device. This would drastically reduce the chance of finding a direct neighbor through a discovery protocol, as both nodes must be actively used at the same time. However, data can still be exchanged reliably by using an always available DTN router as persistent contact on the Internet. The tradeoffs here are throughput, latency, and availability. On the other hand, contact and thus energy utilization are getting better, as contacts are shifted from device-to-device contacts to a persistent device-router contacts, which can always be utilized fully.

13.7.4 Discovery using secondary radios

While the principle that broadcasting and listening for beacons, especially when using high bandwidth radios, consumes a lot of energy cannot be solved, for some applications using a low-bandwidth, low-power radio exclusively for discovery is a viable solution to save energy. One such system has been implemented by the Dieselnet project: a fleet of buses communicates via Wi-Fi with battery-powered roadside units called throwboxes (Banerjee et al., 2010). However, most of the time the Wi-Fi hardware and processing unit will be powered off. Only a small sensor node with a long-range, low-power 900 MHz radio is active. The buses are equipped with a similar radio. When a bus comes in range the processing hardware and Wi-Fi interface of the throwbox will be powered up and ready once the bus comes into Wi-Fi communication range. A flexible IEEE 802.15.4-based power management module for solar-powered outdoor DTN nodes that can wake up the Wi-Fi-powered hardware using a ruleset based on requests, priorities, and current battery status was presented and experimentally evaluated by Doering et al. (2011b).

13.8   Conclusions

Performance is of major importance for all communication systems and, thus, has always been an active research area. Although many generic issues regarding the performance of protocols and overall communication systems apply to DTNs as well, due to the specific characteristics of DTNs, further aspects have to be considered.

In this chapter, we discussed the meaning of the performance of DTN systems and issues which have an impact on this. First, we identified metrics that are relevant for the performance of DTN systems. In addition to standard items like throughput, we presented DTN-specific metrics such as contact utilization. Afterward, we analyzed processing steps and storage-induced costs. The latter is of particular importance due to the widely varying size of data bundles exchanged between DTN nodes.

We studied throughput in a DTN from two perspectives: (i) hop-by-hop as well as (ii) end-to-end throughput; the second, for which we also introduced a capacity model, depends on the first, which again hinges on local mechanisms. In a DTN, latency could be considered as a concern of secondary order. Yet, here also we would like to lower it as much as possible. Thus, we reviewed aspects such as contact utilization and scheduling of data exchange. Since the performance of a DTN system depends on the contact utilization, it is important that nodes discover each other quickly. However, this may come, among other things, at the expense of energy.

Clearly, the routing method used in a DTN has a significant impact on the achievable performance. Multicopy approaches may put a high load on the DTN Engines in the DTN nodes. For some routing schemes, a lot of (signaling) information must be exchanged between nodes before actual data flow can begin. Therefore, the local node mechanisms and the protocol engine are of utmost importance to achieve a good overall performance. Depending on the peculiarities of the routing approaches, different local methods might be of advantage for implementation on a DTN node. Specific optimizations can be applied if the concrete deployment scenario is well known; otherwise, generic approaches should be used which provide good performance in typical cases.

Throughout this chapter, we used examples from existing DTN implementations such as DTN2, ION, and especially our own implementation IBR-DTN (Schildt et al., 2011). Nevertheless, we believe that most of the findings are independent of specific implementations but relevant for DTNs in general. Future research should investigate which approaches lead to good performance for major DTN applications and whether more adaptivity can be introduced to improve performance further if the characteristics of the deployment scenario do not match well with default assumptions.

References

Anders, L., Avri, D., Olov, S., January 2004. Probabilistic routing in intermittently connected networks. Sigmobile Mob. Comput. Commun. Rev. 7, 2003.
Bakht, M., Trower, M., Kravets, R.H., August 2012. Searchlight: won't you be my neighbor? In: Proceedings of the 18th Annual International Conference on Mobile Computing and Networking, Mobicom'12, Istanbul, Turkey. ACM, New York, NY, USA, pp. 185–196.

Banerjee, N., Corner, M.D., Levine, B.N., April 2010. Design and field experimentation of an energy-efficient architecture for DTN throwboxes. IEEE/ACM Trans. Network. 18 (2), 554–567.

Burleigh, S., January 2007. Interplanetary overlay network: an implementation of the DTN bundle protocol. In: 4th IEEE Consumer Communications and Networking Conference, 2007 (CCNC 2007), Las Vegas, NV, USA, pp. 222–226.

Burleigh, S., May 2011. Compressed Bundle Header Encoding (CBHE). RFC 6260 (Experimental).

Burleigh, S., October 2012. Delay-Tolerant Networking LTP Convergence Layer (LTPCL) Adapter. Internet-Draft http://tools.ietf.org/html/draft-burleigh-dtnrg-ltpcl-05.

Cerf, V., Burleigh, S., Hooke, A., Torgerson, L., Durst, R., et al., April 2007. Delay-Tolerant Networking Architecture. RFC 4838 (Informational).

Demmer, M., Brewer, E., Fall, K., Jain, S., Ho, M., et al., December 2004. Implementing delay tolerant networking. Technical report, IRB-TR-04-020.

Doering, M., Wolf, L., March 2012. Work-in-progress: evaluation of generic bundle transmission scheduling strategies in vehicular disruption tolerant networks. In: Proceedings of the 4th Extreme Conference of Communication (ExtremeCom 2012), Zurich, Switzerland.

Doering, M., Pögel, T., Pöttner, W.-B., Wolf, L.C., 2010a. A new mobility trace for realistic large-scale simulation of bus-based DTNs. In: ACM MobiCom 2010 Workshop on Challenged Networks (CHANTS 2010), Chicago, USA, September. ACM, New York, NY, USA.

Doering, M., Pögel, T., Wolf, L., 2010b. DTN routing in urban public transport systems. In: CHANTS'10: Proceedings of the 5th ACM Workshop on Challenged Networks, Chicago, USA, September. ACM, New York, NY, USA.

Doering, M., Pöttner, W.-B., Pögel, T., Wolf, L., 2011a. Impact of radio range on contact characteristics in bus-based delay tolerant networks. In: Eighth International Conference on Wireless On-Demand Network Systems and Services (WONS 2011), Bardonecchia, Italy, January, pp. 195–202.

Doering, M., Rottmann, S., Wolf, L., 2011b. Design and implementation of a low-power energy management module with emergency reserve for solar powered DTN-nodes. In: Proceedings of the 3rd Extreme Conference on Communication The Amazon Expedition – ExtremeCom'11, Zurich, September. ACM, New York, NY, USA, pp. 1–6.

Dutta, P., Culler, D., 2008. Practical asynchronous neighbor discovery and rendezvous for mobile sensing applications. In: Proceedings of the 6th ACM Conference on Embedded Network Sensor Systems, SenSys'08, Raleigh, North Carolina. ACM, New York, NY, USA, pp. 71–84.

Hui, P., Chaintreau, A., Gass, R., Scott, J., Crowcroft, J., et al., 2006. Pocket switched networking: challenges, feasibility and implementation issues. LNCS 3854, 1–12. https://doi.org/10.1007/11687818.

IEEE, 2007. 802.15.4: Wireless Medium Access Control (MAC) and Physical Layer (PHY) Specifications for Low-Rate Wireless Personal Area Networks (WPANs). IEEE Standard for Information Technology. IEEE Computer Society, New York, NY, USA.

Lee, U., Lee, K.-W., Oh, S.Y., Gerla, M., 2007. Understanding the capacity and delay scaling laws of delay tolerant networks: a unified approach. Technical report: Tr-070020, UCLA CSD.

Lee, U., Oh, S.Y., Lee, K.-W., Gerla, M., 2009. Scaling properties of delay tolerant networks with correlated motion patterns. In: Proceedings of the 4th ACM Workshop on Challenged Networks, CHANTS'09, Beijing, China. ACM, New York, NY, USA, pp. 19–26.

Morgenroth, J., Schildt, S., Wolf, L., August 2012. A bundle protocol implementation for android devices. In: Proceedings of the 18th Annual International Conference on Mobile Computing and Networking – Mobicom'12, Istanbul, Turkey. ACM, New York, NY, USA, p. 443.

Patterson, D.A., Hennessy, J.L., 2005. Computer Organization and Design: The Hardware/Software Interface. Morgan Kaufmann.

Pöttner, W.-B., Morgenroth, J., Schildt, S., Wolf, L., 2011a. An Empirical Performance Comparison of DTN Bundle Protocol Implementations. Informatikbericht 2011–08, September, Technische Universität Braunschweig.

Pöttner, W.-B., Morgenroth, J., Schildt, S., Wolf, L., 2011b. Performance comparison of DTN bundle protocol implementations. In: Proceedings of the 6th ACM Workshop on Challenged Networks, CHANTS'11. ACM, New York, NY, USA, pp. 61–64.

Pöttner, W.-B., Büsching, F., von Zengen, G., Wolf, L., October 2012. Data elevators: applying the bundle protocol in delay tolerant wireless sensor networks. In: The Ninth IEEE International Conference on Mobile Ad-hoc and Sensor Systems (IEEE MASS 2012), Las Vegas, Nevada, USA.

Ronan, J., Walsh, K., Long, D., 2010. Evaluation of a DTN convergence layer for the ax.25 network protocol. In: Proceedings of the Second International Workshop on Mobile Opportunistic Networking (MobiOpp 2012), MobiOpp'10, Zurich. ACM, New York, NY, USA, pp. 72–78.

Schildt, S., Morgenroth, J., Pöttner, W.-B., Wolf, L., January 2011. IBR-DTN: a lightweight, modular and highly portable bundle protocol implementation. Electron. Commun. EASST 37, 1–11.

Scott, K., Burleigh, S., November 2007. Bundle Protocol Specification. RFC 5050 (Experimental).

Using emulation to validate applications on opportunistic networks

Gwilherm Baudic[a], Antoine Auger[a], Victor Ramiro[a], and Emmanuel Lochin[b]
[a]ISAE-SUPAERO, University of Toulouse, Toulouse, France, [b]ENAC, University of Toulouse, Toulouse, France

14.1 Introduction

Opportunistic networks are a special case of DTNs (Fall, 2003), where nodes systematically exploit their mobility to benefit from contacts as a communication opportunity to forward messages. This mobility introduces delays when a node cannot forward its message, keeping it in its own buffer. This allows routing protocols to exploit opportunistic contacts, in the absence of stable end-to-end path, as a means to create a temporal path for delivery. The store-carry-forward paradigm allows nodes to exploit spatiotemporal paths created by contact opportunities in order to deliver messages overtime. Opportunistic networks are also suitable for communications in pervasive environments that are saturated by other devices. The ability to self-organize using the local interactions among nodes, added to mobility, leads to a shift from legacy packet-based communications toward a message-based communication paradigm.

However, dealing with the dynamics of opportunistic networks is complicated (Conti and Giordano, 2014). Let us consider the number of parties involved in the network, for instance, n inhabitants in a medium dense city. First, the number of interactions between them grows as $O(n^2)$. Second, the continuously changing topology, due to the nodes mobility and its interactions, leads to an explosion of the number of states needed to characterize the behavior for any algorithm to be deployed. The number and density of nodes, their interactions, their mobility, the different routing strategies impact on the delay, packet lost, retransmissions, etc. Different ways have been proposed to study those dynamics. The main focus of research up to now was to define an optimal routing strategy to deliver some domain-specific information, but they do not consider the final application development.

With the current trend on connected devices, the idea of *opportunistic applications* (i.e., applications running over opportunistic networks) is getting closer to being a reality. However, there are still several obstacles remaining before we see a massive deployment over this paradigm. Indeed, to conceive applications working on these networks remains one of the biggest challenges to overcome. This can be explained by the complexity to undertake a performance evaluation of an opportunistic application before a real-world deployment.

Advances in Delay-tolerant Networks (DTNs). https://doi.org/10.1016/B978-0-08-102793-6.00014-X

Simulations offer a fast and lightweight way of getting insight in the behavior of the network (Chang, 1999; Henderson et al., 2008; Keränen et al., 2009). Unfortunately, they do not give a simple way of thinking in terms of real applications, and typically cannot derive metrics related to QoE because they usually focus on purely network-related performance. Test beds (Beuran et al., 2012, 2013; Li et al., 2015; Giordano et al., 2012; Yoon et al., 2009; Morgenroth et al., 2010; Bittencourt et al., 2013; Zhou et al., 2006; Maeda et al., 2008; Liu et al., 2007; Zhang et al., 2009) are effective in terms of an almost real-world feedback, but they are really expensive to deploy in the middle of the development process.

Developers of opportunistic applications must not only deal with network characterization, but also with its impact on the application. Filling this gap between network characterization and development should be supported by current developments tools. However, even the ability to test the applications conformity to a simple DTN messaging protocol is missing today. *We need to better integrate how developers consider network metrics obtained from the characterization phase into the development process of opportunistic applications.*

The rest of the chapter is structured as follows. We first introduce the challenges related with opportunistic networks in Section 14.2. It allows us to highlight the gap between network characterization and application development. In order to cope with this challenge, we define important requirements presented in Section 14.3. Finally, we conclude and give some directions for future work in Section 14.4.

14.2   Development challenges of opportunistic applications

In this section, we discuss the challenges when developing opportunistic applications. We see those challenges from two perspectives: the way developers deal with the network characterization and the way they assess the network impact on the application. Finally, we highlight the existence of a gap between these two perspectives.

14.2.1   Dealing with opportunistic networks characterization

As we said, dealing with the dynamics of opportunistic networks is complicated. In this section, we discuss the main challenges when dealing with opportunistic networks from a developer's point of view. We discuss the current available alternatives (and their drawbacks) to characterize opportunistic networks.

14.2.1.1   Analytical modeling

Several analytical models have been presented to characterize opportunistic networks. The main goal of analytical models is to provide a closed formula for specific characteristics. Others propose algorithms to approach reality (Groenevelt et al., 2005; Ramiro et al., 2015; Haas and Small, 2006; Zhang et al., 2007). However, most of these models assume unrealistic hypotheses or they cannot scale. Indeed, the number of states needed to model DTNs increases with the interactions parties have in the network. The number of interactions and the changes they follow makes this problem a highly combinatorial one. Most of them belong to the NP-class.

14.2.1.2   DTN simulators

Opportunistic networks simulators (Keränen et al., 2009; Chang, 1999; Henderson et al., 2008) are mainly focused on nodes mobility and efficient routing issues. Indeed, most of the research on DTNs has been focused on the message routing problem as the application.

On the one hand, we find many tailored simulators for specific cases. On the other hand, we find an effort to standardize the results based on the ONE (Keränen et al., 2009) simulator. Nevertheless, current simulators do not allow developers to plug real devices to interact with them. For instance, the ONE simulator provides a simulated network stack. This stack includes a network interface layer, a connection layer (constant and variable bit rates), a routing layer (with several routing algorithms) and, finally, an application layer on the top. However, this application layer is just a basic handler class for messages passed by from the simulated routing protocol. For a real developer, the impossibility to think in terms of a concrete user application, independently of all those complexities, is still a huge problem.

14.2.1.3   Traces collection

Another effort to better understand opportunistic networks has been the development of devices and applications to collect peer contact traces. The characterization of the contact and intercontact time distribution allows to understand the dynamics behind the network. Ideally, this abstraction in terms of contact should be independent from the link layer. However, this is not the case in reality (Baudic et al., 2014). Indeed, current communication layers lack the characteristics needed to really deploy opportunistic applications. This makes the collected traces to be less representative than we need, and therefore hard to replay in order to help application developers.

14.2.1.4   DTN emulators and test beds

Emulation naturally provides a bridge between simulation and real-world testing (Beuran, 2013). It consists of putting together real and simulated components in a single system. In existing proposals, real parts are most of the time the application and underlying operating system, while the network is simulated. For instance, KauNet (Garcia et al., 2007) is a pattern-based link emulator for mobile and wireless systems. Pérennou et al. (2011) extend it to support opportunistic networks, thanks to trigger mechanisms.

More generally, emulation is seen as a convenient way to get closer to the realism of field trials, while at the same time offering the repeatability and scalability of simulations. A good emulation system must also be completely transparent to the real part and should offer sufficient flexibility to be used with various mobility models. However, scalability is often achieved by using test beds, which are hard and costly to setup.

14.2.1.5   Middlewares and real DTN stacks

Most of DTN middlewares (e.g., MaDMAN (Petz et al., 2010) or the solution of Jiang et al. (2011)) focus on integration and architectural considerations, arguing that human mobility will address connectivity issues and that applications will keep working correctly within a DTN context. Although these solutions can be considered as proofs of concept, they do not provide any useful tools nor metrics to potential applications developers. Plus, the proper functioning of these middlewares is only assessed on the field, by looking at the number of packets delivered. Finally, the results are only valid for a given application and for a given context.

Implementations of the DTN stack are still in early stage. We count among them ION (Burleigh, 2007), DTN2 (Demmer et al., 2004), or IBR-DTN (Schildt et al., 2011). Even though they exist, they are not yet massively deployed in any final user platform.

14.2.2   Dealing with opportunistic networks impact

Within an opportunistic network context, end-to-end delays, delivery ratio, and drop ratio are very important factors that developers want to study before deploying their applications. However, this set of network-related metrics is not sufficient to correctly describe the impact of those networks for the end user.

Quality of experience (QoE) is defined by the ITU-T as "the overall acceptability of an application or service, as perceived subjectively by the end-user" (ITU-T, n.d.). As a matter of fact, QoE has objective and subjective dimensions. Although subjective dimension refers to human components (such as emotions, ease of use, etc.), objective dimension relates to more measurable and quantifiable factors (such as network parameters for instance). Developers need to test their applications from a user viewpoint (subjective dimension) while having an overview of some key metrics in real time (objective dimension).

In line with the vision of Brooks and Hestnes (2010), we believe that it is relevant to express QoE as a combination of user experience and technical measurements. In an emulator, we assume that technical measurements should refer to network-related metrics, while user experience relies on the following question: *Is my application working as I expected?* Please note that several measures of user experience exist and can also be envisioned, including for mobile use cases (Park et al., 2013).

14.2.3   Reconciling perspectives

We argue that there is a gap between opportunistic network characterization and applications development.

On the one hand, network characterization provides metrics to better understand the underlying behavior of the opportunistic network. Indeed, opportunistic and DTN networks have led to a change from a connection oriented to a message-based paradigm. Opportunistic applications rely on the store-and-forward paradigm to correctly work. Some properties of these networks, such as the nonguarantee of end-to-end paths, make impossible to ignore the characterization phase when developing opportunistic applications.

On the other hand, developers may not have the adequate knowledge or resources to take advantage of the network characterization. Instead, they want to know, possibly in a quick and simple way, if *their* applications will *still* work within an opportunistic use case. Making hypotheses on the underlying network (end-to-end delay, drop ratio, etc.) often leads to over provisioning and resources waste.

Hence, there is a gap in the way developers deal with network metrics obtained from network characterization. In the following section, we motivate the need to use emulation to reconcile both.

14.3   Requirements for opportunistic network emulation

As presented in Section 14.2, there is a gap between network characterization and application development. We argue that an opportunistic network emulator can fill this gap. In the following, we distill the main requirements for opportunistic network emulation. Emulation provides an environment to deploy real applications, helping developers to craft opportunistic applications.

14.3.1   Link layer requirements

One problem of opportunistic networks is the lack of a clear communication technology that can deal with the evaluation of real applications. Indeed, current link layer technologies, such as Bluetooth, Wi-Fi Direct, or other ad hoc communication technologies lack many of the assumptions made by opportunistic networks algorithms.

Nevertheless, we do not see this as a real problem but rather as an opportunity. Indeed, based on the idea of traces collection, we propose to focus on the contact and intercontact time characteristics of the network over a proper link layer. Contacts between nodes can easily be emulated by adding some local parameters to represent delays or packet losses on a pair-wise basis, allowing in some sort of a worst-case scenario analysis. Hence, we define two requirements that we assume as hypotheses to be met by any communication technology to support opportunistic applications:

C1 *Connection bidirectionality*: Algorithms proposed over opportunistic networks can assume connections to be bidirectional. Consider for example the summary vector exchange in epidemic routing, which is impossible if the connection is not bidirectional.

C2 *Multicast communication*: Nodes can connect with multiple devices at the same instant of time over opportunistic networks.

14.3.2   Connection-oriented vs. contact-oriented emulation

Simulators are usually implemented in a connection-oriented basis: whenever a contact between two nodes occurs, a global state is set to denote an open connection between nodes. While the connection is open, the simulator will try to send and receive data. This open state is updated at each step of the simulation until a disconnection occurs. At this moment, any nonfinished transfer will be dropped.

Instead, we propose a contact-oriented approach: since we know beforehand the duration of the contacts, we can precalculate which messages may be actually transferred along the duration of the connection. This allows us to optimize the number of events we will generate, while keeping the same behavior, instead of keeping the state of the connection at each step of the emulation.

14.3.3   Opportunistic emulation requirements

In order to help developers to design, develop, and test opportunistic applications, we need to provide a tool that can be easily integrated into the development process. Developers interact with IDE tools where usually testing and debugging tasks are executed to polish applications. We propose the use of network emulation to fill this gap. In order to accomplish the challenges defined before, we define the following basic requirements for such an emulator:

E1 *Real-time emulation*: The applications being tested need to operate at the same time scale as in normal operation, hence the need to attach a noncomplex real-time emulator. Notice that this differs from a simulator or a test bed in the sense that we focus on the application development process and not on the network characterization.

E2 *Contact-oriented emulation*: In order to simplify the view of the system, we think that a contact-oriented emulator must be put in place. Contacts are easy to understand in terms of behavior, while hiding the complexity of nodes' mobility. They also allow to know the contact duration beforehand, thus greatly simplifying routing decisions and buffer management. Since we do not rely on a clear link layer technology to deploy opportunistic applications and based on requirements (C1) and (C2), we can abstract the network and still conceive fully functional applications.

E3 *Real-time tuning*: Since the parameters of opportunistic algorithms can be cumbersome, we argue that a real-time parameterization is needed to better understand the impact of changes in the emulated opportunistic networks. This lets developers better plan for changes and adapt their applications accordingly.

E4 *Real-time monitoring*: In order to tune parameters, the need to observe network and application behaviors, along with their reaction to changes, is very important. Ideally, this observation should be possible during the experiment, and not only afterwards as with simulators. For that, we argue that any opportunistic network emulator needs a real-time monitoring system.

E5 *Transparency*: The application that is being tested should not be aware that it runs over an emulator. This also means that we should ideally be able to run the software unmodified. Indeed, to obtain meaningful insights, we need to make sure that the emulation platform itself does not bias application operation.

E6 *Repeatability*: To successfully debug an application, we need to exactly reproduce the conditions for an error to happen. Consequently, an opportunistic emulator has to support exact reproduction of simulated network conditions over several distinct runs if necessary for the end user. This includes routing decisions, message generation, and contacts. It also means that the emulator needs to have a deterministic behavior, so any part requiring randomness (like synthetic mobility models) should be left out, for example, by using precomputed traces.

E7 *Availability*: Unlike a test bed, the emulator should ideally be able to run on more limited hardware resources like a single computer, just like other IDE tools. It cannot afford either to require the booking of resources (hardware and time) in advance, because developer needs may be unpredictable.

14.4 Conclusions

This chapter highlights the gap between network characterization and opportunistic applications development. While opportunistic networking is a promising alternative, its inherent complex dynamics makes application design very challenging. Usually, a phase of network characterization is needed to better understand the challenges we will face later during development. Unfortunately, current tools such as DTN simulators or DTN test beds do not provide any integration with development. Hence, developers must not only deal with network characterization, but also with its impact on the application.

In this positioning chapter, we advocate the use of emulation to reconcile both aspects. We address the limitations of current network characterization approaches. This allows us to distill important requirements for an opportunistic network emulator.

References

Baudic, G., Pérennou, T., Lochin, E., 2014. Revisiting pitfalls of DTN datasets statistical analysis. In: ACM CHANTS '14, pp. 73–76.

Beuran, R., 2013. Introduction to Network Emulation. Pan Stanford Publishing.https://doi.org/10.1201/b13256.

Beuran, R., Miwa, S., Shinoda, Y., 2012. Making the best of two worlds: a framework for hybrid experiments. In: ACM WiNTECH '12.

Beuran, R., Miwa, S., Shinoda, Y., 2013. Network emulation testbed for DTN applications and protocols. In: IEEE Conference on Computer Communications Workshops (INFOCOM WKSHPS), April.

Bittencourt, D., Mota, E., Silva, E.N., Souza, C., 2013. Towards realism in DTN performance evaluation using virtualization. In: IFIP Wireless Days (WD), 2013, November.

Brooks, P., Hestnes, B., 2010. User measures of quality of experience: why being objective and quantitative is important. IEEE Netw. 24, 8–13.

Burleigh, S., 2007. Interplanetary overlay network: an implementation of the DTN bundle protocol. In: IEEE CCNC, January 2007, pp. 222–226.

Chang, X., 1999. Network simulations with OPNET. In: Inproceedings of the 31st Conference on Winter Simulation: Simulation—A Bridge to the Future, vol. 1ACM.

Conti, M., Giordano, S., 2014. Mobile ad hoc networking: milestones, challenges, and new research directions. IEEE Commun. Mag. 52, 85–96.

Demmer, M., Brewer, E., Fall, K., Jain, S., Ho, M., Patra, R., 2004. Implementing Delay Tolerant Networking. Intel Research, Berkeley, CA.

Fall, K., 2003. A delay-tolerant network architecture for challenged internets. In: Proceedings of the 2003 Conference on Applications, Technologies, Architectures, and Protocols for Computer Communications. ACM.

Garcia, J., Conchon, E., Pérennou, T., Brunstrom, A., 2007. KauNet: improving reproducibility for wireless and mobile research. In: Proceedings of the 1st International Workshop on System Evaluation for Mobile Platforms. ACM.

Giordano, E., Codecà, L., Geffon, B., Grassi, G., Pau, G., Gerla, M., 2012. MoViT: the mobile network virtualized testbed. In: ACM VANET '12.

Groenevelt, R., Nain, P., Koole, G., 2005. The message delay in mobile ad hoc networks. Perform. Eval. 62, 210–228.

Haas, Z.J., Small, T., 2006. A new networking model for biological applications of ad hoc sensor networks. IEEE/ACM Trans. Netw. 14, 27–40.

Henderson, T.R., Lacage, M., Riley, G.F., Dowell, C., Kopena, J., 2008. Network simulations with the ns-3 simulator. In: SIGCOMM Demonstration, vol. 14.

ITU-T, n.d. G.1080: quality of experience requirements for IPTV services, recommandations G.1080 (12/08), approved in 2008-12-07, status : in force. https://www.itu.int/rec/T-REC-G.1080-200812-I.

Jiang, P., Bigham, J., Bodanese, E., Claudel, E., 2011. Publish/subscribe delay-tolerant message-oriented middleware for resilient communication. IEEE Commun. Mag. 49, 124–130.

Keränen, A., Ott, J., Kärkkäinen, T., 2009. The ONE simulator for DTN protocol evaluation. In: Proceedings of the 2nd International Conference on Simulation Tools and Techniques, ICST.

Li, Y., Hui, P., Jin, D., Chen, S., 2015. Delay-tolerant network protocol testing and evaluation. IEEE Commun. Mag. 53, 258–266.

Liu, J., Mann, S., Van Vorst, N., Hellman, K., 2007. An open and scalable emulation infrastructure for large-scale real-time network simulations. In: IEEE INFOCOM 2007, May.

Maeda, K., Nakata, K., Umedu, T., Yamaguchi, H., Yasumoto, K., Higashinoz, T., 2008. Hybrid testbed enabling run-time operations for wireless applications. In: PADS '08, June.

Morgenroth, J., Schildt, S., Wolf, L., 2010. HYDRA: virtualized distributed testbed for DTN simulations. In: ACM WiNTECH '10ACM.

Park, J., Han, S.H., Kim, H.K., Oh, S., Moon, H., 2013. Modeling user experience: a case study on a mobile device. Int. J. Ind. Ergon. 43, 187–196.

Pérennou, T., Brunstrom, A., Hall, T., Garcia, J., Hurtig, P., 2011. Emulating opportunistic networks with Kaunet triggers. EURASIP J. Wireless Commun. Netw. (EURASIP JWCN) 2011, 1–15.

Petz, A., Bednarczyk, A., Paine, N., Stovall, D., Julien, C., 2010. Madman: A Middleware for Delay-Tolerant Mobile Ad-Hoc Networks. University of Texas at Austin, Austin, TX.

Ramiro, V., Dang, D.K., Baudic, G., Pérennou, T., Lochin, E., 2015. A Markov chain model for drop ratio on one-packet buffers DTNs. In: IEEE WoWMoM Workshop on Autonomic and Opportunistic Communications (AOC), June.

Schildt, S., Morgenroth, J., Pöttner, W.B., Wolf, L., 2011. IBR-DTN: a lightweight, modular and highly portable bundle protocol implementation. Electron. Commun. EASST 37, 1–10.

Yoon, H., Kim, J., Ott, M., Rakotoarivelo, T., 2009. Mobility emulator for DTN and MANET applications. In: ACM WINTECH '09, pp. 51–58.

Zhang, X., Neglia, G., Kurose, J., Towsley, D., 2007. Performance modeling of epidemic routing. Comput. Netw. 51 (10), 2867–2891.

Zhang, Z., Jin, Z., Chen, H., Shu, Y., Zhao, C., 2009. Design and implementation of a delay-tolerant network emulator based in QualNet simulator. In: WiCom '09, September, pp. 1–4.

Zhou, J., Ji, Z., Bagrodia, R., 2006. TWINE: a hybrid emulation testbed for wireless networks and applications. In: IEEE INFOCOM 2006, April.

The quest for a killer app for delay-tolerant networks (DTNs)☆

A.G. Voyiatzis
Industrial Systems Institute, RC "Athena", Patras, Greece

15.1 Introduction

Delay-tolerant networks (DTNs) were introduced in 2003 for fighting the enormous delays involved in deep space communications (in the order of minutes, hours, or even days). Such delays cannot be handled with existing (terrestrial) networking technologies and thus must be handled at the application level. Extending the concept of delay tolerance, it may be possible with the same design to handle long-term disruptions as well.

Disruptions (intentional or unintentional) can be considered as unplanned delays. Depending on the context and application scenario, a DTN may be a "delay-tolerant," a "disruption-tolerant," or even a "delay- and disruption-tolerant" network.

More than a decade after its introduction, DTN is still lacking a killer application that will help it unveil its potential and result in wide adoption. This is a point of concern for the research community (Lindgren and Hui, 2009). However, a killer application is not a panacea; the now successful Transmission Control Protocol/Internet Protocol (TCP/IP) suite of network protocols did not have a killer application for more than 11 years (Crowcroft et al., 2008).

In this chapter, we survey the breadth of applications in which DTN is already tested, solving actual, real-world problems related to intermittent connectivity and harsh operational environments around our planet (Section 15.2). Characteristic examples are described and, in many cases, the reader may consult the respective chapters of this book for more details. The wide applicability of DTN in such diverse environments indicates that DTN is, in fact, an enabling network protocol stack and technology for building next-generation applications and apps for smart mobile devices that can cope transparently with long disconnections (Section 15.3). The chapter concludes with an outlook for the future (Section 15.4).

15.2 The quest for a problem

The origins of DTN can be traced back to the late 1990s and the exploration of extending the Internet into space in the context of the Interplanetary (IPN) Internet project (Fall, 2003; Wood et al., 2009a,b). As in the case of many space technologies, it was

---

☆ This chapter is a reprint of the chapter originally published in the first edition of Advances in Delay-Tolerant Networks (DTNs): Architecture and Enhanced Performance.

Advances in Delay-tolerant Networks (DTNs). https://doi.org/10.1016/B978-0-08-102793-6.00015-1

realized that deep space communications have much in common with terrestrial applications, especially opportunistic wireless networks. What differentiates most terrestrial applications from space applications is the stochastic nature of disconnections in the former while the latter exhibit (normally) scheduled connectivity periods. Next, DTN terrestrial applications are presented. The applications are organized into three major categories: implementations for remote areas without network infrastructure (Section 15.2.1); low-cost alternatives to satellite-assisted communications for sea application (Section 15.2.2); and smart networking applications (Section 15.2.3).

15.2.1   Isolated and reduced-connectivity areas

Internet connectivity in isolated areas is a challenging task. In many cases, the network infrastructure may not be available, the cost of operation may be prohibitively high, or it may not be possible to install anything due to environmental protection regulations. In such cases, a DTN stack and moving objects with short-range wireless connectivity can be the only viable option.

DakNet is an example of real-world DTN deployed in rural villages of India and Cambodia (Pentland et al., 2004). DakNet uses moving objects, such as bicycles and motorcycles, to transfer information from one disconnected village to another. These so-called "data mules" were enhanced with network services, such as addressing and routing, forming a "mechanical backhaul" infrastructure (Seth et al., 2006). KioskNet improved the design of the prototype device based on observations of a pilot deployment (Guo et al., 2011).

Saami herders near the Arctic Circle were provided with basic Internet services based on DTN within the context of project SNC (Saami Network Connectivity) (Lindgren et al., 2008). Short-range infrastructure for providing connectivity and services in disconnected areas, such as the Padjelanta National Park in Sweden and Dharamsala in the Indian Himalaya, was further implemented in the context of project N4C (Networking for Communications Challenged Communities: Architecture, Test Beds and Innovative Alliances) (http://www.n4c.eu/) and the series of ExtremeComm scientific workshops (Lindgren and Hui, 2011).

Wildlife and open-space environment monitoring is an application area that can benefit from DTN. The ZebraNet project implemented small devices based on DTN principles that were mounted on zebras. The device sensors monitored the social behavior and the movement of the zebras in a large park in Kenya (Zhang et al., 2004). Other researchers monitored the pollution of an Irish lake using boats as "data mules" (Farrell and Cahill, 2006). This was a cheap alternative solution compared with transmitting the information using cellular networks. Later, an autonomous system was designed for monitoring a lake in Greenland for a whole year (Bonnet and Chang, 2010). The severely harsh and unapproachable environment of a frozen lake provided the researchers with several insights regarding the system design and installation while proving the applicability and value of a DTN approach. LUSTER (Light Under

Shrub Thicket for Environmental Research) is an environmental wireless sensor network (WSN) utilizing in-network storage like a DTN (Selavo et al., 2007). LUSTER also acts as a bridge between two wireless network protocols, namely, IEEE 802.11 and IEEE 802.15.4. It was tested in a forested area and was successfully deployed in Virginia, United States.

About 4000 mines are operating worldwide. Mines are a dynamic system, with new paths being developed as mining progresses deeper and (maybe) some old ones being destroyed. The environment in a mine can be very harsh, with increased humidity and significant signal attenuation and reflections. Also, it is not efficient to lay a wired infrastructure, as it may need to be moved following the mining progress. People or trucks carrying laptops can be used as the "data mules" for carrying information back and forth to and from the mine using DTN technology. A successful demonstration of the DTN applicability took place in a chromium mine in Finland (Ginzboorg et al., 2010).

The need to feed an ever-increasing population leads to the optimization of food production on diminishing arable lands. Precision agriculture relates to the use of information and communication technology (ICT) for monitoring crops and environmental conditions to maximize production. A large set of sensors can be deployed in rural areas and collect valuable information relating to crop parameters such as humidity and soil quality. The information must be transferred to back end systems for deeper processing and combination with other sources of information, such as satellite imagery, to advise and support the farming decisions. Cellular technologies, such as General Packet Radio Service (GPRS), have been used extensively. However, the related subscription costs are not always affordable and in many cases, there is insufficient radio coverage for these technologies. Tractors and farmers, or even low-altitude unmanned aerial vehicles (UAVs), can be ideal "data mules" for applying the DTN concepts and transferring back the collected information. The University of Tokyo is already performing field experiments with as many as 39 DTN nodes based on the Linux operating system, observing delivery success rates as high as 99.8% (Ochiai et al., 2010).

Most, if not all, of the previous scenarios, exploit DTN but are not based on the standardized Bundle Protocol (BP) as defined in IETF RFC 5050.

The provision of health services, especially in developing areas and regions, is a topic of utmost importance. In many cases, there exists a low-bandwidth Internet connection that allows some basic information exchange between a remote area and a healthcare post. In these environments, DTN can be an ideal technology for transferring high-volume information (e.g., high-resolution images), which do not have real-time transmission requirements. An early survey among health workers in low-resource settings revealed a sufficient interest in DTN-supported services (Syed-Abdul et al., 2011). A teleconsultation service based on DTN and Diaspora professionals has already been demonstrated in Ghana (Luk et al., 2007, 2008, 2009). Other researchers have demonstrated a sentinel surveillance application in Tanzania (Ntareme and Domancich, 2011).

15.2.2   Sea applications

Sailing the seas leaves only a few communication options, mainly using satellite links. Even when near the coast, such as in a harbor, ships may have a hard time connecting with landline wireless network infrastructure due to channel congestion and signal attenuation and scattering due to waves and other ships. In this setting, an end-to-end path may not always be available. DTN is a profound and cost-effective solution to this problem compared with satellite links. Additionally, DTN routing is shown to achieve a higher packet delivery ratio compared with regular routing schemes (Lin et al., 2010).

Underwater communication cannot be realized with radio frequency (RF) technologies. Instead, acoustic signals must be used. Acoustic signals have low propagation (1500 m/s) and high attenuation, resulting in low-rate communications through special "acoustic modems." An underwater convergence layer (UCL) was developed that allowed acoustic modems to be interfaced with the DTN2 Reference Implementation (Merani et al., 2011). Furthermore, the University of Porto, in collaboration with NATO Underwater Research Centre, performed a field test in Tuscan Archipelago, Italy, that showcased how a DTN can connect heterogeneous networks (acoustic and RF) below and above the sea level (NURC, 2011).

Naval and airborne networks include multiple links with dynamic topologies for a multitude of reasons. The Marine Corps CONDOR (C2 on-the-move network digital over-the-horizon relay) is a proof-of-concept system built around a Cisco router with a ported version of the DTN2 implementation (Parikh and Durst, 2005). The system has multiple connection links for the field and even a satellite protocol for coping with disconnections and disruptions in the most efficient manner.

15.2.3   Smart networking

Urban areas have, in many cases, rich connectivity through broadband connections at home and at work, while cellular technologies offer sufficient coverage for mobile devices and people on the move. In many cases, it is necessary to collect information about the environment and the surrounding space. However, it is not necessary to provide real-time information. Furthermore, the wealth of connectivity options provides the opportunity to use available networking technologies in a power- and cost-effective manner. DTN offers significant advantages in this direction, especially for an open-space environment with a large number of information sources, such as in the case of a WSN.

The Environmental Monitoring in Metropolitan Areas (EMMA) project deployed a wide-area measurement infrastructure for collecting information about air pollution at the city level (Schildt et al., 2011). Rather than deploying and maintaining a grid of wireless sensors in the field, EMMA utilizes the available public transport vehicles (e.g., trams, trains, and buses). These vehicles continuously collect data while roaming across the city. Here, DTN architecture is used to reduce the transmission costs incurred by cellular data connections. Vehicles exchange information through short-range wireless links as they meet and "unload" information to predefined points with backbone connections.

The mobility patterns of vehicles can provide valuable information for car maintenance, commuting, and city planning. There is no need to utilize cellular connections for transmitting these data, assuming some in-vehicle storage and opportunistic Wi-Fi connectivity. CarTel is a system built on the DTN principles and successfully tested on a metropolitan area for road traffic analysis, Wi-Fi measurements, and automotive diagnostics (Hull et al., 2006). DieselNet is a testbed deployed in Amherst, Massachusetts, United States that involves 40 buses equipped with DTN nodes (Burgess et al., 2006). The testbed allows studying connectivity patterns and testing DTN routing algorithms that exploit opportunistic connectivity. BikeNet collects information regarding bicycle rides and routes across a city (Eisenman et al., 2009). DTN is used for transparently handling both cellular and short-range communications. The cellular traffic is kept to a minimum; the connection is only used for transmitting urgent signals, such as when a cyclist is in danger.

Bulk transfer of storage media disks can be the fastest and most cost-effective means to transfer large amounts of information between two remote places. In this case, DTN can be used to move the information transparently for the application layer. The idea was initially explored at a global scale using airports, air carriers, and flight schedules (Keränen and Ott, 2009). TrainNet is a similar idea but on a smaller geographic scale, exploring trains and stations (Zarafshan-Araki and Chin, 2010).

Information and content sharing on the go are also explored as an application area for DTN technology. The key idea here is to better serve spontaneous crowds that move across the city, such as in the case of train commuters at rush hours. Example proposals include drive-thru access to the Internet (Ott and Kutscher, 2004), web search (Balasubramanian et al., 2007), popular media sharing (McNamara et al., 2008), and access to social media (Zaragoza et al., 2011).

Uninterrupted service provision while on the move requires significant investments for covering large areas. The need is more apparent in the case of video streaming, where high bandwidth and fast handoff between access points are required for providing streamlined user experience. One costly solution is to invest in upgrading and condensing the access points throughout the motorways to increase the network capacity and sustain a specific quality of service for moving cars. The Smart Caching proposal takes a more cost-effective approach (Goebbels, 2010). It exploits high-speed connectivity episodes for proactively caching content on customer devices or the in-car entertainment system. Smart Caching forwards the content to the access points that the vehicle will probably connect to in the future. This allows storage capacity and bandwidth requirements for access points to be reduced. Even in the case where network coverage is sufficient and, cellular data access is cheap, offloading cellular traffic through Wi-Fi networks may have significant advantages since the latter offers an order of magnitude more bandwidth. Real mobility traces from the city of San Francisco showed that half of the cellular data traffic could be offloaded using only a few hundred Wi-Fi access points for an area of more than 300 km^2 (Dimatteo et al., 2011). Not all motorways are free of obstacles, as tall buildings and tunnels may disrupt communications. DTN is proposed for supporting the seamless connectivity of land mobile satellites (LMS) in such environments as well (Caini et al., 2010).

A user-provided network is a network formed by sharing a user's broadband connection to nearby passing mobile users. Such networks service the needs of the home users on preference, but on a best-effort case may also serve the needs of mobile users. This can happen, for example, when the connection is lightly utilized or at specific times of the day. In these cases, mobile users benefit by accessing the Internet more affordably and by enjoying high-speed transfers compared with cellular networks. A home access point (HAP) equipped with some storage capability can provide transparent network access based on the DTN principles without user intervention (Koutsogiannis et al., 2011).

DTN has found a way even in the data centers, where plenty of bandwidth is available. NetStitcher utilizes the DTN principles for supporting big volume data transfers between data centers that are dispersed around the globe (Laoutaris et al., 2011). The key idea of NetStitcher is to move data temporarily in intermediate data centers during nonpeak hours and then forward traffic based on predictions regarding future bandwidth availability. A fivefold bandwidth saving compared with alternative means was observed on simulations and when deployed in a content delivery network used in production.

e-Health applications in urban settings, such as patient monitoring, are an area of increasing interest. DAPHNE (Delay-tolerant Application Proxy for e-Health Network Environments) is a DTN proxy for e-health services that transparently overcomes network instabilities, incompatibilities, or even absent for a long duration (Spanakis and Voyiatzis, 2013). DAPHNE decouples network connectivity management from the e-health application, allowing the continuous collection of monitored parameters even when a network connection is not available. DTN is also proposed for the biomedical monitoring of marathon runners (Benferhat et al., 2011, 2012). A WSN monitors the cardiac activity of athletes and the collected signals are "unloaded" in sparsely placed base stations for uploading to a central system. As a marathon race involves a large number of athletes and covers a large area, a DTN approach can be very cost-effective compared with alternative options. A similar approach can be used for biomedical monitoring of nonhospitalized subjects while walking outdoors (Guidec et al., 2012).

One of the major strengths of the DTN architecture and the BP defined in IETF RFC 5050 is the ability to interface a DTN with practically any network protocol stack. Implementations of so-called "convergence" layers already exist for a wide variety of network protocols and stacks, including TCP; User Datagram Protocol (UDP); Ethernet; Negative ACKnowledgment (NACK)-Oriented Reliable Multicast (NORM); AX.25; underwater convergence layer (UCL) for acoustic modems; Licklider Transmission Protocol (LTP); and IEEE 802.15.4. This variety offers multiple connectivity options and the ability to interface transparently multiple networks: terrestrial, underwater, aerial, and satellite. In this sense, DTN can be an ideal technology for emergency relief operations, where infrastructures may not be available and the exchange of information is vital by any possible means. Furthermore, DTN has by design the capability to cope with the enormous delays and disruptions that are commonplace in such situations. DTN over AX.25 is already tested for these scenarios (Ronan et al., 2010, 2011). AX.25 is a link-layer protocol for low-rate packet radio networking over ultra-high frequency (UHF), very high frequency (VHF), and high frequency (HF) links. Furthermore, researchers are already exploring the idea of providing situational awareness in emergencies by using DTN and any available citizen infrastructure (Fall et al., 2010).

Sensor networks, especially WSN can benefit from DTN technology. Sensor systems are usually limited by available power; thus it is very important to optimize computation and communications to an absolute minimum to achieve energy efficiency and long-term operation. Short-range communication technologies and network protocols have been developed for coping with these limitations. Examples include IEEE 802.15.4, ZigBee, and Bluetooth Low Energy. Although there are attempts, such as IPv6 over 6LoWPAN, to run "standard" Internet protocols over sensor systems, still most sensor systems are engineered to run a specialized low-power protocol. In this setting, it is necessary to deploy gateway systems, which transport information between the two networks (WSN and Internet). Further, these short-range protocols have, in many cases, scalability and reliability issues. DTN routing can exploit the node mobility in a Building Energy Management System (BEMS) to increase reliability and decrease overheads (Buranachokphaisan et al., 2013). The BP is criticized as being "heavy" for WSN applications. Pöttner et al. (2012) showed how BP can be implemented in 8- and 16-bit microprocessors and presented an IEEE 802.15.4 convergence layer, achieving interoperability with a BP-compliant DTN implementation for personal computers. The researchers evaluated the technology using a lift as a "data mule" to transfer measurements from the roof of a 15-story building to a computer on the third floor.

15.3 DTN as an enabling technology

DTN was designed for interplanetary networking. A wide variety of terrestrial application scenarios in which DTN can be applied and provides advantages over conventional networking technologies were presented in the previous section. Communication disruption and intermittent connectivity are becoming a norm even in the urban environment due to the increased mobility of nodes. DTN tries to exploit this mobility for increasing network capacity and reach rather than extending the infrastructure for wider coverage.

A key concept for DTN is the provision of "in-network storage." This storage is provided transparently for the application layer and the application itself. This is an important differentiator, as the network storage is not exposed to the applications. Hence, the application programmers need not integrate the functionality to their implementation. More importantly, they need not handle all the accompanying networking issues, such as discovering available storage nodes, handling credentials, coping with reconnections, and retransmissions. Rather, all the details are automatically handled by the DTN stack.

The second key concept of DTN is that it is network-agnostic. The "convergence layer" concept allows integration with any kind of network or network protocol stack. As soon as a convergence layer is implemented, DTN can cross through networks with quite different and incompatible characteristics. Carrier examples and proposals include acoustic signals; RF signals; boats and ferries; bicycles; even pigeons (Scholl and Lindgren, 2012), human networks (Zhuo et al., 2011), and crowds (Tournoux et al., 2010).

15.3.1 DTN implementations

It is a challenging task to provide the necessary tools for developing DTN applications that can cope with so many and different assumptions. DTN2 is an open-source implementation of the BP that evolved over an older approach named DTN-RI. DTN2 is considered as the reference implementation of the protocol. It is available in C++ and works on Linux and Mac OS operating systems. DTN2 offers many routing schemes and defines an extensible architecture that allows interested parties to develop and test new ones using XML messaging. Among the available implementations, DTN2 also offers the largest number of convergence layers, being able to interface different networks, such as IP-based and LTP-based (used in space environments).

ION (Interplanetary Overlay Network) is a software implementation in C of the BP and the required functionality for DTN into space. Its main shortcoming for implementing terrestrial applications is that ION supports only scheduled contacts at predefined times. This may be sufficient for simple, static scenarios, like train communications, but it cannot support opportunistic ones, like buses or human mobility. ION is distributed with an open-source license and is available for Linux, Mac OS X, Solaris, FreeBSD, and even real-time operating systems, such as VxWorks and RTEMS. The interoperability of ION and DTN2 is confirmed (Jenkins et al., 2010; de Frescheville et al., 2011; Nichols et al., 2010).

POSTELLATION is a licensed implementation written in C and targeting embedded systems. It runs on Linux, RTEMS, Mac OS X, and Microsoft Windows. POSTELLATION offers an HTTP proxy for web browsing over a DTN.

IBR-DTN is an implementation of DTN in C++ (Doering et al., 2008; Schildt et al., 2011). It is very portable, slim, and extensile. IBR-DTN runs on Linux and embedded systems, with emphasis on devices running the OpenWRT firmware. As such, it can support a wide range of systems, from laptops down to low-budget wireless access points, such as the Fonera FON2200.

Portable devices carried by human beings, such as smartphones, are a very interesting option for realizing DTN nodes. These devices have sufficient processing, power, and storage capacity for transferring large amounts of data from the field. Also, they offer multiple connectivity options, such as cellular, wireless, USB, and even short-range communications such as Bluetooth, Bluetooth Low Energy, and Near Field Communication (NFC). In many cases, they also integrate navigation components, such as a global navigation satellite system (GNSS) receiver or a digital compass. The availability of a DTN implementation for smartphones is thus very useful.

Bytewalla is a DTN implementation in Java for Google Android smartphones (Domancich, 2010). It was originally conceived for DTN communication in African rural villages using humans carrying a mobile phone as "data mules."

DASM is an implementation for Symbian phones, tested with Nokia Communicators 9300i and 9500 devices (Mukhtar and Ott, 2006). DASM is superseded by DTNS60, which is a complete rewrite for newer, Maemo-based Symbian S60 devices (Turkulainen, 2010).

IBR-DTN is ported for Google Android smartphones, and example applications are already available. The development follows advances in main IBR-DTN and retains interoperability with the DTN2 reference implementation. This is important for freely exchanging information through infrastructure that may be realized with another DTN implementation.

WSNs can be a target for DTN applications, especially in scenarios where real-time communication is not critical. The BP specification is considered "heavy" for implementation in sensor nodes since it introduces significant overheads (bundle headers) that may not fit well with sensor network protocols. For example, an IEEE 802.15.4 can carry a packet of up to 115 bytes (128 bytes with the necessary headers), which is already smaller than the bundle header. Thus, it requires fragmentation and reassembly at the edge of the network.

DTNLite is an implementation following the DTN architecture on the TinyOS sensor network operating system running on Mica motes (Nedenshi and Patra, 2004). Due to platform limitations, DTNLite does not implement BP. ContikiDTN is an attempt to run DTN on the Contiki operating system. ContikiDTN is shown to be compatible with the DTN2 reference implementation running on a computer (Loubser, 2006). IBR-DTN is the most complete approach till date. It is shown to run on an iMote2 sensor and supports the IEEE 802.15.4 convergence layer (Schildt et al., 2011).

15.3.2   DTN tools and performance

Once implementation is available, it is necessary to have various tools for debugging, monitoring, and reasoning about a DTN. The DTNperf_2 tool follows the design of the Iperf network performance analysis tool and can be used for DTN performance evaluation (Caini et al., 2009). A BP dissector is available in the popular Wireshark network protocol analyzer (http://www.wireshark.org/). The dissector can be very helpful for analyzing DTN network traffic, testing for interoperability, and debugging implementations of convergence layers. Tools for monitoring the status of a DTN2 network at the BP layer have also been developed (Papalambrou et al., 2011).

One point of concern for DTN is its performance in the sense of latency resulting from bundle storage and processing overhead due to BP complexity. The effect of storage back end technology for IBR-DTN and DTN2 implementations leads to the conclusion that per bundle processing is the limiting factor for attainable throughput rather than the storage bandwidth (Oliver and Falaki, 2007; Doering et al., 2008; Schildt et al., 2011). For high-end systems, the performance improves significantly and storage bandwidth becomes the bottleneck (Pöttner et al., 2011a,b). The Bundle Security Protocol (BSP) is defined in IETF RFC 6257 for bundle security. Preliminary analysis of the performance of a BSP implementation on a smartphone platform revealed a small transmission overhead but a large effect on battery due to heavy use of public-key cryptography (Domancich, 2010). A series of works have explored the best bundle size, and the findings confirm the assumption that there is no global optimal size; rather, the selection depends on the specifics of the convergence layers and the operation environment parameters, such as the bit error rate (Ivancic et al., 2008; Ronan et al., 2008, 2010; Ronan and O'Connor, 2010; Samaras and Tsaoussidis, 2010, 2011; Magaia et al., 2011; Morgenroth et al., 2011).

The operation conditions of a DTN must be studied and accounted for ahead of time. Availability of simulation and emulation tools is a necessity, especially for DTN routing protocols, as they allow the network behavior to be accurately studied before costly deployments. General-purpose network simulators have been enhanced to incorporate some DTN functionality; namely, NS2 and OMNET++. Specialized simulators with increased capabilities include DTNSim and ONE (Liu et al., 2011). The latter can also support the vehicular DTN (VDTN) architecture (Soares et al., 2009). Emulator tools can provide a more accurate representation of the network behavior but are harder to implement. Available emulators include VDTN@Lab for VDTN (Dias et al., 2009); DOME for moving nodes on scheduled routes, like buses (Soroush et al., 2009); SPICE and EGGS for space communications (Samaras et al., 2009; de Cola et al., 2009); and UTMesh for wireless mesh networking (Ochiai et al., 2011).

15.3.3 Technology adoption

In the previous sections, a wide range of application scenarios are described and a large variety of implementations and tools are presented. Even more, developments are in progress, and there is a significant experience in deploying DTN even in harsh environments. DTN has proven its ability to cope with long-term disconnections and disruptions while retaining good performance in "normal" network conditions. It can no longer be claimed that DTN is a niche solution for "exotic" applications only. There are real-world examples of its applicability in urban environments and even in data centers.

The technology adoption of DTN is still limited. Despite the availability of open-source implementations and tools, DTN is not considered in general as a first choice, even for scenarios that include mobility and communication disruptions. Rather, they are handled at the application level. The programmers integrate the necessary logic for session handling and data buffering within the application. A DTN implementation can handle these situations more efficiently and transparently at the expense of using a slightly different application programming interface (API) compared with the one most commonly used for developing networking applications (e.g., sockets).

The resistance to change is justified at this stage. Most, if not all, DTN-enabled systems until now are designed and deployed for handling information related to just one application. The cost to integrate the DTN technology in an application only for handling opportunistic connectivity remains higher than the expected benefit. If more than one or even all applications installed in a device can exploit this functionality, then it could become affordable. Till date, no such cases have been reported. Furthermore, poor connectivity due to increased mobility in urban settings can be confronted by upgrading cellular and wireless infrastructures. Although these upgrades can be very costly, if decided by network operators in the first place, they constitute a strong argument against investing in DTN technology. However, even in an always-on world, DTN has proven its advantages, increasing network capacity, and reducing transmission costs (Dimatteo et al., 2011; Laoutaris et al., 2011).

15.4 Conclusions and future trends

The true value of DTN is not a single killer application. Rather, it is the ability to seamlessly and transparently support so many different scenarios and network technologies, and, further, the ease of development for recurring application tasks such as handling disconnections. This trend becomes even more important as mobility and free movement in open space become the norm rather than the exception. In such a dynamic environment, the assumptions of uninterrupted connectivity through a long-range communication infrastructure and the existence of an end-to-end path are rather weak. Opportunistic connectivity characterized by various degrees of uncertainty and availability of bandwidth is a more accurate representation of the operational state. Legacy protocols designed with the inherent assumption of end-to-end connectivity cannot fit this world.

In this context, we argue that a bundle of mobile applications (apps), each supporting a different application scenario and operating in an opportunistic connectivity environment, can be the "killer app" for DTN technology. Mobile devices, such as tablets and smartphones, carried by people, always on the go, can benefit most from an embedded networking layer that handles intermittent connectivity most efficiently. Given the multiple connectivity options and the diversity of mobile ecosystems, driving application development through DTN can both benefit developers, and spread technology adoption.

Research in DTN is far from complete. As adoption advances, new problems must be solved. Opportunistic routing in DTN is an area of significant attention. Efficient storage techniques and optimized data delivery while minimizing network traffic are also areas of interest. Even more, problems arise as DTN systems and nodes are engineered; energy efficiency and appropriate system architectures for specific application domains are always a concern for real-life deployments. Bundle security and privacy, especially for data transfers involving or relating information about humans, remain unexplored until now and should be a top priority for DTN.

15.5 Sources of further information and advice

The best source of information for DTN developments is the website of the DelayTolerant Research Group (DTNRG) of the Internet Research Task Force (IRTF). The website moved to https://sites.google.com/site/dtnresgroup/home in October

2013. Interested parties can follow the progress of DTN standards, join the mailing lists, download available code of the DTN2 reference implementation, and access available documents including the latest Internet drafts, scientific papers, tutorials, and presentations.

References

Balasubramanian, A., Zhou, Y., Croft, W., Levine, B., Venkataramani, A., 2007. Web search from a bus. In: Proceedings of the Second ACM Workshop on Challenged Networks (CHANTS '07). ACM, New York, NY, USA, pp. 59–66.

Benferhat, D., Guidec, F., Quinton, P., 2011. Disruption-tolerant wireless biomedical monitoring for marathon runners: a feasibility study. In: 1st International Workshop on Opportunistic and Delay/Disruption-Tolerant Networking (WODTN'11), Brest, France, IEEE Xplore (October), pp. 1–5.

Benferhat, D., Guidec, F., Quinton, P., 2012. Disruption-tolerant wireless sensor networking for biomedical monitoring in outdoor conditions. In: 7th International Conference on Body Area Networks (BODYNETS'12), September. ACM Digital Library, Oslo, pp. 1–7.

Bonnet, P., Chang, M., 2010. Monitoring in a high-arctic environment: some lessons from Mana. IEEE Pervasive Comput. 9 (4), 16–23.

Buranachokphaisan, M., Ochiai, H., Esaki, H., 2013. Disruption tolerant wireless sensor network for building energy management system. In: Poster Presentation in Internet Conference 2013 (IC 2013), 24–25 October, Tokyo, Japan.

Burgess, J., Gallagher, B., Jensen, D., Levine, B., 2006. MaxProp: routing for vehicle-based disruption-tolerant networks. In: INFOCOM 2006. 25th IEEE International Conference on Computer Communications. Proceedings. vol. 6. IEEE, pp. 1–11.

Caini, C., Cornice, P., Firrincieli, R., Livini, M., 2009. DTNperf 2: a performance evaluation tool for delay/disruption tolerant networking. In: Proceedings of the International Conference on Ultra Modern Telecommunications, ICUMT 2009, 12–14 October 2009. IEEE, St Petersburg, Russia, pp. 1–6.

Caini, C., Firrincieli, R., Livini, M., 2010. DTN bundle layer over TCP: retransmission algorithms in the presence of channel disruptions. J. Commun. 5 (2), 106–116.

Crowcroft, J., Yoneki, E., Hui, P., Henderson, T., 2008. Promoting tolerance for delay tolerant network research. SIGCOMM Comput. Commun. Rev. 38 (5), 63–68.

de Cola, T., Ronga, L., Pecorella, T., Barsocchi, P., Chessa, S., et al., 2009. Communications and networking over satellites: SatNEx experimental activities and testbeds. Int. J. Satell. Commun. Netw. 27 (1), 1–33.

de Frescheville, F.B., Martin, S., Policella, N., Patterson, D., Aiple, M., et al., 2011. Set-up and validation of METERON end-to-end network for robotic experiments. In: ESA/ESTEC, Noordwijk, The Netherlands, 12–14 April 2011.

Dias, J., Isento, J., Silva, B., Soares, V., Ferreira, P., et al., 2009. Creation of a vehicular delay tolerant network prototype. In: Conferncia de Engenharia, 5, Covilh, 25 Novembro 2009—Inovao e Desenvolvimento, Universidade da Beira Interior, Covilh, Portugal, pp. 1–5.

Dimatteo, S., Hui, P., Han, B., Li, V., 2011. Cellular traffic offloading through WiFi networks. In: Proc. of IEEE MASS, Valencia, Spain, 17–22 October. IEEE, pp. 192–201.

Doering, M., Lahde, S., Morgenroth, J., Wolf, L., 2008. IBR-DTN: an efficient implementation for embedded systems. In: Proceedings of the Third ACM Workshop on Challenged Networks (CHANTS '08). ACM, New York, NY, USA, pp. 117–120.

Domancich, S., 2010. Security in Delay Tolerant Networks for the Android Platform (Master's thesis). Royal Institute of Technology (KTH), Sweden.

Eisenman, S., Miluzzo, E., Lane, N., Peterson, R., Ahn, G., et al., 2009. Bikenet: a mobile sensing system for cyclist experience mapping. ACM Trans. Sens. Netw. 6 (1), 6:1–6:39.

Fall, K., 2003. A delay-tolerant network architecture for challenged internets. In: Proceedings of the 2003 Conference on Applications, Technologies, Architectures, and Protocols for Computer Communications (SIGCOMM '03), Karlsruhe, Germany, 25–29 August. ACM, pp. 27–34.

Fall, K., Iannaccone, G., Kannan, J., Silveira, F., Taft, N., 2010. A disruption-tolerant architecture for secure and efficient disaster response communications. In: Proceedings of the 7th International Conference on Information Systems for Crisis Response and Management, Seattle, Washington, USA, 2–5 May.

Farrell, S., Cahill, V., 2006. Delay- and Disruption-Tolerant Networking. Artech House, Norwood, MA, USA.

Ginzboorg, P., Krkkinen, T., Ruotsalainen, A., Andersson, M., Ott, J., 2010. DTN communication in a mine. In: Extreme Workshop on Communication—The Himalayan Expedition (ExtremeCom 2010), Dharamsala, India, 4–10 November, pp. 1–6.

Goebbels, S., 2010. Disruption tolerant networking by smart caching. Int. J. Commun. Syst. 23 (5), 569–595.

Guidec, F., Benferhat, D., Quinton, P., 2012. Biomedical monitoring of non-hospitalized subjects using disruption-tolerant wireless sensors. In: Proceedings of MobiHealth '12, Number 61 in Springer LNICST, Paris, France, November, pp. 11–19.

Guo, S., Derakhshani, M., Falaki, M.H., Ismail, U., Luk, R., et al., 2011. Design and implementation of the KioskNet system. Comput. Netw. 55 (1), 264–281.

Hull, B., Bychkosky, V., Zhang, Y., Chen, K., Goraczko, M., et al., 2006. CarTel: a distributed mobile sensor computing system. In: Proc. ACM SENSYS, Boulder, Colorado, USA, 31 October–3 November. ACM.

Ivancic, W., Eddy, W., Wood, L., Stewart, D., Jackson, C., et al., 2008. Delay/disruption-tolerant network testing using a LEO satellite. In: 8th Annual NASA Earth Science Technology Conference (ESTC 2008). University of Maryland University College, Maryland, USA, pp. 1–7. 24–26 June. Paper A4P2, Proceedings. Available at: http://esto.nasa.gov/conferences/estc2008/index.html.

Jenkins, A., Kuzminsky, S., Gifford, K., Pitts, R., Nichols, K., 2010. Delay/disruption-tolerant networking: flight test results from the International Space Station. In: 2010 IEEE Aerospace Conference Proceedings. IEEE, Piscataway, NJ, USA, pp. 1–8.

Keränen, A., Ott, J., 2009. DTN over aerial carriers. In: Proceedings of the 4th ACM Workshop on Challenged Networks (CHANTS '09). ACM, New York, NY, USA, pp. 67–76.

Koutsogiannis, E., Mamatas, L., Psaras, I., 2011. Storage-enabled access points for improved mobile performance: an evaluation study. In: Proceedings of the 9th IFIP TC 6 International Conference on Wired/Wireless Internet Communications (WWIC'11). Springer-Verlag, Berlin, Heidelberg, pp. 116–127.

Laoutaris, N., Sirivianos, M., Yang, X., Rodriguez, P., 2011. Inter-datacenter bulk transfers with NetStitcher. In: Proceedings of the ACM SIGCOMM 2011 Conference on SIGCOMM, SIGCOMM '11. ACM, New York, NY, USA, pp. 74–85.

Lin, H.-M., Ge, Y., Pang, A.-C., Pathmasuntharam, J., 2010. Performance study on delay tolerant networks in maritime communication environments. In: IEEE OCEANS 2010, pp. 1–6, Seattle, Washington, USA, 20–23 September.

Lindgren, A., Hui, P., 2009. "The quest for a killer app for opportunistic and delay tolerant networks" (invited paper). In: Proceedings of the 4th ACM Workshop on Challenged Networks (CHANTS '09). ACM, New York, NY, USA, pp. 59–66.

Lindgren, A., Hui, P., 2011. ExtremeCom: to boldly go where no one has gone before. SIGCOMM Comput. Commun. Rev. 41 (1), 54–59.

Lindgren, A., Doria, A., Lindblom, J., Ek, M., 2008. Networking in the land of northern lights: two years of experiences from DTN system deployments. In: Proceedings of the 2008 ACM Workshop on Wireless Networks and Systems for Developing Regions (WiNS-DR '08). ACM, New York, NY, USA, pp. 1–8.

Liu, M., Yang, Y., Qin, Z., 2011. A survey of routing protocols and simulations in delay-tolerant networks. In: Proceedings of the 6th International Conference on Wireless Algorithms, Systems, and Applications (WASA '11). Springer-Verlag, Berlin, Heidelberg, pp. 243–253.

Loubser, M., 2006. Delay tolerant networking for sensor networks. In: Technical Report T2006:01, Swedish Institute of Computer Science, Kista, Sweden.

Luk, R., Ho, M., Aoki, P.M., 2007. A framework for designing teleconsultation systems in Africa. In: Proc. Int'l Conf. on Health Informatics in Africa (HELINA), Bamako, Mali, January 2007, pp. 1–5.

Luk, R., Ho, M., Aoki, P.M., 2008. Asynchronous remote medical consultation for Ghana. In: Proceedings of the Twenty-Sixth Annual SIGCHI Conference on Human Factors in Computing Systems (CHI '08). ACM, New York, NY, USA, pp. 743–752.

Luk, R., Zaharia, M., Ho, M., Levine, B., Aoki, P., 2009. ICTD for healthcare in Ghana: two parallel case studies. In: 2009 International Conference on Information and Communication Technologies and Development (ICTD), Doha, Qatar, 17–19 April, pp. 118–128.

Magaia, N., Pereira, P., Casaca, A., Rodrigues, J., Dias, J., et al., 2011. Bundles fragmentation in vehicular delay-tolerant networks. In: 7th Euro-NF Conference on Next Generation Networks (NGI2011), Kaiserslautern, Germany, 27–29 June. IEEE, pp. 27–29.

McNamara, L., Mascolo, C., Capra, L., 2008. Media sharing based on colocation prediction in urban transport. In: MobiCom '08: Proceedings of the 14th ACM International Conference on Mobile Computing and Networking. ACM, New York, NY, USA, pp. 58–69.

Merani, D., Berni, A., Potter, J., Martins, R., 2011. An underwater convergence layer for disruption tolerant networking. In: Proceedings of the 2011 Baltic Congress of Future Internet Communications (BCFIC 2011), Riga, Latvia, 16–18 February. IEEE, pp. 103–108.

Morgenroth, J., Poegel, T., Heitz, R., Wolf, L., 2011. Delay-tolerant networking in restricted networks. In: Proceedings of the 6th ACM Workshop on Challenged Networks (CHANTS '11). ACM, New York, NY, USA, pp. 53–56.

Mukhtar, O., Ott, J., 2006. Backup and bypass: introducing DTN-based ad-hoc networking to mobile phones. In: Proceedings of the 2nd International Workshop on Multi-hop Ad hoc Networks: From Theory to Reality (REAL-MAN '06). ACM, New York, NY, USA, pp. 107–109.

Nedenshi, S., Patra, R., 2004. DTNLite: a reliable data transfer architecture for sensor networks. In: 8th International Conference on Intelligent Engineering Systems, Cluj-Napoca, Romania, 19–21 September.

Nichols, K., Holbrook, M., Pitts, R.L., Gifford, K.K., Jenkins, A., et al., 2010. DTN implementation and utilization options on the International Space Station. In: SpaceOps 2010 Conference, Von Braun Center, Huntsville, Alabama, USA, 25–30 April. Conference Proceeding Series, AIAA.

Ntareme, H., Domancich, S., 2011. Security and performance aspects of Bytewalla: a delay tolerant network on smartphones. In: IEEE 7th International Conference on Wireless and Mobile Computing, Networking and Communications (WiMob 2011), Shanghai, China, 10–12 October, pp. 449–454.

NURC, 2011. The Centre Quarterly. URL http://www.cmre.nato.int/research/publications/the-centre-quarterly/doc_download/10-december-2011. (Accessed 15 July 2014).

Ochiai, H., Ishizuka, H., Kawakami, Y., Esaki, H., 2010. A DTN-based sensor data gathering for agricultural applications. IEEE Sensors J. 11 (11), 2861–2868.

Ochiai, H., Matsuo, K., Matsuura, S., Esaki, H., 2011. A case study of UTMesh: design and impact of real world experiments with Wi-Fi and Bluetooth devices. In: 11th International Symposium on Applications and the Internet (SAINT), 2011 IEEE/IPSJ. IEEE Computer Society, Washington, DC, USA, pp. 433–438.

Oliver, E., Falaki, H., 2007. Performance evaluation and analysis of delay tolerant networking. In: Proceedings of the 1st International Workshop on System Evaluation for Mobile Platforms (MobiEval '07). ACM, New York, NY, USA, pp. 1–6.

Ott, J., Kutscher, D., 2004. Drive-thru Internet: IEEE 802.11b for "automobile" users. In: INFOCOM 2004. 23rd IEEE International Conference on Computer Communications. Proceedings. vol. 1. IEEE, pp. 362–373.

Papalambrou, A., Voyiatzis, A.G., Soufrilas, P., Serpanos, D.N., 2011. Monitoring of a DTN2 network. In: 2011 Baltic Congress on Future Internet Communications (BCFIC Riga), pp. 116–119.

Parikh, S., Durst, R., 2005. Disruption tolerant networking for marine corps CONDOR. In: Military Communications Conference, 2005. MILCOM 2005, Atlantic City, NJ, USA, 17–21 October. vol. 1. IEEE, pp. 325–330.

Pentland, A.S., Fletcher, R., Hasson, A., 2004. DakNet: rethinking connectivity in developing nations. Computer 37 (1), 78–83.

Pöttner, W.-B., Morgenroth, J., Schildt, S., Wolf, L., 2011a. An empirical performance comparison of DTN bundle protocol implementations. Informatikbericht 2011–08, Technische Universitaat Braunschweig.

Pöttner, W.-B., Morgenroth, J., Schildt, S., Wolf, L., 2011b. Performance comparison of DTN bundle protocol implementations. In: Proceedings of the 6th ACM Workshop on Challenged Networks (CHANTS '11). ACM, New York, NY, USA, pp. 61–64.

Pöttner, W.B., Busching, F., Von Zengen, G., Wolf, L., 2012. Data elevators: applying the bundle protocol in delay tolerant wireless sensor networks. In: 9th International Conference on Mobile Adhoc and Sensor Systems (MASS) 2012, Las Vegas, NV, USA, 8–11 October. IEEE, pp. 218–226.

Ronan, J., O'Connor, C., 2010. A comparison of different TCP/IP and DTN protocols over the D-Star Digital Data Mode. In: 29th ARRL and TAPR Digital Communications Conference, ARRL, 225 Main Street, Newington, CT 06111–1494, USA, pp. 134–138.

Ronan, J., Walsh, K., Darren, L., 2008. Experiments with delay and disruption tolerant networking in AX.25 and D-Star networks. In: 28th ARRL and TAPR Digital Communications Conference, ARRL, 225 Main Street, Newington, CT 06111–1494, USA, pp. 97–106.

Ronan, J., Walsh, K., Long, D., 2010. Evaluation of a DTN convergence layer for the AX.25 network protocol. In: Proceedings of the Second International Workshop on Mobile Opportunistic Networking (MobiOpp '10). ACM, New York, NY, USA, pp. 72–78.

Ronan, J., Darren, L., Walsh, K., 2011. More experiments with delay and disruption tolerant networking over AX.25 networks. In: 30th ARRL and TAPR Digital Communications Conference, ARRL, 225 Main Street, Newington, CT 06111–1494, USA, pp. 65–71.

Samaras, C., Tsaoussidis, V., 2010. Design of delay-tolerant transport protocol (DTTP) and its evaluation for Mars. Acta Astronaut. 67 (7–8), 863–880.

Samaras, C.V., Tsaoussidis, V., 2011. Adjusting transport segmentation policy of DTN Bundle Protocol under synergy with lower layers. J. Syst. Softw. 84, 226–237.

Samaras, C., Komnios, I., Diamantopoulos, S., Koutsogiannis, E., Tsaoussidis, V., et al., 2009. Extending internet into space—ESA DTN testbed implementation and evaluation. In: Granelli, F., Skianis, C., Chatzimisios, P., Xiao, Y., Redana, S., et al. (Eds.), Mobile Lightweight Wireless Systems. Vol. 13 of Lecture Notes of the Institute for Computer Sciences, Social Informatics and Telecommunications Engineering, Springer, Berlin, Heidelberg, pp. 397–404.

Schildt, S., Morgenroth, J., Poettner, W.-B., Wolf, L., 2011. IBR-DTN: a lightweight, modular and highly portable Bundle Protocol implementation. Electr. Commun. EASST 37, 1–11.

Scholl, J., Lindgren, A., 2012. Considering pigeons for carrying delay tolerant networking based internet traffic in developing countries. Electr. J. Inf. Syst. Dev. Ctries. 54 (4), 1–19.

Selavo, L., Wood, A., Cao, Q., Sookoor, T., Liu, H., et al., 2007. LUSTER: wireless sensor network for environmental research. In: Proceedings of the 5th International Conference on Embedded Networked Sensor Systems (SenSys '07). ACM, New York, NY, USA, pp. 103–116.

Seth, A., Kroeker, D., Zaharia, M., Guo, S., Keshav, S., 2006. Low-cost communication for rural Internet kiosks using mechanical backhaul. In: Proceedings of the 12th Annual International Conference on Mobile Computing and Networking (MobiCom '06). ACM, New York, NY, USA, pp. 334–345.

Soares, V., Farahmand, F., Rodrigues, J., 2009. A layered architecture for vehicular delay tolerant networks. In: IEEE Symposium on Computers and Communications (ISCC 2009), Sousse, Tunisia, 5–8 July. IEEE, pp. 122–127.

Soroush, H., Banerjee, N., Balasubramanian, A., Corner, M., Levine, B., et al., 2009. DOME: a diverse outdoor mobile testbed. In: Proceedings of the 1st ACM International Workshop on Hot Topics of Planet-Scale Mobility Measurements. ACM, New York, NY, USA, pp. 2:1–2:6.

Spanakis, E.G., Voyiatzis, A.G., 2013. DAPHNE: "A disruption-tolerant application proxy for e-health network environments". In: Proceedings of the 3rd International Conference on Wireless Mobile Communication and Healthcare, Paris, France, 21–23 November. Springer-Verlag, pp. 88–95.

Syed-Abdul, S., Scholl, J., Lee, P., Jian, W.-S., Liou, D.-M., et al., 2011. Study on the potential for delay tolerant networks by health workers in low resource settings. Comput. Methods Prog. Biomed. 107 (3), 557–564.

Tournoux, P., Lochin, E., Leguay, J., Lacan, J., 2010. Robust streaming in delay tolerant networks. In: 2010 IEEE International Conference on Communications (ICC), pp. 1–5.

Turkulainen, A., 2010. Delay-Tolerant Networking for the Open Mobile Software Platform: An Application Case Study (Master's thesis). Aalto University, Finland.

Wood, L., Eddy, W., Holliday, P., 2009a. A bundle of problems. In: 2009 IEEE Aerospace Conference Proceedings. IEEE, Piscataway, NJ, USA, pp. 1–17.

Wood, L., Holliday, P., Floreani, D., Eddy, W., 2009b. Sharing the dream: the consensual networking hallucination offered by the Bundle Protocol. In: 2009 International Congress on Ultra Modern Telecommunications and Control Systems and Workshops (ICUMT)', St Petersburg, Russia, 12–14 October. IEEE, pp. 1–2.

Zarafshan-Araki, M., Chin, K., 2010. TrainNet: a transport system for delivering non real-time data. Comput. Commun. 33 (15), 1850–1863.

Zaragoza, K., Thai, N., Christensen, T., 2011. An implementation for accessing Twitter across challenged networks. In: Proceedings of the 6th ACM Workshop on Challenged Networks (CHANTS '11). ACM, New York, NY, USA, pp. 71–72.

Zhang, P., Sadler, C.M., Lyon, S.A., Martonosi, M., 2004. Hardware design experiences in ZebraNet. In: Proceedings of the 2nd International Conference on Embedded Networked Sensor Systems (SenSys '04). ACM, New York, NY, USA, pp. 227–238.

Zhuo, X., Li, Q., Cao, G., Dai, Y., Szymanski, B., et al., 2011. Social-based cooperative caching in DTNs: a contact duration aware approach. In: Proc. of IEEE MASS, Valencia, Spain, 17–22 October. IEEE, pp. 92–101.

Index

Note: Page numbers followed by *f* indicate figures and *t* indicate tables.

Printed in the United States
By Bookmasters